THE PHYSICS AND EVOLUTION OF ACTIVE GALACTIC NUCLEI

Research into active galactic nuclei (AGNs) – the compact, luminous hearts of many galaxies – is at the forefront of modern astrophysics. Understanding these objects requires extensive knowledge in many different areas: accretion disks, the physics of dust and ionized gas, astronomical spectroscopy, star formation, and the cosmological evolution of galaxies and black holes. This new text by Hagai Netzer, a renowned astronomer and leader in the field, provides a comprehensive introduction to the theory underpinning our study of AGNs and the ways that we observe them. It emphasizes the basic physics underlying this phenomenon, the different types of active galaxies and their various components, and the complex interplay between them and other astronomical objects. Recent developments regarding the evolutionary connections between active and dormant black holes and star-forming galaxies are explained in detail. Both graduate students and researchers will benefit from Netzer's authoritative contributions to this exciting field of research.

PROFESSOR HAGAI NETZER is a world-class expert in the area of active galaxies and super-massive black holes. He has been at Tel Aviv University since 1977, was the director of the Wise Observatory and the Chair of the Department of Astronomy and Astrophysics, was a visiting professor at numerous other universities, and won several prizes and awards. He has written several advanced and introductory texts in astronomy, as well as a popular science book about the search for life in the universe.

THE PHYSICS AND EVOLUTION
OF ACTIVE GALACTIC NUCLEI

HAGAI NETZER

Tel Aviv University

Shaftesbury Road, Cambridge CB2 8EA, United Kingdom

One Liberty Plaza, 20th Floor, New York, NY 10006, USA

477 Williamstown Road, Port Melbourne, VIC 3207, Australia

314–321, 3rd Floor, Plot 3, Splendor Forum, Jasola District Centre, New Delhi – 110025, India

103 Penang Road, #05–06/07, Visioncrest Commercial, Singapore 238467

Cambridge University Press is part of Cambridge University Press & Assessment,
a department of the University of Cambridge.

We share the University's mission to contribute to society through the pursuit of
education, learning and research at the highest international levels of excellence.

www.cambridge.org
Information on this title: www.cambridge.org/9781107021518

First published 2013
Reprinted 2019

A catalogue record for this publication is available from the British Library

Library of Congress Cataloging-in-Publication data
Netzer, Hagai.
The physics and evolution of active galactic nuclei / Hagai Netzer,
Tel Aviv University.
pages cm
Includes bibliographical references and index.
ISBN 978-1-107-02151-8 (hardback)
1. Active galactic nuclei. I. Title.
QB858.3.N48 2013
523.1´12–dc23 2012047526

ISBN 978-1-107-02151-8 Hardback

To my students and colleagues who helped me appreciate
the beauty and complexity of Astronomy.

Contents

Color plates following page 178

Preface

The field of active galactic nuclei (AGNs) is exploding. From a narrow discipline
dealing with massive active black holes (BHs) and their immediate surroundings,
it now includes the host galaxies of such BHs, the correlated evolution of BHs and
galaxies, and the physics of extremely energetic phenomena like γ-ray jets. More
than 1000 articles are being published in refereed journals every year about this
topic, and the numbers are still growing. The equivalent number in the mid-1970s
was about 200.

This book, *The Physics and Evolution of Active Galactic Nuclei*, is an attempt
to cover most of the central topics in this large field in a way that emphasizes the
basic physics and the complex connections between AGNs and other astronomical objects. It grew from a graduate-level course taught at Tel Aviv University
over many years and from numerous international schools on this topic where I
participate as a lecturer, and it contains three main themes. The first is a comprehensive description of the more important physical processes associated with
AGNs: the physics of photoionized gas; dust in AGNs; nonthermal processes;
and various modes of accretion onto BHs, including accretion disks and accretion
flows. The second is a detailed description of various subgroups of AGNs and the
main components in individual sources. These include radio-loud and radio-quiet
AGNs, type-I and type-II sources, LINERs, blazars, broad and narrow emission
line regions, broad and narrow absorption lines, megamasers, dusty tori, and X-ray-
emitting gas near the BH. The third part deals with the various connections, evolutionary and others, between BHs and their host galaxies, including star-forming
galaxies.

The book is meant to be self-contained, an almost impossible mission given the
nature of the field. There is no way to pay justice to all the papers that were used in
writing this text – there are simply too many of them and, no doubt, others that are
as important but escaped my notice. Instead, I chose to give itemized references
at the end of each chapter to help guide the reader to other books, to important

review articles, and to central papers that were used in the writing. These lists are "references to references" and should be viewed as shortcuts to a more thorough study of this vast area of modern research.

Hagai Netzer
April 2012

1

Observations of active galactic nuclei

The names "active galaxies" and "active galactic nuclei" (AGNs) are related to the main feature that distinguishes these objects from inactive (normal or regular) galaxies: the presence of supermassive accreting black holes (BHs) in their centers. As of 2011, there were approximately a million known sources of this type selected by their color and several hundred thousand by basic spectroscopy and accurate redshifts. It is estimated that in the local universe, at $z \leq 0.1$, about 1 out of 50 galaxies contains a fast-accreting supermassive BH, and about 1 in 3 contains a slowly accreting supermassive BH.

Detailed studies of large samples of AGNs, and the understanding of their connection with inactive galaxies and their redshift evolution, started in the late 1970s, long after the discovery of the first quasi-stellar objects (hereinafter quasars or QSOs) in the early 1960s. Although all objects containing active supermassive BHs are now referred to as AGNs, various other names, relics from the 1960s, 1970s, and even later, are still being used. Some of the names that appear occasionally in the literature, such as "Seyfert 1 galaxies" and "Seyfert 2 galaxies," in honor of Carl Seyfert, who observed the first few galaxies of this type in the late 1940s (see Chapter 6 for a detailed discussion of the various groups), are the result of an early confusion between different sources that are now known to have similar properties. The main observational difference between Seyfert 1 and Seyfert 2 galaxies is their different optical–ultraviolet (UV) spectra. Seyfert 1 galaxies show strong, very broad (2000–10,000 km s^{-1} if interpreted as Doppler broadening) permitted and semiforbidden emission lines, whereas the broadest lines in Seyfert 2 galaxies have widths that do not exceed $\sim$1200 km s^{-1}. Such differences are now interpreted as arising from different viewing angles to the centers of such sources and from a large amount of obscuration along the line of sight. The common nomenclature used throughout this book is *type-I AGNs* for those sources with unobscured lines of sight to their centers and *type-II AGNs* for objects with heavy obscuration along the line of sight that extincts basically all the optical–UV radiation from the inner parsec (pc).

Another example related to the nature of various types of AGNs was the tendency to separate AGNs according to their luminosity. The name "Seyfert galaxies" was reserved for the lower-luminosity, mostly low-redshift AGNs, whereas QSOs were considered to be the more luminous members of the family. In fact, the dividing line between Seyfert galaxies and quasars was never defined properly and is not very precise; some researchers suggest that the line should be drawn at about $L_{bol} = 10^{45}$ erg s^{-1}, where L_{bol} is the bolometric luminosity of the central source. Others prefer a redshift-based division, for example, in some papers, all AGNs with $z \geq 0.2$ are considered quasars.

Several other names have been proposed over the years: "N-galaxies," "broad-line radio galaxies" (BLRGs), "narrow-line radio galaxies" (NLRGs), "narrow emission-line X-ray galaxies" (NLXGs), "BL-Lac objects," "optically violently variable QSOs" (OVVs), and "low-ionization nuclear emission-line regions" (LINERs), among others. The preference in this book is to use the generic name "AGNs" and to distinguish objects by their basic physical properties such as L_{bol}, the level of ionization of the line-emitting gas, the width of emission and/or absorption lines, and the intensity of the nonthermal radiation source. A detailed discussion of the various subgroups, and general ways of classification, is given in Chapter 6.

The definition of nuclear activity in galaxies can be based either on the physical mechanism involved or on the observational signature of this activity. The physical definition is simple: an extragalactic object is considered to be an AGN if it contains a massive accreting BH in its center. The observational classification is, in many cases, not so clear because of observational limitations, because of source obscuration, and because the term *activity* covers many orders of magnitude in accretion rate. The definition adopted here is purely observational – an object is classified as an AGN if at least one of the following is fulfilled:

1. It contains a compact nuclear region emitting significantly beyond what is expected from stellar processes typical of this type of galaxy.
2. It shows the clear signature of a nonstellar continuum emitting process in its center.
3. Its spectrum contains strong emission lines with line ratios that are typical of excitation by a nonstellar radiation field.
4. It shows line and/or continuum variations.

This very broad definition extends the limits of the AGN phenomenon to objects that are not considered by all researchers to be part of this family. Moreover, given this definition, an object can leave the AGN family, and new members can be born. The former happens when the luminosity drops below a certain limit; the later occurs when a nonactive source undergoes a sudden burst of activity. The current physical model explains such cases as reflecting large changes in the accretion rate onto the BH. The subclassification of AGNs into various groups is directly related to one of the preceding four indicators. For example, radio galaxies are AGNs because of point 2, LINERs are AGNs because of point 3, and blazars are AGNs because of point 4. All these subgroups, and others, are described in Chapter 6.

1.1. AGNs across the electromagnetic spectrum

The characteristic spectral signatures of AGNs are easily distinguished in several wavelength bands. It is customary to refer to the spectral energy distribution (SED) and describe it in terms of the monochromatic luminosity per unit frequency (L_ν erg s^{-1} Hz^{-1}), per unit energy (L_E erg s^{-1} erg^{-1}), or per unit wavelength (L_λ erg s^{-1} Å^{-1}). The equivalent monochromatic fluxes (F_ν, F_E, or F_λ) contain an additional unit of cm^{-2} and are used to describe the observed properties. Photon flux, typically used by X-ray astronomers, is discussed later. The conversion between frequency and wavelength is obtained from simple energy conservation considerations:

$$L_\nu d\nu = L_\lambda d\lambda. \tag{1.1}$$

The SED of many AGNs can be described, over a limited energy range, as

$$L_\nu \propto \nu^{-\alpha} \tag{1.2}$$

or

$$L_\lambda \propto \lambda^{-\beta}, \tag{1.3}$$

where α is the frequency spectral index, β is the wavelength spectral index, and $\beta = 2 - \alpha$. For example, the observed 1200–6000 Å continuum of many luminous AGNs is described, adequately, by $F_\nu \propto \nu^{-0.5}$ or $F_\lambda \propto \lambda^{-1.5}$. This single power-law approximation clearly fails for wavelengths below 1200 Å or above $\sim$6000 Å. Another example is the 2–20 keV range, where, to a good approximation, $F_E \propto E^{-0.9}$ for a large number of type-I and type-II AGNs. Examples of various SEDs covering the entire range from radio to X-ray frequencies are shown in Figure 1.1.

1.1.1. Optical–UV observations of AGNs

Optical images of luminous type-I AGNs show clear signatures of pointlike central sources with excess emission over the surrounding stellar background of their host galaxy. The nonstellar origin of these sources is determined by their SED shape and by the absence of strong stellar absorption lines. Type-II AGNs do not show such excess.

The luminosity of the nuclear, nonstellar source relative to the host galaxy luminosity can vary by several orders of magnitude. In particular, many AGNs in the local universe are much fainter than their hosts, and the stellar emission

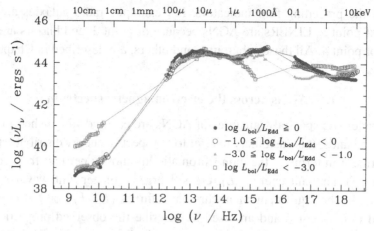

Figure 1.1. Broadband spectral energy distributions (SEDs) for various types of AGNs (from Ho, 2008; reproduced by permission of ARAA). The SEDs are normalized and do not reflect the very large range in intrinsic luminosity between different objects and different redshifts.

can dominate their total light. For example, the V-band luminosity of a high-stellar-mass AGN host can approach 10^{44} erg s^{-1}, a luminosity that far exceeds the luminosity of many local type-I AGNs. This must be taken into account when evaluating AGN spectra obtained with large-entrance-aperture instruments. The relative AGN luminosity increases with decreasing wavelength, and contamination by stellar light is not a major problem at UV wavelengths.

The optical–UV spectra shown in Figures 1.2 and 1.3 represent typical spectra of high-ionization luminous type-I and type-II AGNs. The added "high-ionization" is

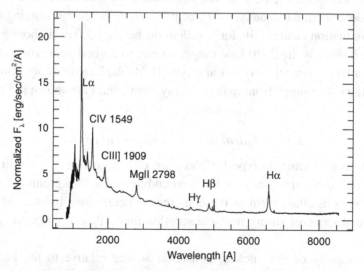

Figure 1.2. The average optical–UV SED of several thousand high-luminosity type I AGNs (adapted from data in Vanden Berk, 2001).

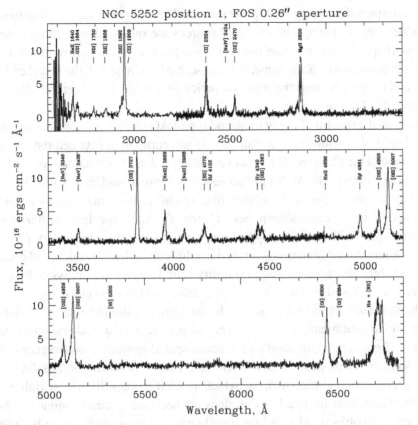

Figure 1.3. The spectrum of the low-luminosity, low-redshift type-II AGN NGC 5252 (courtesy of Zlatan Tsvetanov).

needed to distinguish such sources from low-ionization type-I and type-II sources described later. The type-I spectrum is a composite composed of several thousand spectra of different redshift AGNs. This is done to illustrate the entire rest wavelength range of 900–7000 Å using only ground-based observations. The data used to obtain this composite at $\lambda > 5000$ Å are based on spectra of lower luminosity, low-redshift objects, and the SED at those wavelengths is affected by host galaxy contamination. The type-II spectrum covers a similar range, but this time, the spectrum is a combination of a ground-based optical spectrum with a spaceborne (Hubble Space Telescope; HST) UV spectrum.

The striking differences between the high-ionization type-I and type-II spectra, which were the reason for the early classification into Seyfert 1 and Seyfert 2 galaxies, are the shape and width of the strongest emission lines. Type-II AGNs show only narrow emission lines with typical full-width-at-half-maximum (FWHM) of 400–800 km s^{-1}. In type-I spectra, all the permitted line profiles, and a few semi-forbidden line profiles, indicate large gas velocities, up to 5000–10,000 km s^{-1}

when interpreted as owing to Doppler motion. The line ratios and line widths of the forbidden lines in the spectra of type-I sources are very similar to those observed in type-II spectra and indicate that the basic physics in the narrow line-emitting region of both classes is the same. As is described in Chapter 7, the broad emission lines can be used to map the gas kinematics very close to the central BH and to measure the BH mass.

Study of the spectra of many thousand type-I AGNs shows a considerable range in optical–UV continuum slope but little if any correlation between the slope and L_{bol}. Some of the observed differences are attributed to a small amount of reddening in the host galaxy of the AGN or other sources of foreground dust. More discussion about the origins of the SED and the effect of dust is given in Chapters 7 and 9.

Although the spectra shown here clearly illustrate the large differences in emission-line widths between type-I and type-II sources, a cautionary note is in order. Observational limitations can make it difficult to detect weak broad emission lines. Slightly obscured or low-luminosity type-I AGNs are occasionally classified as type II based on their stellarlike continua and narrow emission lines. This can be the result of reddening of the broad wings of the Hβ line or a relatively strong stellar continuum, especially in large-aperture, low-spatial-resolution observations. Higher signal-to-noise (S/N), better-spatial-resolution observations of the same sources reveal, in some cases, very broad wings in one or more Balmer lines.

The term *broad emission lines*, which is used to describe the permitted and semiforbidden lines in type-I AGNs, does not necessarily mean similar widths for all lines in all objects. The various broad emission lines show typically different widths, and in general, the width reflects the level of ionization of the gas, the source luminosity, and the mass of the central BH. Historically, it was found that broad emission lines in some type-I AGNs are narrower than narrow emission lines in type-II sources. A well-known example is the subgroup of narrow-line Seyfert 1 galaxies (NLS1s).[1] This class of objects was historically defined as those Seyfert 1 galaxies with FWHM(Hβ) < 2000 km s^{-1}. In many such sources, FWHM(Hβ) < 1000 km s^{-1}, similar to the width of Hβ in many type-II AGNs. Evidently, the distinction between type-I and type-II sources requires other criteria, such as the presence of a nonstellar continuum; strengths of emission lines typical of type-I sources such as FeII emission lines; or the presence of a strong, unobscured X-ray continuum. There are also differences in the shape and even the velocity of the same line in different objects. Some examples are shown in Figure 1.4, and more discussion is given in Chapters 5, 6, and 7.

A similar remark should be made about the width of the narrow emission lines. For example, the width of the strong [O III] λ5007 line can depend on the mass

---

[1] Because of these, some authors chose also to use the name "broad-line Seyfert 1 galaxies" (BLS1s) to distinguish them from the NLS1s.

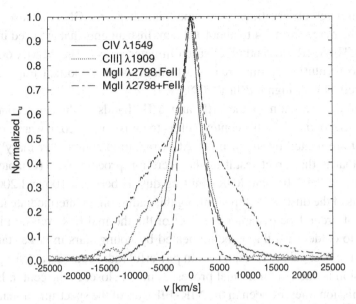

Figure 1.4. Comparison of different broad-line profiles in a typical type-I AGN.

of the central BH (or, more accurately, the mass of the bulge; see Chapter 8). Thus [O III] $\lambda5007$ lines with FWHM ≥ 1000 km s^{-1} are commonly observed in high-redshift, large-M_{BH}, large-L_{bol} AGNs.

1.1.2. Infrared–submillimeter observations of AGNs

The earlier infrared (IR) observations of AGNs provided broadband photometry in the near-IR (NIR) bands, J ($\sim$1.2 μm), H ($\sim$1.6 μm), K ($\sim$2.2 μm), and L ($\sim$3.5 μm). Advances in building IR detectors and launch of IR-dedicated experiments allowed the extension of the measurements in wavelength and spectral resolution. Intermediate-resolution ($E/\Delta E = 1000$–3000) J, H, and K spectroscopy of AGNs is now a standard procedure. Ground-based N-band ($\sim$10 μm) imaging is commonly performed, and mid-IR (MIR) spectroscopy, mostly by ISO (launched in the 1990s) and Spitzer (launched in 2003), have provided good-quality spectra of hundreds of sources, some at redshift as high as 3.

The far-IR (FIR) band of thousands of AGNs has been observed by IRAS, with limited spatial resolution, and by Spitzer, with much improved resolution. The 2009 launch of Herschel is the most recent development in this area. Broadband images with much improved spatial resolution are now available between 70 and 500 μm. Systematic surveys have already produced high-quality photometry of hundreds of AGNs and their host galaxies, up to redshift of 5 and beyond. Lower-sensitivity, high-resolution spectroscopy over the FIR range is also provided by the Herschel instruments.

Submillimeter observations of a small number of AGNs are also available, covering the range from 0.4 to about 1.2 μm. Instruments that are used in this area include SCUBA, IRAM, and other submillimeter arrays. The ALMA observatory is likely to revolutionize this area of astronomy with important implications for local as well as very high redshift AGNs.

Most of the emission in the NIR and MIR bands is due to secondary dust emission. "Secondary" in this context refers to emission by cold, warm, or hot dust grains that are heated by the primary AGN radiation source. "Primary" refers to radiation that is the direct result of the accretion process itself (Chapter 4). The temperature of the NIR- and MIR-emitting dust is between 100 and 2000 K. The dimensions of the dusty structure emitting this radiation, in intermediate luminosity AGNs, is of order 1 pc (Chapter 7). Most of the thermal FIR emission is thought to be due to colder dust that is being heated by young stars in large star-forming regions in the host galaxy (Chapter 8). In powerful radio sources, at least part of the FIR emission is due to nonthermal processes much closer to the center. Broad and narrow emission lines are seen in the NIR–FIR part of the spectrum of many AGNs. They are thought to originate in the broad- and narrow-line regions discussed in Chapter 7.

Figure 1.5 shows a composite 0.3–30 μm spectrum of intermediate-luminosity type-I AGNs. The emission longword of 1 μm is due primarily to secondary radiation from dust. The dip at 1 μm is due to the decline of the disk-produced continuum (Chapter 4) on the short-wavelength side and the rise of the emission due to hot dust on the other side.

1.1.3. X-ray observations of AGNs

X-ray images of AGNs are usually not very interesting; a point source at all X-ray energies in type-I sources and a point source in hard X-rays only in type-II AGNs. Low-resolution X-ray spectra of AGNs are available since the late 1970s. They cover the energy range from about 0.5 keV to 10 keV with a spectral resolution typical of proportional counters and CCD detectors ($\Delta E \sim 100$ eV). Using optical band terminology, these can be described as broad- or intermediate-band photometry. The situation is somewhat improved at higher energies, close to the strong 6.4 keV iron Kα line, where the resolution approaches that of low-dispersion optical spectroscopy. The Chandra and XMM-Newton missions that were launched in 1999 improved this situation dramatically by providing grating spectroscopy of nearby AGNs. The resolution of these instruments, below about 1 keV, is of order $E/\Delta E \simeq 1000$. This has revolutionized X-ray studies of AGNs and resulted in the identification of hundreds of previously unobserved emission and absorption lines. Present-day X-ray instruments like Suzaku and Swift/BAT extend the low-resolution observations to 100 keV and even beyond.

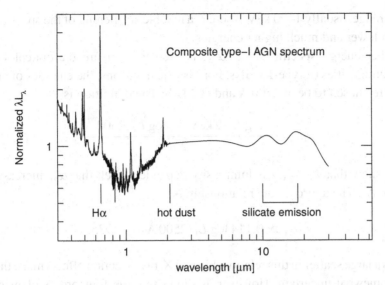

Figure 1.5. A composite spectrum of type-I AGNs covering the range 0.3–40 μm. The observations were obtained by several ground-based telescopes and Spitzer and were normalized to represent a typical intermediate-luminosity source. Note the dip at around 1 μm and the two silicate emission features around 10 and 18 μm.

The soft X-ray spectrum of many type-I AGNs is dominated by a plethora of narrow absorption lines superimposed on a strong X-ray continuum. This must represent material along the line of sight to the source. Narrow emission lines are often associated with the strongest absorption lines. X-ray spectroscopy of type-II AGNs, those with an obscured soft X-ray continuum, shows the narrow emission lines more clearly because of the attenuated central continuum. A common feature near 6.4 keV is the iron Kα line. In type-I AGN, this line is relatively weak and very broad with equivalent width (EW) of 100 eV or less. In type-II sources that are totally obscured at 6.4 keV (line-of-sight column density of $N_H = 10^{24}$ cm^{-2} or larger), the EW is much larger, 1–2 keV, and the line is narrow and barely resolved. The issue of very broad, relativistic iron-K lines is discussed in Chapter 7.

A single power law of the type $L_\nu \propto \nu^{-\alpha_X}$ fits well the intrinsic spectrum of many type-I and type-II AGNs over the energy range 0.2–20 keV. In many cases, $\alpha_X \sim 1$, but there is a clear tendency for larger α_X in objects with narrower broad emission lines. A clear example is the X-ray spectrum of NLS1s defined earlier as those type-I sources with FWHM(Hβ) < 2000 km s^{-1}.

X-ray astronomers prefer, in many cases, the use of "photon index" over "energy index." This index, Γ, is defined such that $N(E) \propto E^{-\Gamma}$, where $N(E)$ is the number of emitted photons per unit time and energy. Obviously $\Gamma = \alpha_X + 1$. The X-ray spectral index is changing with energy, and the preceding slope corresponds to a limited range, usually 1–20 keV. There is a notable steepening of the spectrum both at much lower and much higher energies.

Another energy spectral slope, α_{ox}, is used to compare the optical–UV and X-ray luminosities of type-I AGNs. For historical reasons, the energies of comparison were chosen to be at 2500 Å and at 2 keV. The definition is

$$\alpha_{ox} = -\frac{\log L_\nu(2\,\mathrm{keV}) - \log L_\nu(2500\,\text{Å})}{2.605}. \tag{1.4}$$

Studies show that L_X/L_{UV} is luminosity dependent such that α_{ox} increases with $L_\nu(2500\,\text{Å})$. The approximate relationship is

$$\alpha_{ox} \simeq 0.114 \log L_\nu(2500\,\text{Å}) - 1.975. \tag{1.5}$$

There is a large scatter in this relationship and X-ray selection effects make the exact slope somewhat uncertain. However, it shows that the four orders of magnitude difference between low- and high-luminosity AGNs correspond to about a factor of 500 in L_X/L_{UV}. Thus the X-ray luminosity of the most luminous AGNs is only a small fraction of their total (bolometric) luminosity.

1.1.4. Radio observations of AGNs

The discovery of radio galaxies preceded the optical discovery of AGNs. It goes back to the late 1940s and the early 1950s (except, of course, for the famous paper by Seyfert from 1943). Many of these sources were later shown to have optical–UV spectra that are very similar to the various types of optically discovered AGNs. The main features of many such sources are single- or double-lobe structures with dimensions that can exceed those of the parent galaxy by a large factor and strong radio cores and/or radio jets in some sources that coincide in position with the nucleus of the optical galaxy.

Like optically classified AGNs, there are broad-line radio galaxies (BLRGs), the equivalent of the type-I sources; narrow-line radio galaxies (NLRGs), the spectroscopic equivalent of type-II AGNs; and even weak-line radio galaxies (WLRGs), the equivalent of LINERs (see Chapter 6).

While most AGNs show some radio emission, there seems to be a clear dichotomy in this property. It is therefore customary to define the "radio loudness" parameter, R, which is used to separate radio-loud from radio-quiet AGNs. R is a measure of the ratio of radio (5 GHz) to optical (B-band) monochromatic luminosity,

$$R = \frac{L_\nu(5\,\mathrm{GHz})}{L_\nu(4400\,\text{Å})} = 1.36 \times 10^5 \frac{L(5\,\mathrm{GHz})}{L(4400\,\text{Å})}, \tag{1.6}$$

where $L(5\text{ GHz})$ and $L(4400\text{ Å})$ represent the value of λL_λ at those energies. The dividing line between radio-loud and radio-quiet AGNs is usually set at $R = 10$. Statistics of a large number of AGNs show that about 10 percent of the sources are radio loud, with some indication that the ratio is decreasing with redshift.

Much of the radio emission in radio-loud AGNs originates in a pointlike radio core. The spectrum of such core-dominated radio sources suggests emission by a self-absorbed synchrotron source (Chapter 2). Except for the self-absorption low-frequency part, the spectrum is represented well by a single power law, $F_\nu \propto \nu^{-\alpha_R}$. Sources with $\alpha_R < 0.5$ are usually referred to as *flat-spectrum radio sources*, and those with $\alpha_R > 0.5$ are *steep-spectrum radio sources*. There is a clear connection between the radio structure and the radio spectrum of such sources. Steep-spectrum radio sources show lobe-dominated radio morphology and are also less variables. Flat-spectrum sources have in general higher luminosity cores, larger amplitude variations, and weak or undetected lobes. This dichotomy is interpreted as a dependence on the viewing angle to the core. In steep-spectrum sources, one is looking away from the direction of the nuclear radio jet, and the radio emission is more or less isotropic. In flat-spectrum sources, we are looking at a small angle into the core. The intensity is boosted due to the relativistic motion of the radio-emitting particles, and the variations are amplified. In many cases, there is evidence for superluminal motion in such sources. These and other properties of radio-loud AGNs are further discussed in Chapters 6 and 7.

1.1.5. Gamma-ray observations of AGNs

Observations at energies above 100 keV show that most AGNs are weak high-energy emitters. However, this is not the case for a small fraction (less than 10%) of the population that are strong γ-ray emitters. All these AGNs are also powerful, core-dominated radio-loud sources. They are highly variable, at all wavelength bands, and it is thought that their γ-ray emission is highly collimated. If correct, the apparent high-energy luminosity is large, but the isotropic γ-ray emission is not a large fraction of L_{bol}. The physical mechanisms responsible for the high-energy emission are discussed in Chapters 2 and 7.

Because of the low spatial resolution of present-day γ-ray instruments, all AGNs detected at such energies appear as point sources. The most advanced γ-ray observatory as of 2011, the Fermi Gamma-Ray Space Telescope, allows us to probe their extremely high-energy emission up to about 300 GeV. The typical SED of one such source is shown in Figure 1.6. It is made of two "humps" or peaks, one at energies well below 1 MeV and one at the γ-ray range. The γ-ray part appears as a power-law source with a typical energy spectral index of about 1–1.5. In some sources, there is a clear break in the slope at the highest energies, and a broken power law is a better description of the data.

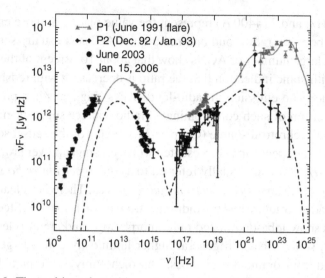

Figure 1.6. The multiepoch, multiwavelength spectrum of the blazar 3C 279 (from Böttcher et al., 2007; reproduced by permission of the AAS) showing the two characteristic peaks at low and high energies and the long-term variations of the source.

1.2. AGN variability

Most AGNs are variable. The variability of type-I sources is observed at all wavelength bands with a tendency for faster variations to occur at higher energies. Hard X-ray variations are also seen in low-obscuration, Compton thin type-II AGNs. The correlation of variability amplitude and variability time scale with the frequency band is not very clear. For example, the time scale for large-amplitude soft X-ray variations is, in many cases, shorter than the one for hard X-ray variations.

Variability in different frequency bands is correlated, which provides important clues about the physics of the central radiation source. For example, X-ray variations are proposed to be the result of instabilities in the disk corona (Chapter 4), while optical continuum variations are thought to be related to instabilities in the more extended, colder and optically thick parts of the same disk. The simplistic assumption is that the X-ray and optical variations will be well correlated. This is true in some but not all objects. A comparison of optical and X-ray light curves in a number of sources shows that in some of these, the optical light curve "leads" the X-ray light curve, whereas in others, one finds the opposite behavior.

The multiband light curves of many type-I AGNs indicate variability time scales, and variability amplitudes, that seem to be inversely correlated with source luminosity. This is clearly seen in the X-ray and optical bands, where most such studies have been conducted. There are some indications that the driver of the variability amplitude is the BH mass. Indeed, AGNs hosting low-mass BHs show larger amplitude variations at all wavelengths compared with objects containing larger

BHs. Like several other claims, it is difficult to separate the BH mass dependence from the luminosity dependence since L_{bol} and M_{BH} are strongly correlated.

The most systematic attempts to define the variability pattern in a large AGN sample are based on a study of many thousands of Sloan Digital Sky Survey (SDSS) AGNs. The assumption is that the variability in a certain photometric band can be described by a "structure function," $V(\Delta t_{i,j})$, which connects the variability, in magnitude, between epochs i and j ($\Delta m_{i,j}$), and the errors on the photometric measurements, σ_i and σ_j, in the following way:

$$V(\Delta t_{i,j}) = |\Delta m_{i,j}| - \sqrt{(\sigma_i^2 + \sigma_j^2)}. \qquad (1.7)$$

The study shows that a power-law structure function,

$$V(\Delta t_{i,j}) = A(\Delta t)^y, \qquad (1.8)$$

where A is the amplitude in magnitude and Δt is the time in years, gives a reasonable approximation for the variability of most sources. Typical numbers are $A = 0.2$ mag. and $y = 0.4$, but the range of properties is large, and y tends to be larger when A is smaller.

Another aspect of the large-scale optical studies is the very weak dependence of the optical continuum slope on the variability amplitude and time scale. In fact, most AGNs retain the same "color" during high and low stages of activity. This uniformity, together with the very similar optical–UV SEDs of high- and low-luminosity AGNs, must be related to the nature of the central power house, probably an optically thick, geometrically thin accretion disk (Chapter 4).

Unlike the optical and X-ray variations, radio variability, and also γ-ray variability, are thought to arise in jet-related processes. Here theory and observation are not always in good agreement. Figure 1.6 is an example of correlated, very large energy band variations in a type-I AGN classified as a blazar (Chapter 6) over a period of 15 years. In this case, and others, the overall shape of the SED is retained, but some bands are more variable than others.

Broad-emission-line variations are seen in all type-I sources that have been monitored for long enough periods of time. One such case is shown in Figure 1.7. These are thought to be the direct result of the UV continuum variations and are discussed in § 7.1. NIR (mostly K-band) variability is seen also in several nearby type-I sources. This is interpreted as the result of the time-dependent heating of the nuclear dust by the variable, primary source of radiation (§ 7.5).

1.3. Discovering AGNs

There are various methods to discover AGNs, some almost as old as the subject itself. The more recent methods have resulted in the largest number of new sources and the most uniform samples. The most important techniques, in terms of the number of newly detected sources, are as follows.

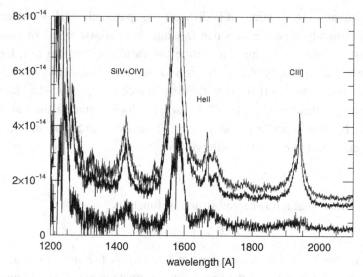

Figure 1.7. UV variability in the spectrum of NGC 5548. Two HST spectra taken 30 days apart are shown on the top, and their difference is shown on the bottom. The flux scale is in units of erg s^{-1} cm^{-2} Å^{-1}. The difference spectrum indicates both broad-emission-line and continuum variations.

1.3.1. Discovery by optical–UV properties

As explained, typical AGN SEDs are different in several ways from stellar SEDs. They cover a broader energy range and do not resemble a single-temperature blackbody. This difference provides a simple and efficient way of discovering AGNs using broadband multicolor photometry. Several color combinations, based on three-band and five-band photometry, are useful in separating AGNs from stars by their color. Earlier methods were based on a UVB photometry in large areas of the sky. This three-band system is useful for discovering low-redshift sources but fails to detect many high-redshift objects because the spectrum gets effectively red and resembles the colors of nearby stars. In addition, even the low-redshift AGNs can be confused with the local population of hot white dwarfs.

Some AGNs are intrinsically red, or reddened by dust, which results in colors that are not very different from those of stars. Moreover, intrinsically blue, high-redshift AGNs are "effectively red" due to the absorption of their short-wavelength radiation by intergalactic gas. This is illustrated in Figure 1.8, which shows the $u-g$ colors of a large number of quasars at higher redshifts observed in the Sloan Digital Sky Survey. The blue AGNs (small values of $u-g$) are seen at all $z < 2.5$, but the color is much redder at higher redshifts. More color combinations and other techniques are needed to find the high-redshift sources.

The more sophisticated five-band systems overcome most of these difficulties. They use a combination of several colors and are very efficient in detecting AGNs

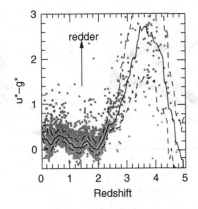

Figure 1.8. The u–g color of a large number of SDSS AGNs with various redshifts (adopted from Richards et al., 2009; reproduced by permission of the AAS).

up to $z \simeq 6$. The SDSS system, which produced the largest number (so far) of AGN candidates, is a color-color system based on five photometric bands: u (0.35 μm), g (0.48 μm), r (0.62 μm), i (0.76 μm), and z (0.91 μm). The system is very efficient for low-redshift AGNs because of the blue color of such sources. The additional bands help to separate AGNs from white dwarfs. The five-band system, with its many color combinations, is also very efficient in discovering high-redshift AGNs. An illustration of the method as adopted by the SDSS survey is shown in Figure 1.9. Such methods have been shown to produce flux-limited AGN samples that are complete to a level of 90 percent and even higher. The total number of type-I AGNs discovered in this way, as of 2011, is more than 100,000.

Type-I AGNs can be directly discovered by their spectrum, because of the large contrast between the strong broad emission and absorption lines and the underlying continuum. This method was used in the 1970s and resulted in several large high-redshift samples.

This spectroscopic method is based on objective prism surveys that produce a small, low dispersion spectrum, instead of a single point, for every object in the field. This was eventually supplemented by high-resolution follow-up spectroscopy. The method is most useful in detecting broad-absorption-line (BAL) AGNs (Chapter 7). In fact, several such surveys gave the erroneous impression that BAL AGNs are more common than they really are because they are difficult to miss in objective prism surveys. The method is very inefficient in discovering type-II AGNs, with their relatively weak emission lines and strong stellar continuum.

1.3.2. Discovery by radio properties

About 10 percent of all AGNs are core-dominated radio-loud sources. This provides an additional way to identify AGNs in deep radio surveys by correlating their radio

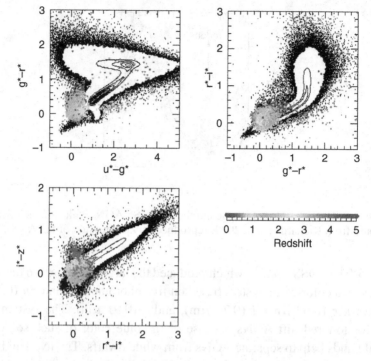

Figure 1.9. Discovering AGNs by their broadband colors. The plots show the AGN location in various color-color diagrams using the SDSS bands u, g, r, i, z. Black points and contours are stars of different types. Colors specify the redshift of the AGN (from Richards et al., 2004; reproduced by permission of the AAS). (See color plate)

and optical positions. Stars are extremely weak radio sources, and hence an optical point source that is also a strong radio source is likely to be a radio-loud AGN. The positional accuracy of optical and radio telescopes is one arcsec or better, and there is hardly any problem in verifying that the radio and optical emitters are one and the same source. Most of the early AGN samples were discovered in this way. A well-known example is the 3C radio sample, which includes some of the most powerful radio-loud, early-discovered AGNs such as 3C 48 and 3C 273.

1.3.3. Discovery by X-ray properties

Almost all AGN are strong X-ray emitters. This property can be used to discover AGNs by conducting deep X-ray surveys. An example of a sample that resulted in the detection of numerous new AGNs is the ROSAT all-sky survey. The most sensitive deepest X-ray surveys tend to pick bright soft X-ray sources with strong 0.5–2 keV emission. Type-II AGN with obscuring column densities of 10^{22} cm^{-2} or larger are more difficult to detect. Recent (since 1999) deep surveys are those conducted with Chandra and XMM-Newton. The Chandra surveys are extremely deep because of the superb resolution of this instrument (about 1 arcsec). Unlike the low-energy ROSAT survey, both Chandra and XMM-Newton can observe at higher energies, up to about 10 keV. However, these missions have only covered a small fraction of the sky. Recent hard X-ray all-sky surveys include those from Swift Burst Alert Telescope (BAT) and INTEGRAL missions at energies from 15 to 150 keV. The amount of obscuration at high energies is much smaller, which results in X-ray discovery of many type-II AGNs. Needless to say, optical follow-up spectroscopy is needed to confirm those detections.

X-ray observations are not very efficient in discovering very high redshift AGN because of the limited sensitivity of the X-ray instruments and the sharp drop of X-ray luminosity of such sources (Equation 1.5).

1.3.4. Discovery by IR properties

Several recent IR surveys, most notably by Spitzer, have been used to search for AGNs using their unique IR properties. This requires the use of at least two IR bands or a combination of one IR band (e.g., the Spitzer 24 μm band) with X-ray or optical observations.

A very important aspect of such techniques is the ability to detect highly obscured (Compton thick; see later) AGNs. A large fraction of such objects, especially at high redshift, do not show detectable X-ray emission, and being type-II sources, their optical spectrum is completely dominated by the host galaxy. Such sources would not be classified as AGNs based on their optical and X-ray continuum properties. However, their mid-IR (MIR) spectrum is dominated by warm dust emission, the result of the heating of the central torus by the central source (§ 7.5). A luminosity ratio like L(24 μm)/L(R), where R is the red optical band, will be much larger in such sources compared with inactive galaxies because the AGN light is heavily obscured at the R-band. Spectroscopic follow-up of such objects can be used to look for the unique emission-line spectrum of the AGN. Indeed, systematic searches in uniformly scanned Spitzer fields reveal a large number of Compton thick AGNs.

1.3.5. Discovery by continuum variability

This is an independent method based on the fact that optical variability is very common in type-I AGNs. The few available variability studies of large AGN samples show that their typical structure function (Equation 1.7) is very different from the structure function of variable stars. Thus, the method can clearly distinguish the two types of sources. Also, it does not suffer from the limitation of the color selection methods in various redshift bands where AGN colors are similar to those of some stars, for example, at $2.5 < z < 3$. The method requires at least two visits per field and follow-up spectroscopy. It is expected that this way of discovering type-I AGNs will become dominant when the Large Synopsis Survey Telescope (LSST) becomes operational around 2017.

All discovery methods discussed so far have their own band-dependent biases. In general, flux-limited samples miss more faint sources in the band in question. The X-ray selection methods discover a larger fraction of the X-ray luminous AGNs, and IR selection is likely to to be biased toward heavily obscured AGNs. Thus the mean optical/X-ray luminosity ratio in X-ray-selected AGN samples will be smaller than the same ratio in optically selected samples. Understanding the selection biases, and combining surveys of different wavelength bands in a proper statistical manner, is a must for revealing the underlying population properties.

1.4. AGN samples and census

AGNs are found by a combination of techniques at different wavelength bands. There are several dedicated AGN surveys and many more general surveys that can be used to find AGNs. The following is an *incomplete* list of some of the surveys and their capabilities:

Radio surveys

The Cambridge 178 MHz radio survey: A northern hemisphere survey with a flux limit of 2 Jy ($1\,\mathrm{Jy} = 10^{-23}\,\mathrm{erg\,s^{-1}\,Hz^{-1}cm^{-2}}$). The results are listed in the third (3C), revised (3CR), and fourth (4C) Cambridge catalogs.

The PKS 408 MHz, 1410 MHD, and 2650 MHz survey: A southern hemisphere survey. The high-frequency part reaches a flux limit of 0.3 Jy.

The NRAO VLA sky survey (NVSS): A northern hemisphere 1.4 GHz survey with a flux limit of 2.5 mJy.

The faint images of the radio sky at 21 cm (FIRST) survey: A 1.4 GHz survey with a flux limit of 1 mJy at higher angular resolution than NVSS.

Optical–UV surveys

The first Byurakan objective prism survey: A 17,000 square degrees survey. Resulted in a list of about 1500 ultraviolet excess Markarian galaxies.

The Tololo survey: A southern hemisphere objective prism survey. Resulted in several hundred high-luminosity, high-redshift AGNs.

The large bright quasar survey (LBQS): Northern hemisphere survey combining Schmidt plates, colors, and objective prism observations ($\sim$1000 bright QSO with $m_B < 19$ mag.).

The Palomar QUEST survey: Northern hemisphere variability (transient) survey. Already proven to be useful in detecting blazars and other large-amplitude variables.

The Palomar Green survey: Color detection of UV excess sources. About 5 percent are type-I AGNs. Provided a list of 87 PG quasars that have been observed at basically all available wavelengths.

The two degree field (2DF) survey: A southern hemisphere survey combining imaging and spectroscopy.

The Sloan digital sky survey (SDSS): A northern hemisphere ~8000 square degree survey combining five-color (ugriz) photometry and follow-up spectroscopy. Resulted in more than 150,000 type-I AGN at all redshifts.

The galaxy evolution explorer (GALEX) survey: A two broadband (1500 and 2300A) UV survey. The shallow, all-sky survey detected more than 50 percent of the type-I SDSS AGNs up to redshift 2.

Infrared surveys

The infrared astronomical satellite (IRAS) survey: Four MIR and FIR (12.5, 25, 60, and 100 μm) all-sky, low-spatial-resolution survey.

The 2 micron all-sky survey (2MASS): Imaging in the JHK bands.

The wide-field infrared survey explorer (WISE) survey: All-sky NIR-MIR survey with four bands at 3.4, 4.6, 12, and 22 μm.

X-ray surveys

ROentgen satellite (ROSAT) all-sky survey: A 5" resolution survey in two bands covering the energy range 0.1–2 keV.

Chandra surveys: Several areas with 1" resolution in the 0.3–9 keV range. Some small fields are very deep (2 and 4 Ms) in the north (CDN) and in the south (CDS).

XMM-Newton surveys: Several 10" resolution 0.3–10 keV surveys across the sky.

Swift burst alert telescope (BAT) survey: A 17' resolution 15–150 keV shallow X-ray survey.

Multiwavelength surveys

These are several preselected areas that were imaged by many ground-based and space-borne instruments including VLA, HST, Spitzer, Herschel, Chandra, XMM, VLT, Subaru, and more. They include, among others, the GOODS north and south fields, the COSMOS field, and the Lockman hole field.

AGN census

What is the the number of detected AGN per unit area on the sky, $N(\text{AGN})$? The answer depends on the wavelength band and the survey depth. Using the large data set provided by SDSS, we find $N(i < 19.1\,\text{mag.}) \sim 20$ type-I AGN with i magnitude less than 19.1 per square degree. Going deeper, we find

$N(i < 21.3 \text{ mag.}) \sim 120$. The number of type-II sources is larger by a factor of 2–3, but the detection of such sources at $z > 0.1$ is incomplete. The largest numbers are obtained from deep X-ray surveys conducted by Chandra. Results from the deepest, 4 Ms survey indicates $N(\text{AGN}) \propto L(2 - 10 \,\text{keV})^{-1.6}$. The current number of X-ray-detected point sources (which are almost exclusively AGN) is approaching 10,000 per square degree.

The fraction of high-ionization AGN among local galaxies is about 2 percent. A more complete census, arranged by BH mass, is given in Chapter 9, where it is shown that for some galaxy and BH mass, the fraction can approach 10 percent. The fraction of low-ionization AGN, like LINERs in the local universe, is larger, ~25 percent.

1.5. AGN terminology

Several decades of study have also produced many terms and abbreviations for the different classes of sources, Seyfert galaxies, quasars, QSOs, blazars, BLRGs, NLRGs, and more. In this book, there are only two categories: "type-I AGN" (those objects showing broad, strong optical–UV emission lines in their spectrum) and "type-II AGN" (those showing prominent narrow emission lines, very faint, if any, broad emission lines, and a large X-ray-obscuring column). Type-I or type-II sources can have low or high ionization lines, strong or weak radio sources, strong or weak X-ray sources, and so on.

The study has also resulted in a dictionary full of names, abbreviations, and acronyms for the various AGN components. Only the most common abbreviations will be used in this book, for example, the narrow-line region (NLR); the extended narrow-line region (ENLR); the broad-line region (BLR), which is occasionally referred to also as the broad-emission-line region (BELR); the broad absorption line (BAL); the narrow absorption line (NAL); the highly ionized gas (HIG); and the black hole (BH). More complete descriptions, based on physical properties, are given in Chapters 6 and 7.

1.6. Useful cosmological relationships

We follow standard astronomical conventions and measure distances in parsec (pc), kilo-parsec (kpc) and Mega parsec (Mpc). Most of the numbers used in this book are based on a Λ cold dark matter (ΛCDM) cosmology and the present somewhat rounded values of the cosmological parameters, $\Omega_m \simeq 0.3$, $\Omega_\Lambda \simeq 0.7$, and $H_0 = 70$ km s^{-1} Mpc^{-1}. In the following, we assign the latter "e" (emitted) to quantities in the source frame and the latter "o" (observed) to those in the local observer's frame.

The redshift of a source is defined by

$$z = \frac{\lambda_o - \lambda_e}{\lambda_e}. \tag{1.9}$$

A nonzero redshift may be the result of random velocity, the expansion of the universe, or both. The relativistic Doppler shift is

$$1 + z = \left[\frac{1 + v/c}{1 - v/c}\right]^{1/2}, \tag{1.10}$$

which, for $v \ll c$, gives

$$z \simeq \frac{\Delta v}{c} = \frac{v_2 - v_1}{c}. \tag{1.11}$$

Known cosmological relationships are used to convert from monochromatic luminosity at the rest frame of the source, L_ν (in erg s^{-1} Hz^{-1}), to observed flux, F_ν (in erg s^{-1} Hz^{-1}cm^{-2}), using the luminosity distance D_L. Assume

$$F = \int_{\nu_1/(1+z)}^{\nu_2/(1+z)} F_\nu d\nu \quad L = \int_{\nu_1}^{\nu_2} L_\nu d\nu; \tag{1.12}$$

then D_L satisfies

$$F = \frac{L}{4\pi D_L^2}. \tag{1.13}$$

The luminosity distance D_L depends on the photon path and thus on the curvature of space-time. For an empty universe with no cosmological constant,

$$D_L = \frac{cz}{H_0}\left(1 + \frac{z}{2}\right), \tag{1.14}$$

and for a matter-dominated universe with no cosmological constant,

$$D_L = \frac{c}{H_0 q_0^2}\left\{q_0 z + (q_0 - 1)\left[(2q_0 z + 1)^{1/2} - 1\right]\right\}, \tag{1.15}$$

where $q_0 = \Omega_m/2$ is the present value of the deceleration parameter.

For the currently preferred cosmology, there is no simple analytical form for the luminosity distance and the calculation involves an integral that contains both Ω_m and Ω_Λ. The angular diameter distance is obtained from D_L via

$$D_\Theta = \frac{D_L}{(1 + z)^2}. \tag{1.16}$$

The conversion to magnitudes depends on the wavelength and the color system. A practical and useful system is the AB magnitude system, which is designed to give $m_{AB} = V$ for a flat spectrum source. The constant in this system is defined by

$$m_{AB} = -2.5 \log F_\nu - 48.60, \tag{1.17}$$

where F_ν is the observed monochromatic flux. The absolute magnitude is then

$$M_{AB} = m_{AB} - 5 \log D_L + 5. \tag{1.18}$$

The conversion to the commonly used UBVR colors involves some constants, for example, $B = B_{AB} + 0.16$, $V = V_{AB} + 0.044$, $R = R_{AB} - 0.055$.

An equivalent method, based on F_λ, provides the following wavelength-based magnitudes:

$$m_\lambda = -2.5 \log F_\lambda - 21.175. \tag{1.19}$$

The two methods can differ substantially, by different amounts at different wavelengths, depending on the shape of the SED. For example, if we use L_{5100} to specify the value of λL_λ at $5100\,\text{Å}$, we find that for an AGN at $z = 1$, with $L_{5100} = 10^{46}$ erg s^{-1}, and a standard $L_\nu \propto \nu^{-0.5}$ SED, the observed quantities at $5100(1 + z)\text{Å}$ are $m_{10200} = 18.1$ mag. and $m_{AB,10200} = 16.82$ mag. The values at observed wavelength $2000(1 + z)\text{Å}$ are $m_{4000} = 16.57$ mag. and $m_{AB,4000} = 17.32$ mag.

1.7. Further reading

General observations and SEDs: There are thousands of papers describing AGN observations of different types. Comprehensive reviews on observations at all redshifts until 1997 can be found in Peterson (1997) and Robson (1996). Reviews and references for more recent observations of high-redshift AGN are found in Fan (2006) and Fan et al. (2006). Optical–UV SEDs of type-I AGN are described in Vanden Berk (2001), Shang (2011), and Richards (2006). Broader-band, low-resolution SEDs are summarized in Ho (2008). References to Herschel observations of AGN and their host galaxies, with connection to submillimeter galaxies, can be found in Bonfield et al. (2011), Santini et al. (2012), and Rosario et al. (2012). For Fermi observations, see catalog and references in Ackermann et al. (2011).

Discovery methods: Older techniques, including color-based methods, the objective prism method, and radio methods, are described in Peterson (1997) and Robson (1996). More recent five-color selections of SDSS and 2dF AGN are described in Richards et al. (2004) and Croom et al. (2008).

AGN variability: See Peterson (1997), Robson (1996), Villforth et al. (2010), McHardy (2010), Schmidt et al. (2012), and references therein.

2

Nonthermal radiation processes

Much of the electromagnetic radiation emitted by AGNs is very different from a simple blackbody emission or a stellar radiation source. The general name adopted here for such processes is *nonstellar emission*, but the term *nonthermal emission* is commonly used to describe such sources. There are several types of nonstellar radiation processes, and this chapter addresses the ones that are more relevant to AGNs and to the physical conditions in the vicinity of a massive BH.

2.1. Basic radiative transfer

Describing the interaction of radiation with matter requires the use of three basic quantities: the first is the specific intensity I_ν, which gives the local flux per unit time, frequency, area, and solid angle everywhere in the source. The second quantity is the monochromatic absorption cross section, κ_ν (cm^{-1}), which combines all loss (absorption and scattering) processes. The third quantity is the volume emission coefficient, j_ν, which gives the locally emitted flux per unit volume, time, frequency, and solid angle. The three are combined into the equation of radiative transfer,

$$\frac{dI_\nu}{ds} = -\kappa_\nu I_\nu + j_\nu, \tag{2.1}$$

where ds is a path length interval. The first term on the right in this equation describes the radiation loss due to absorption, and the second gives the radiation gain due to local emission processes.

It is customary to define the optical depth element, $d\tau_\nu = \kappa_\nu ds$, and to divide the two sides of the equation by κ_ν to obtain

$$\frac{dI_\nu}{d\tau_\nu} = -I_\nu + S_\nu, \tag{2.2}$$

where

$$S_v = j_v/\kappa_v \tag{2.3}$$

is the *source function*.

The formal solution of the equation of transfer depends on geometry. For a slab of thickness τ_v in a direction perpendicular to the slab, it is

$$I_v(\tau_v) = I_v(0)e^{-\tau_v} + \int_0^{\tau_v} e^{-(\tau_v - t)} S_v(t) dt. \tag{2.4}$$

For any other direction θ, both τ_v and dt must be divided by $\cos\theta$.

The general equation of radiative transfer is difficult to solve and requires numerical techniques. However, there are simple cases in which the solution is straightforward. In particular, the case of a slab and a constant source function that is independent of τ_v allows a direct integration and gives the following solution for Equation 2.4:

$$I_v = I_v(0)e^{-\tau_v} + S_v(1 - e^{-\tau_v}). \tag{2.5}$$

For an opaque source in full thermodynamical equilibrium (TE), the optical depth is large, and both I_v and S_v approach the Plank function

$$B_v(T) = \frac{2hv^3/c^2}{\exp(hv/kT) - 1}. \tag{2.6}$$

This chapter describes several continuum emission processes that operate in AGN gas. The treatment of these processes is relatively simple, and the approximate solutions of the radiative transfer equation are adequate, in most cases. The situation is different when dealing with optically thick atomic transitions. These require additional approximations and are described in Chapter 5.

2.2. Synchrotron radiation

2.2.1. Emission by a single electron in a magnetic field

Consider an electron of energy E that is moving in a uniform magnetic field B of energy density $u_B = B^2/8\pi$. The energy loss rate, $-dE/dt$, which is also the power emitted by the electron, P, is given by

$$P = 2\sigma_T c\gamma^2 \beta^2 u_B \sin^2\alpha, \tag{2.7}$$

where σ_T is the Thomson cross section, c is the speed of light, $\gamma = E/mc^2$ is the Lorentz factor, and $\beta = v/c$, where v is the speed of the electron. The angular term $\sin^2\alpha$ reflects the direction of motion, where α is the pitch angle between the direction of motion and the magnetic field. Averaging over isotropic pitch angles gives

$$\overline{P} = (4/3)\sigma_T c \gamma^2 \beta^2 u_B. \tag{2.8}$$

The radiation emitted by a single electron is beamed in the direction of motion. The spectral energy distribution (SED) of this radiation is obtained by considering the gyro frequency of the electrons around the field lines ($\omega_B = eB/\gamma m_e c$) and the mean interval between pulses ($2\pi/\omega_B$). The calculation of the pulse width is obtained by considering the relativistic time transformation between the electron frame and the observer frame. This involves an additional factor of γ^2. Thus the pulse width is proportional to γ^{-3} or, expressed with the Larmor angular frequency, $\omega_L = eB/m_e c$ (which differ from ω_B by a factor of γ), to γ^{-2}. Fourier transforming these expressions gives the mean emitted spectrum of a single electron, $\overline{P_\nu}(\gamma)$, which peaks at a frequency near $\gamma^2 \omega_L$.

2.2.2. *Synchrotron emission by a power-law distribution of electrons*

Assume now a collection of electrons with an energy distribution $n(\gamma)d\gamma$ that gives the number of electrons per unit volume with γ in the range $\gamma - (\gamma + d\gamma)$. The emission coefficient due to the electrons is obtained by summing $\overline{P_\nu}(\gamma)$ over all energies:

$$j_\nu = \frac{1}{4\pi} \int_1^\infty \overline{P_\nu}(\gamma) n(\gamma) d\gamma. \tag{2.9}$$

There is no general analytical solution to this expression since $n(\gamma)$ can take various different forms. However, there are several cases of interest where $n(\gamma)$ can be presented as a power law in energy:

$$n(\gamma)d\gamma = n_0 \gamma^{-p} d\gamma. \tag{2.10}$$

The additional assumption that all the radiation peaks around a characteristic frequency, $\gamma^2 \nu_L$, where ν_L is the Larmor frequency,[1] gives the following solution for j_ν:

$$4\pi j_\nu = \frac{2}{3}\sigma_T n_0 u_B \nu_L^{-1} \left(\frac{\nu}{\nu_L}\right)^{-\frac{p-1}{2}}. \tag{2.11}$$

As a practical example, consider $p = 2.5$, which is expected in various important cases, for example, if the particles are accelerated by relativistic shocks. This gives $j_\nu \propto \nu^{-0.75}$. To obtain the monochromatic luminosity of an optically thin medium

---

[1] Note the slight difference from the earlier used expression $\gamma^2 \omega_L$.

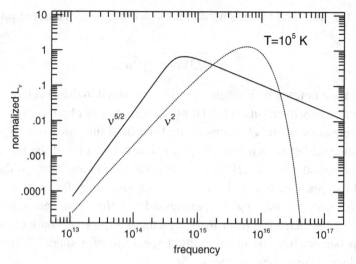

Figure 2.1. A comparison of a synchrotron source with $p = 2.5$ (solid line) and a 10^5 K blackbody source (dotted line).

emitting synchrotron radiation, we must integrate over the volume of the source,

$$L_\nu = \int_V j_\nu dV \propto \nu^{-0.75}. \qquad (2.12)$$

This is similar to the observed continuum slope of many AGNs at radio, optical–UV, and X-ray energies.

2.2.3. Synchrotron self-absorption

The source of fast electrons can be opaque to its own radiation. This results in a significant modification of the emergent spectrum especially at low frequencies, where the opacity is the largest. It can be shown that in this case,

$$\kappa_\nu \propto \nu^{-\frac{p+4}{2}}, \qquad (2.13)$$

that is, the largest absorption is at the lowest frequencies. Using the equation of radiative transfer (Equation 2.2) for a uniform homogeneous medium, and substituting j_ν and κ_ν into S_ν, we get the solution at the large optical depth limit, $I_\nu \propto \nu^{5/2}$, which describes the synchrotron SED at low energies. Note that this function drops faster toward low energies than the low-energy drop of a blackbody spectrum ($I_\nu \propto \nu^2$). The overall shape of such a source is shown, schematically, in Figure 2.1.

Table 2.1. *Synchrotron sources in AGNs*

Source	B (G)	ν (Hz)	γ	t_{cool} (yr)	E (erg)
Extended radio sources	10^{-5}	10^9	10^4	10^7	10^{59}
Radio jets	10^{-3}	10^9	10^3	10^4	10^{57}
Compact jets	10^{-1}	10^9	10^2	10^1	10^{54}
BH magnetosphere	10^4	10^{18}	10^4	10^{-10}	10^{47}

2.2.4. Polarization

Synchrotron radiation is highly linearly polarized. The intrinsic polarization can reach 70 percent. However, what is normally observed is a much smaller level of polarization, typically 3 to 15 percent. This indicates a mixture of the highly polarized synchrotron source with a strong nonpolarized source. For AGNs, especially radio-loud sources, this polarization is clearly observed. There is also a correlation between high-percentage polarization and large-amplitude variations. AGNs showing such properties go under the name *blazars* (Chapter 6). In the NIR–optical–UV spectrum of radio-loud AGNs, the region around 1 μm shows most of the polarization. The percentage polarization seems to drop toward shorter wavelengths, in contrast to what is expected from a pure synchrotron source. This is interpreted as an indication of an additional thermal, nonpolarized source at those wavelengths.

2.2.5. Synchrotron sources in AGNs

It is thought that most of the nonthermal radio emission in AGNs is due to synchrotron emission. There are various ways to classify such radio sources using the slope, $(p - 1)/2$, and the break frequency below which it is optically thick to its own radiation. Table 2.1 gives a summary of the properties of several observed and expected synchrotron sources in AGNs. It includes the typical strength of the magnetic field, B (in gauss), the Lorentz factor, γ, and the total energy generated in the source, E, which is obtained by integrating u_B over the volume of such sources. The table also shows the typical cooling time of the source, t_{cool}, which is a characteristic lifetime defined by

$$t_{cool} = \frac{\gamma m_e c^2}{\overline{P}} \simeq 5 \times 10^8 B^{-2} \gamma^{-1} \text{ sec.} \tag{2.14}$$

2.3. Compton scattering

The interaction between an electron and a beam of photons is described by the classical Compton scattering theory. For stationary or slow electrons, one uses energy and momentum conservation to obtain the relationship between the frequencies of the coming (v') and scattered (v) photons. If $\vec{n}_v$ and $\vec{n}_{v'}$ are unit vectors in the directions of these photons, and $\cos\theta = \vec{n}_v.\vec{n}_{v'}$, we get

$$v = \frac{m_e c^2 v'}{m_e c^2 + hv'(1 - \cos\theta)}. \tag{2.15}$$

For nonrelativistic electrons, the cross section for this process is given by

$$\frac{d\sigma}{d\Omega} = \frac{1}{2} r_e^2 \left[1 + \cos^2\theta\right], \tag{2.16}$$

where $r_e = e^2/m_e c^2$ is the classical electron radius. Integrating over angles gives the Thomson cross section, σ_T. In the high-energy limit, the cross section is replaced by the Klein–Nishina cross section, σ_{K-N}, which is normally expressed using $\epsilon = hv/m_e c^2$. The approach to the low-energy limit is given roughly by

$$\sigma_{K-N} \sim \sigma_T(1 - 2\epsilon), \tag{2.17}$$

and for $\epsilon \gg 1$,

$$\sigma_{K-N} \sim \frac{3}{8} \frac{\sigma_T}{\epsilon} \left[\ln 2\epsilon + \frac{1}{2}\right]. \tag{2.18}$$

2.3.1. Comptonization

The term *Comptonization* refers to the way photons and electrons reach equilibrium. The fractional amount of energy lost by the photon in every scattering is

$$\frac{\Delta v}{v} \simeq -\frac{hv}{m_e c^2} = -\epsilon. \tag{2.19}$$

Consider a distance r from a point source of monochromatic luminosity L_v in an optically thin medium where the electron density is N_e. The flux at this location is $L_v/4\pi r^2$, and the heating due to Compton scattering is

$$H_{CS} = \int \frac{L_v}{4\pi r^2} N_e \sigma_T \left[\frac{hv}{m_e c^2}\right] dv. \tag{2.20}$$

The cooling of the electron gas is the result of *inverse Compton scattering*. Like Compton scattering, this process is a collision between a photon and an electron, except that in this case, the electron has more energy that can be transfered to the radiation field. In this case, the typical gain in the photon energy is a factor of γ^2 larger than the one considered earlier. This factor is obtained by first transforming to the electron's rest frame and then back to the laboratory frame.

It is customary to treat Compton cooling by specifying the fraction x of the electron energy kT, which is transferred to the photon,

$$\left\langle \frac{\Delta v}{v} \right\rangle = x \frac{kT_e}{m_e c^2}, \qquad (2.21)$$

where T_e is the electron temperature. Using this terminology, we can write the cooling term for the electron gas as

$$C_{CS} = \int \frac{L_v}{4\pi r^2} N_e \sigma_T \left[\frac{xkT_e}{m_e c^2} \right] dv. \qquad (2.22)$$

A simple thermodynamical argument suggests that if Compton heating and Compton cooling are the only heating–cooling processes, and if the radiation field is given by the Planck function ($L_v = B_v$), the equilibrium requirement, $H_{CS} = C_{CS}$, gives $x = 4$. Because this is a general relation between a physical process and its inverse, the result must also hold for any radiation field.

The radiation field in luminous AGNs can be very intense, and the energy density of the photons normally exceeds the energy density due to electrons. The requirement $H_{CS} = C_{CS}$ gives, in this case, a Compton equilibrium temperature of

$$T_C = \frac{h\overline{v}}{4k}, \qquad (2.23)$$

where the mean frequency, $\overline{v}$, is defined by integrating over the SED of the source,

$$\overline{v} = \frac{\int v L_v dv}{\int L_v dv}. \qquad (2.24)$$

As an example, consider dilute gas near a strong AGN source whose X-ray spectrum between the frequencies $v_{min} = 1 \text{ keV}/h$ and $v_{max} = 1000 \text{ keV}/h$ is given by $L_v = C v^{-\alpha}$ with $\alpha = 0.9$ (the typical spectral index observed in luminous AGNs). The mean frequency in this case is

$$\overline{v} = \frac{\alpha - 1}{2 - \alpha} \left[\frac{v_{min}}{v_{max}} \right]^{\alpha-1} v_{max}. \qquad (2.25)$$

Substituting for the minimum and maximum frequencies gives $T_C \simeq 10^8$ K. An important conclusion is that, for highly ionized stationary gas whose energy balance is entirely due to Compton heating and cooling, the Compton temperature, which is the equilibrium temperature of the electrons, depends entirely on the SED or, in the preceding example, on the slope (spectral index) of the nonthermal radiation field.

2.3.2. The Compton parameter

The emitted spectrum of thermal and nonthermal radiation sources that are embedded in gas with a thermal distribution of velocities is modified due to Compton and inverse Compton scattering. For high-energy electrons, inverse Compton is the dominant process, and the resulting collisions will up-scatter the photon energy. The emergent spectrum is modified, and its spectral shape will depend on the original shape, the electron temperature, and the Compton depth of the source, which determines the number of scattering before escape.

Consider an initial photon energy of $h\nu_i$ and the case of thermal electrons with temperature T_e such that $h\nu_i \ll 4kT_e$. The scattering of such photons by a fast electron will result in energy gain per scattering (inverse Compton scattering). The photon continues to gain energy, during successive scatterings, as long as $h\nu_i \ll 4kT_e$. If the final photon energy is $h\nu_f$, and the number of scattering is N, we get

$$h\nu_f \simeq h\nu_i \exp\left[N\frac{4kT_e}{m_e c^2}\right]. \tag{2.26}$$

For a medium with Compton depth τ_e, the mean number of scatterings is roughly $\max(\tau_e, \tau_e^2)$. Using this, we can define a Compton parameter y,

$$y = \max(\tau_e, \tau_e^2)\left[\frac{4kT_e}{m_e c^2}\right], \tag{2.27}$$

such that

$$h\nu_f \sim h\nu_i \exp(y). \tag{2.28}$$

The factor $\exp(y)$ is an energy amplification factor. For $y > 1$, we are in the regime of *unsaturated inverse Comptonization*. For $y \gg 1$, the process reaches a limit where the average photon energy equals the electron thermal energy. This is referred to as *saturated Compton scattering*.

2.3.3. Inverse Compton emission

An important example is the case of a source whose spectrum is due to scattering of "soft" photons onto relativistic electrons. Like in previous examples, we consider first the typical energy following a single scattering and then average over the energy distribution of the photons and electrons.

A simple way to estimate the power emitted in the preceding process is to consider a beam of photons with number density n_{ph} and mean energy before scattering $\overline{h\nu_0}$. The energy density of these photons is $n_{\text{ph}}\overline{h\nu_0}$, and the energy flux of photons incident on a stationary electron is $cu_{\text{rad}} = n_{\text{ph}}\overline{h\nu_0}c$. As explained earlier, the mean energy after scattering, $\overline{h\nu}$, is larger than the mean energy before scattering by a factor of order γ^2. In the rest frame of the electron, the process can be considered as a simple Thomson scattering with radiated power given by the classical expression $\overline{P} = \sigma_T c u_{\text{rad}}$. Thus the simple $L_\nu \propto \nu^{-(p-1)/2}$ estimate for the laboratory frame emitted power is $\overline{P} = \gamma^2 \sigma_T c u_{\text{rad}}$. A more accurate derivation of the emitted power must take into account the scattering angle and its transformation between frames. The final expression in this case is

$$\overline{P} = (4/3)\sigma_T c \gamma^2 \beta^2 u_{\text{rad}}, \tag{2.29}$$

which differs from the simple estimate by a factor of order unity.

The expression for the power emitted due to inverse Compton (IC) scattering is basically identical to the power emitted by synchrotron radiation, except that the energy density of the magnetic field, u_B, was replaced by the energy density of the radiation field, u_{rad}. Thus, the mean power of the two processes, assuming they take place in the same volume of space, is simply u_B/u_{rad}. Also, for the same volume of space, the energy distribution of the relativistic electrons is given by the same power-law function used in the synchrotron case, $n(\gamma) \propto \gamma^{-p}$. Thus, we also get a similar dependence of the monochromatic luminosity on the parameter p:

$$L_\nu(\text{IC}) \propto \nu^{-(p-1)/2}. \tag{2.30}$$

2.3.4. Synchrotron self-Compton

In a compact synchrotron source, the emitted photons can be inverse Compton scattered by the relativistic electrons that emit the synchrotron radiation. This gives the photon a big boost in energy. The emergent radiation is referred to as *synchrotron self-Compton* (SSC) emission. The flux emitted by this process can be calculated by integrating over the synchrotron radiation spectrum and the electron velocity distribution. To a good approximation, the resulting spectral index is identical to the spectral index of the synchrotron source.

The synchrotron self-Compton process can repeat itself, in the same source, by additional scattering of the emergent photons, which results in an additional boosting, by a factor γ^2, to the photons. The natural limit for the process is when the scattered photon energy extends into the γ-ray and the condition of $h\nu_\gamma \ll m_e c^2$ (the condition for no Compton recoil of the electron) no longer holds. At this limit, the resulting radiation density decreases dramatically.

2.4. Annihilation and pair production

The observations of γ-ray jets in many AGNs suggest that, under some conditions, the density of high-energy photons is large enough to result in efficient pair production and a concentration of both electrons and positrons in some parts of the central source. Under these conditions, energetic γ-ray photons, with energies much above the rest energy of the electron, can react with lower-energy photons to create electron–positron pairs. Short-time-scale variations of the X-ray spectrum, in the lower-luminosity AGN, indicate extremely small dimensions; γ-ray photons that are associated with the X-ray source would not be able to escape these regions and would create electron–positron pairs. Likely locations where such processes take place are in the corona of the central accretion disk (Chapter 4) or inside the γ-ray jet (Chapter 7).

The basic processes discussed here are pair production and its reverse process, electron–positron annihilation:

$$e^+ + e^- \rightleftharpoons \gamma + \gamma. \tag{2.31}$$

Consider the interaction between a γ-ray photon with frequency ν_γ, above the rest mass frequency of the electron, with an X-ray photon of frequency ν_X below this frequency. Using the notation of unit vectors for the photons, we can write the threshold frequency for pair production as

$$\nu_\gamma = \left(\frac{m_e c^2}{h}\right)^2 \frac{2}{\nu_X(1 - \vec{n}_\gamma.\vec{n}_X)}. \tag{2.32}$$

The $\gamma\gamma$ cross section is given by

$$\sigma_{\gamma\gamma} = \frac{3}{16}\sigma_T(1 - \beta^2)\left[(3 - \beta^4)\ln\left(\frac{1+\beta}{1-\beta}\right) - 2\beta(2 - \beta^2)\right], \tag{2.33}$$

where the value of β for the electron and the positron is measured in the center of momentum frame. The typical value of $\sigma_{\gamma\gamma}$ near threshold is $\sim 0.2\sigma_T$, and it declines with frequency as ν_γ^{-1}.

The size of the radiation source, R, plays an important role in determining the optical depth of the source and hence the probability of pair production taking place. It is convenient to describe this dependence by defining a *compactness parameter* for the γ-ray source, l_γ, using the source size and its luminosity, L_γ. There is an equivalent compactness parameter for the X-ray source, l_X. To understand the definition and meaning of the compactness parameter, assume that the typical γ-ray photon energy is $\epsilon = m_e c^2$ and that the photon number density is

$$N_\gamma = \frac{L_\gamma}{4\pi R^2 c\epsilon}. \tag{2.34}$$

The mean free path of the photons for pair production is

$$\lambda_{\gamma\gamma} = (N_\gamma \sigma_T)^{-1} \tag{2.35}$$

and for unit optical depths, $R \approx \lambda$, which gives

$$\frac{L_\gamma \sigma_T}{4\pi m_e c^3 R} \approx 1. \tag{2.36}$$

This leads to the following expression for the compactness parameter:

$$l_\gamma = \frac{L_\gamma \sigma_T}{4\pi m_e c^3 R}, \tag{2.37}$$

which is equivalent to the pair production optical depth of the source.[2] In principle, l_γ can be measured from the variability time scale of the γ-ray source. In reality, however, this is difficult to measure and is occasionally replaced by l_X and the X-ray variability time scale. When $l_X \gg h\nu_X / m_e c^2$, it will be difficult for the γ-rays to escape the source without creating pairs.

The rate of the inverse process, pair annihilation, in the nonrelativistic limit is independent of temperature and is roughly $0.4 N_e \sigma_T c$ per unit volume, where N_e is the combined electron–positron density. In a steady state, pair production is balanced by annihilation,

$$\frac{m_e c^3 l_\gamma}{4\pi \sigma_T R^2 h\nu_\gamma} \approx 0.4 N_e \sigma_T c, \tag{2.38}$$

where l_γ is the compactness parameter for those γ-ray photons for which the source is optically thick to pair production. This equation can be solved for the mean Thomson depth in the source, τ_T. For large τ_T, the electrons and positrons thermalize because their interaction time is short compared with the annihilation time. In AGN gas, where the conditions allow this thermalization, the temperature of the hot, Compton thick pair plasma can reach 10^9 K. Such gas can contribute to the observed high-energy spectrum. It can up-scatter soft (UV) emitted photons and even produce some free–free electron–positron radiation.

2.5. Bremsstrahlung (free–free) radiation

Finally, we consider free–free radiation, which, formally, is thermal radiation. However, in this case, the spectral shape is very different from that of a blackbody, and hence it is described in this chapter.

The free–free emissivity due to ion i of an element of charge Z whose number density is N_i is given by

$$4\pi j_\nu = 6.8 \times 10^{-38} Z^2 T_e^{-1/2} N_e N_i \bar{g}_{ff}(\nu, T_e, Z) e^{-h\nu/kT_e}, \tag{2.39}$$

where $\bar{g}_{ff}(\nu, T, Z)$ is the velocity-averaged Gaunt factor, which accounts for quantum-mechanical effects. This factor is always of order unity and can

---

[2] In some books, l_γ is defined without the factor 4π.

change slightly with frequency, in particular, at X-ray energies $\overline{g}_{ff} \propto \nu^{-0.1}$. The Bremsstrahlung radiation extends over a large range of energies and resembles, over most of this range, a very flat (small spectral index) power law.

We can integrate the free–free emissivity over frequencies to obtain the total energy per unit volume per second, C_{ff}, where we used the symbol C to indicate that this is also the cooling rate due to free–free emission. The integration gives

$$C_{ff} = 1.42 \times 10^{-27} Z^2 T_e^{1/2} g_{ff} N_e N_i \; \mathrm{erg\, s^{-1} cm^{-3}}, \qquad (2.40)$$

where g_{ff} is now the frequency average of the velocity-averaged Gaunt factor. This is typically in the range 1.1–1.5. In Chapter 5, we return to this cooling rate in connection with the thermal equilibrium of photoionized plasma.

2.6. Further reading

The topics discussed in this chapter are very standard and appear in many textbooks. For detailed descriptions, see Rybicki and Lightman (1979), Krolik (1999), and Blandford (1990).

3

Black holes

3.1. Active and dormant black holes in galactic nuclei

BHs of all sizes are very common in the universe. There are numerous known stellar-sized BHs in our galactic neighborhood, and in several nearby galaxies, with masses in the range 3–30 $M_{\odot}$. These are the remnants of core collapse in type-II supernovae (SNs) with very massive progenitors of at least 20 or perhaps even 30 $M_{\odot}$. Such objects are found in binary systems that are also strong X-ray sources, probably the result of accretion onto the BH. A direct causal connection between the SN explosion and the remnant BH has been established in several cases. A well-known example is the strong X-ray source in M 100 with a location that coincides with that of a 1979 SN explosion (SN 1979C). As of 2011, this is the youngest known BH.

Active supermassive BHs, in galactic centers, have been known since the early discovery of QSOs in the 1960s. However, the idea that most galaxies, especially those with dynamically relaxed bulges, contain dormant supermassive BHs in their centers took much longer to develop. Detailed studies of the stellar velocity field and gas motion in about 60 nearby galaxies suggest the existence of such objects. Moreover, there seems to be a strong correlation between the mass of the bulge, its luminosity, and the mass of the BH. Because the physical sizes of the two masses are very different, and the ratio of masses is very large (about 100–4000; see Chapter 8), it is difficult to find a mechanism that will link the two. Understanding these relationships has become an area of intensive research in astronomy and is discussed later in this book.

Systematic and extensive observations of active supermassive BHs show that such systems must have been present very early in the history of the universe. In fact, observations of AGNs at large redshift show that BHs as large as $10^9 M_{\odot}$ have already been in place at $z \simeq 6$. Some of these sources are accreting matter at a very high rate, close to the theoretical upper limits of such processes (see later). Thus the evolutions of supermassive BHs and their parent galaxies have been linked since very early times, perhaps as early as the time of formation of the first galaxies at redshift of about 10. These ideas are detailed in Chapters 8 and 9.

This chapter describes the characteristics of BHs that depend on their mass and spin. They are common to all astrophysical BHs in the mass range of 1–10^{10} $M_\odot$.

3.2. General black hole properties

BHs are extreme cases of curved space-time and are described by general relativity (GR) or alternative gravitational theories. GR is adopted in this book. According to the current theory, mass, charge, and angular momentum are the only properties black holes can possess. This has become known as the "no hair" theorem of BHs.

It is convenient to describe the basic properties of a BH of mass M using the gravitational radius r_g defined as

$$r_g = \frac{GM}{c^2}.$$
(3.1)

To a good approximation,

$$r_g \simeq 1.5 \times 10^{13} M_8 \text{ cm},$$
(3.2)

where M_8 is the BH mass in units of $10^8 M_\odot$. Here we neglect electrically charged BHs and consider only stationary and rotating (Kerr) BHs.

All properties of stationary (Schwarzschild) BHs can be described by using r_g. Regarding rotating or spinning BHs, it is customary to define two other quantities, the angular momentum of the BH,

$$s \sim I\Omega \simeq M r_g^2 \left(\frac{v}{r}\right) \simeq M r_g c,$$
(3.3)

where Ω is the angular velocity at the horizon, and the specific angular momentum of the BH (angular momentum per unit mass), s/M. A related quantity is the specific angular diameter parameter, α, defined such that

$$s/M \equiv \alpha c.$$
(3.4)

It is convenient to define yet another parameter, a, such that $\alpha = a r_g$ or $s/M = a r_g c$. This recovers the more familiar form of the specific angular momentum and shows that a can take all values between -1 and 1, where the plus and minus signs refer to the direction of rotation. Using this notation, and the maximum allowed value of a, we get the following approximation for the specific angular momentum of the black hole:

$$s/M = r_g c \sim 5 \times 10^{23} M_8 \text{ cm}^2 \text{ s}^{-1}.$$
(3.5)

Several important properties of AGN BHs depend on their spin since this determines the maximum energy that can be extracted from the hole during accretion.

For example, in accretion via a thin disk, the subject of the next chapter, the difference in a between 0 and 1 translates to a factor of about 10 in radiation conversion efficiency. Subsequently, we introduce the exact GR expression for the line element ds^2 and its various solutions that depend on a.

The GR line element near a rotating BH was first derived by Kerr in 1963 (hence the name *Kerr BH*). It is given by

$$ds^2 = \left(1 - \frac{2r_g r}{\Sigma}\right) c^2 dt^2 + \frac{4\alpha r_g r \sin^2 \theta}{\Sigma} dt d\phi - \frac{\Sigma}{\Delta} dr^2$$
$$- \Sigma d\theta^2 - \left(r^2 + \alpha^2 + \frac{2r_g r \alpha^2 \sin^2 \theta}{\Sigma}\right) \sin^2 \theta d\phi^2, \qquad (3.6)$$

where

$$\Sigma = r^2 + \alpha^2 \cos^2 \theta \qquad (3.7)$$

$$\Delta = r^2 - 2r_g r + \alpha^2. \qquad (3.8)$$

This expression can be solved for the location of the event horizon of the hole. The simplest solution is for a case of a stationary hole, where $a = 0$ and $\alpha = 0$. In this case, the event horizon of the BH is given by the Schwarzschild radius,

$$r_s = 2r_g. \qquad (3.9)$$

In the case of $a \neq 0$, setting $\Delta = 0$ gives two solutions for the event horizon,

$$r_\pm = r_g \left[1 \pm (1 - a^2)^{1/2}\right]. \qquad (3.10)$$

For extreme BHs with $a = 1$, the outer radius is at $r_+ = r_g$, and for $a = 0$, it is at $r_+ = 2r_g$. Besides the two event horizons, the Kerr metric also features an additional surface of interest called the *static limit*, given by

$$r_0 = r_g \left[1 + (1 - a^2 \cos^2 \theta)^{1/2}\right]. \qquad (3.11)$$

The region of space between r_0 and r_+ is called the *ergosphere*.

GR enables us to calculate the orbits of particles in the vicinity of stationary and rotating BHs. In particular, it allows us to solve for the location of the innermost stable circular orbit (abbreviated to ISCO), also called the *marginal stability radius*, r_{ms}, beyond which the particle loses its orbital motion and falls directly into the event horizon. For a Schwarzschild BH, $r_{ms} = 6r_g$, and for rotating Kerr BHs with $a > 0$, $1 \leq r_{ms} \leq 6r_g$, where the exact value depends on a.

The value of r_{ms} determines the fraction of the gravitational potential energy that is converted to electromagnetic radiation during the accretion. The smaller r_{ms} is, the larger is this fraction. For thin accretion disks (Chapter 4), r_{ms} is taken as the innermost radius of the disk. Thick accretion disks are different, and closer orbits, so-called *marginally bound orbits*, can be achieved. For example, the marginally bound radius of a Schwarzschild BH is at $4r_g$ compared with the stationary thin-disk value of $6r_g$. The one for a spinning BH with $a = 0.5$ is at $2.914r_g$. Theoretical calculations show that for realistic thin accretion disks, the radiation emitted by the disk and swallowed by the hole produces a counteracting torque that prevents spinning up beyond $a = 0.998$. A BH with this spin is sometimes denoted as *maximally rotating BH*, and the value of a is the *canonical spin-up limit*. Here, again, thick disks allow larger values of a and smaller values of r_{ms}.

We should also consider the case of retrograde accretion, where, in contrast to aligned rotation, the accreted inflowing material and the BH are rotating in opposite directions. In this case, $a < 0$, and the marginally stable orbit is farther away from the black hole. For the minimal value of $a = -1$, this is at $r_{ms} = 9r_g$. Obviously, in the beginning of an accretion event around a spinning BH, the BH spin and the angular momentum of the disk material are not necessarily related and can point in completely different directions.

Given the assumption that r_{ms} is the innermost possible orbit, and using the notation $x = r/r_g$ for the normalized distance from the hole, one can solve for η, the efficiency of converting rest mass to electromagnetic radiation, due to the infall of mass m from infinity to r_{ms}. In general, $\eta = [E(\infty) - E(r)]/mc^2$. For a stable circular orbit at the ISCO, this is the equivalent of the Newtonian term $1/2x$. The increase in BH mass due to the subsequent fall of this mass into the BH is $(1 - \eta)m$.

The efficient parameter η depends only on the BH properties, assuming all the gravitational energy is converted to electromagnetic radiation. There are other cases in which part of this energy is advected into the BH, thus decreasing the efficiency of the radiation process. In such cases, we use a different efficiency factor, ϵ_r, to describe the relationship between m and E_G. Obviously, $\epsilon_r \leq \eta$. Various cases of this type are discussed in § 4.2.

GR calculations show that for an ISCO at a normalized radius x,

$$a = \frac{x^{1/2}}{3} \left[4 - (3x - 2)^{1/2} \right]. \tag{3.12}$$

The equivalent relationship between x and η is

$$\eta = 1 - \left[1 - \frac{2}{3x} \right]^{1/2}. \tag{3.13}$$

For example, in a Schwarzschild BH with $x = 6$, $\eta = 0.057$, while the Newtonian approximation gives $1/12$, which is somewhat larger. More examples showing the values of η and r_{ms} for various values of the parameter a are listed in Table 3.1. They illustrate the overall range in the efficiency of the accretion process, between 0.038 and 0.421.

We can compare the specific angular momentum of the BH to the typical specific angular momentum on galactic scales. For example, if $r = 1$ kpc and the rotational velocity $v = 300$ km s^{-1}, we get $rv \sim 10^{29}$ cm^2 s^{-1}, which is many

Table 3.1. *Properties of Schwarzschild and Kerr BHs*

a	r_{ms}/r_g	η
-1.0	9.0	0.038
0	6.0	0.057
0.1	5.67	0.061
0.5	4.23	0.082
0.9	2.32	0.156
0.998	1.24	0.321
1.0	1.00	0.423

orders of magnitude larger than $r_g c$ given in Equation 3.5. If this gas is to be brought to the vicinity of central BH, then there must be an efficient mechanism that gets rid of its excess angular momentum and enables its inflow into the center on time scales that are typical of galactic evolution. It is thought that on very large physical scales, and long time scales, the BH growth depends on mechanisms such as galaxy collision and mergers, bar instability, and other internal ("secular") processes that are capable of bringing far-away gas into distances of 1–100 pc from the BH. The mechanisms that can overcome the "100 pc barrier" (according to some researchers, the "10 pc barrier") and bring the gas to $10^5\,r_g$ or even closer to the BH are still unknown (Chapter 9).

The situation very close to the BH is different. Gas particles at $r = 10^3$–$10^5\,r_g$ are subjected to different types of forces, including the BH gravitational pull, the self-gravity of the gas, the local radiation and gas pressures, stellar winds, SN explosions, and gas and magnetic viscosity. The motion of such particles depends, again, on their angular momentum. Low-angular-momentum material may be subjected to spherical accretion, provided gravity overcomes radiation and gas pressure. High-angular-momentum gas can fall all the way into the BH via a central accretion disk that provides a viscosity-based mechanism to get rid of the excess angular momentum. The case of spherical accretion is discussed in this chapter, and accretion disks are explained in detail in Chapter 4.

3.3. Accretion onto black holes

3.3.1. The Eddington luminosity

Accretion onto massive objects and the associated release of the binding gravitational energy are important sources of radiation in astrophysics. The process is geometry dependent and can proceed in various different routes. In particular, spherical and nonspherical systems can behave in a very different way. Two fundamental quantities that are related to such processes are the *Eddington luminosity* and the *Eddington accretion rate*.

Assume a central point source with mass M, total luminosity L, and monochromatic luminosity L_ν. Assume also fully ionized gas at a distance r from the source. The radiation pressure force acting on a gas particle is

$$f_{\text{rad}} = \frac{N_e \sigma_T}{4\pi r^2 c} \int_0^\infty L_\nu d\nu = \frac{N_e L \sigma_T}{4\pi r^2 c} L, \tag{3.14}$$

where N_e is the electron density and σ_T is the Thomson cross section. The gravitational force per particle is

$$f_g = \frac{GM\mu m_p N_e}{r^2}, \tag{3.15}$$

where μ is the mean molecular weight (mean number of protons and neutrons per electron; about 1.17 for a fully ionized solar composition gas). Spherical accretion of fully ionized gas onto the central object can proceed as long as $f_g > f_{\text{rad}}$. The limiting requirement for accretion, $f_{\text{rad}} = f_g$, leads to the definition of the Eddington luminosity,

$$L_{\text{Edd}} = \frac{4\pi c G M \mu m_p}{\sigma_T} \simeq 1.5 \times 10^{38} (M/M_\odot) \text{ erg s}^{-1}, \tag{3.16}$$

where the factor 1.5×10^{38} depends on the exact value of μ and was calculated for solar metallicity gas. For a pure hydrogen gas, this factor is about 1.28×10^{38}. The value of L_{Edd} defined in this way is the maximum luminosity allowed for objects that are powered, over a long period of time, by a steady-state accretion. Obviously, the luminosity can exceed this limit for a short duration, for example, immediately after an outburst.

The preceding definition of L_{Edd} takes into account only one source of opacity, Compton scattering, which is appropriate for a fully ionized plasma. More realistic situations may involve partly neutral gas and hence much higher opacity. Here the effective L_{Edd} can be significantly smaller than the value defined in Equation 3.16.

Given the definition of L_{Edd}, and the accretion rate $\dot{M} = L/\eta c^2$, where η is, again, the mass-to-luminosity conversion efficiency, we can define several other useful quantities. The first is the Eddington accretion rate, $\dot{M}_{\text{Edd}}$, which is the accretion rate required to produce a total luminosity of L_{Edd},

$$\dot{M}_{\text{Edd}} = \frac{L_{\text{Edd}}}{\eta c^2} \simeq 3 M_8 \left[\frac{\eta}{0.1} \right]^{-1} M_\odot y^{-1}. \tag{3.17}$$

The second is the Eddington time, t_{Edd}, which gives the typical time associated with this accretion rate,

$$t_{\text{Edd}} = \frac{M}{\dot{M}_{\text{Edd}}} \simeq 4 \times 10^8 \eta \text{ yr.} \tag{3.18}$$

The preceding terminology allows us to express the relative accretion rate, which is the accretion rate per unit mass of the BH, by

$$\frac{L}{L_{Edd}} \propto \frac{\dot{M}}{\dot{M}_{Edd}} \propto \frac{\dot{M}}{M}. \tag{3.19}$$

The normalized Eddington accretion rate,

$$\dot{m} = \frac{\dot{M}}{\dot{M}_{Edd}} = \frac{\dot{M}}{L_{Edd}/\eta c^2} = \frac{\eta \dot{M} c^2}{L_{Edd}}, \tag{3.20}$$

is used, occasionally, to define a scale-free relationship that involves the Eddington luminosity. Unfortunately, the normalized accretion rate is used in an inconsistent way in various works. In some articles, it includes the efficiency factor η (as in this book), whereas in others, η is not included (i.e., $\dot{m} = c^2 \dot{M}/L_{Edd}$). Moreover, in practice (e.g., Chapter 7), we normally use the measured quantity L/L_{Edd} as a replacement for the normalized accretion rate. Obviously, this is not a direct measure of the accretion since it involves η.

We note again that $\dot{M}$ used here and later in the text refers to the mass inflow rate into the BH and is thus the *infall accretion rate*. The BH growth rate is $\dot{M}(1 - \eta)$.

3.3.2. Spherical accretion

The simplest possible case of accretion is spherical accretion, also known as *Bondi accretion*. The accretion in this case starts at a large distance from the center, where the gas is at rest. We then follow the gas radial motion under the combined force of gravity and radiation pressure force. We can make an estimate of the spherical accretion rate of hot gas, with constant temperature and no radiation pressure force, by considering the microphysics in the flow. The sound speed in the gas is

$$v_s = \sqrt{\gamma kT/\mu m_p}, \tag{3.21}$$

where μ is the mean atomic weight per particle (about 0.62 for a solar composition fully ionized gas, not to be confused with the mean molecular weight used earlier with the same symbol) and γ is the adiabatic index of the gas ($\sim$5/3). We can now write a simple mass conservation equation,

$$\dot{M}_{Bondi} = 4\pi \lambda v_s \rho r_A^2, \tag{3.22}$$

where r_A is an accretion radius within which the gravitational potential of the BH dominates over the thermal energy of the surrounding gas and λ is a correction factor of order 0.1 that depends on γ. This is also the location where the sound speed equals the free fall speed and is given by

$$r_A = \frac{2GM}{v_s^2}. \tag{3.23}$$

Detailed calculations show that the optical depth of such gas is very large. For example, if the gas is completely ionized, its electron scattering optical depth is

$$\tau_e \simeq \frac{\dot{M}/\dot{M}_{Edd}}{[r/r_g]^{1/2}}, \tag{3.24}$$

which, for a critical accretion rate of $\dot{M} = \dot{M}_{Edd}$, whose inflow velocity is the free fall velocity, gives a Compton depth of order unity.

The mass-to-radiation efficiency factor in spherical accretion is much smaller than the value of η derived earlier for BHs. The reason is that in such cases, most of the radiation is emitted by two-body collisional processes in the ionized plasma. In low-density gas, such processes are relatively inefficient compared with the release of gravitational energy, and the gas does not have enough time to cool before it is accreted onto the central object. As a result, most of the gravitational energy is advected into the central object, and the radiated luminosity is only a small fraction of the released gravitational energy. This means that $\epsilon_r \ll \eta$.

A good estimate for the luminosity of a spherical accretion event is

$$\frac{L}{L_{Edd}} \simeq 10^{-4} \left[\frac{\dot{M}}{\dot{M}_{Edd}} \right]^2. \tag{3.25}$$

In terms of radiative efficiency, this is equivalent to $\epsilon_r \sim 10^{-4}$.

Although the luminosity and L/L_{Edd} in spherical accretion onto a supermassive BH are very small compared with accretion via a disk, the actual energy produced by the process can be large. An interesting recent suggestion, which is discussed in Chapter 4, is that Bondi-type accretion of hot gas onto the central BH in cD and other galaxies results in the conversion of gravitational energy into kinetic outflow energy. This is manifested in a form of a powerful radio jet and/or high-temperature X-ray-emitting shells of gas (X-ray bubbles). Such jets and bubbles are suspected to influence the final evolutionary stages of large elliptical galaxies in clusters of galaxies by inhibiting further accretion and by preventing the cooling of the surrounding gas through cooling flows.

3.4. Further reading

General BH properties: Several comprehensive books and review articles explain this topic in much greater detail. The approach used here is based on Blandford (1990). Narayan and Quataert (2005) is a general review of astrophysical BHs with connection of BH spin to jets.

Eddington luminosity and time and spherical accretion: A comprehensive discussion of these topics is given in Krolik (1999).

4

Accretion disks

Some of the more efficient accretion processes in astrophysics are associated with the presence of accretion disks. Such disks are found in protostars; in various types of binary stars, including low- and high-mass X-ray binaries; in dwarf novae or cataclysmic variables; and in classical novae. They are also believed to be present in galactic nuclei, around the central supermassive BH, during periods of fast accretion.

Accretion disks in galactic centers are naturally formed by infalling gas that sinks into the central plane of the galaxy while retaining most of its angular momentum. The assumption is that the viscosity in the disk is sufficient to provide the necessary mechanism to transfer outward the angular momentum of the gas and to allow it to spiral into the center, losing a considerable fraction of its gravitational energy on the way. The energy lost in the process can be converted into electromagnetic radiation with extremely high efficiency, from about 4 percent and up to 42 percent. It can also be converted to kinetic energy of gas, which is blown away from the disk, or in other cases, it can heat the gas to very high temperatures, which causes much of the energy to be advected into the BH.

AGN disks, and accretion disks in general, are classified according to their shape into thin, slim, and thick disks. Each one of these can be optically thin or thick, depending on the column density (or surface density) and the level of ionization of the gas. The optical depth of AGN disks, during periods of fast accretion, is very large. The disks that receive most attention are optically thick, geometrically thin accretion disks. Such systems are easier to treat analytically and numerically. The next section describes the basic assumptions and the analytical solution of such disks. A full solution of this type can be used to calculate the emergent disk spectrum and to compare it with observations. The additional sections address other types of accretion disks and accretion flows.

4.1. Optically thick, geometrically thin accretion disks

4.1.1. Basic disk parameters

Optically thick, geometrically thin accretion disks are likely to form over a large range of conditions in galactic nuclei during times when large amounts of cold gas are falling into the center. The fundamental parameters that govern the properties of such systems are the accretion rate through the disk, the BH mass, and the BH spin. These parameters determine the geometry of the disk, the gas temperature everywhere in the disk, the overall luminosity, and the emitted spectrum.

As explained in the previous chapter, for practical reasons, it is convenient to measure the accretion rate in units of L/L_{Edd}, which is referred to as the *normalized accretion rate*. This is not a very accurate definition because the accretion rate is related to the observed luminosity via the efficiency factor, η, and therefore the exact relation between L/L_{Edd} and the accretion rate depends on the BH spin. Most of the general properties described here refer to spinning BHs with $\eta \sim 0.1$.

It is thought that "standard" optically thick, geometrically thin accretion disks around massive BHs, that is, those systems that are adequately described by the theory presented subsequently, are characterized by $10^{-2} \lesssim L/L_{Edd} \lesssim 0.3$. Both those limits are not well determined. The exact upper limit depends on the disk shape and its deviation from a simple thin structure. We take the definition of *thin* to indicate a thickness over radius that is well below 0.1. The lower limit is determined by the disk viscosity and the efficiency of converting gravitational potential energy to electromagnetic radiation.

We express all the important accretion disk properties as functions of the distance from the central BH, r. The fundamental parameters are the gas density in grams per cubic cm (gr/cm), $\rho(r)$, the local angular momentum, $s(r)$, the net torque associated with this angular momentum, $N(r)$, the azimuthal rotational velocity, $v_\phi(r)$, and the radial drift velocity, $v_r(r)$. All these quantities are time dependent and are determined by the accretion rate. Time-dependent calculations of such systems are complicated, and most of the discussion here is restricted to time-independent, stationary disks.

An important ingredient of the accretion disk model is viscous dissipation. The derivation of the temperature and emissivity in such cases may be done by using the Navier–Stokes equations for a noncompressible fluid. The derivation used here is different and involves the assumption that gravitational potential energy and the work done by torque in the disk can be treated separately. This somewhat simplistic treatment leads to the same results as the result obtained by solving the complete set of disk equations.

Consider a narrow ring in the disk with radius r, radial thickness dr, and a total mass dm given by

$$dm = 2\pi r \Sigma dr, \tag{4.1}$$

where Σ is the surface density of the disk per square centimeter,

$$\Sigma(r) = \rho(r)H(r), \tag{4.2}$$

and $H(r)$ is the disk height at radius r, to be defined later. The continuity equation in its differential form requires that

$$\frac{\partial \rho}{\partial t} + \vec{\nabla}.(\rho\vec{v}) = 0. \tag{4.3}$$

The application of this equation to the gas in the ring can be written as

$$2\pi r \frac{\partial \Sigma}{\partial t} + \frac{\partial}{\partial r}(2\pi r \Sigma v_r) = 0, \tag{4.4}$$

where we have used the polar coordinate form of the divergence and assume that $H(r)$ changes slowly with radius. In a steady-state disk, $\rho(r)$, $H(r)$, and $\Sigma(r)$ are independent of time ($\partial/\partial t = 0$). This allows us to identify $2\pi r \Sigma v_r$ as a location-independent constant, which is the radial mass inflow rate through the cross section of the disk. We can write this as

$$\dot{M} = -2\pi r \Sigma v_r = \text{const.}, \tag{4.5}$$

where $\dot{M}$ is the mass inflow through the disk and the minus sign allows for the direction of v_r.

Next we consider the angular velocity of the ring, the angular momentum of the ring material, and the time and radial dependences of the net torque. A major assumption in standard thin-disk models is that the accreted gas moves inward slowly while retaining its circular orbital motion. For most cases of interest, $v_r(r) \ll v_\phi(r)$ and $v_\phi(r) \simeq v_K(r)$, where $v_K(r)$ is the Keplerian velocity. The associated angular velocity can be approximated by the Keplerian angular velocity $\Omega_K(r)$:

$$\Omega \simeq \Omega_K = \frac{v_K}{r} = \left[\frac{GM}{r^3}\right]^{1/2}. \tag{4.6}$$

The angular momentum of the gas in a ring of unit radial thickness is

$$s = 2\pi r \Sigma r^2 \Omega. \tag{4.7}$$

A major assumption of the disk model is that the motion of a particle in the ring is coupled to the motion of gas particles just outside and just inside its location through some kind of friction or viscosity. This results in viscous torque that is exerted on the exterior disk. The viscosity is discussed in more detail in § 4.1.3. Similarly, the ring at distance r is subjected to the torque exerted by the gas inside r, which we mark by N. The net radial change of torque across the ring must equal the rate of change of the angular momentum of the ring. This can be written as

$$\frac{\partial N}{\partial r} = \frac{\partial}{\partial t}(2\pi r \Sigma r^2 \Omega). \tag{4.8}$$

Differentiating with respect to t and noting again the steady-state condition $(\partial \Sigma / \partial t = 0)$, we get

$$\frac{\partial N}{\partial r} = 2\pi r \Sigma v_r \frac{\partial}{\partial r}(r^2 \Omega) = -\dot{M}\frac{\partial}{\partial r}(r^2\Omega), \tag{4.9}$$

where we have used the definition of $\dot{M}$ from Equation 4.5. The integration of dN across the ring gives $N_2 - N_1$, where $N_2 = N(r+dr)$ and $N_1 = N(r)$. This allows us to write an expression for the radial dependence of the torque,

$$N(r) = \dot{M}[GMr]^{1/2} + \text{const.}, \tag{4.10}$$

where Ω is obtained from Equation 4.6.

To identify the constant in Equation 4.10, we carry the calculations inward over all rings down to the innermost radius of the disk. At this location, already mentioned in the discussion about BH and normally referred to as the innermost stable circular orbit (ISCO), there is no viscosity and torque from inside ($N_1(r_{in}) = 0$). This translates to the following torque equation at all radii:

$$N(r) = \dot{M}\left([GMr]^{1/2} - [GMr_{in}]^{1/2}\right) = \dot{M}[GMr]^{1/2}f(r), \tag{4.11}$$

where

$$f(r) = 1 - \left(\frac{r_{in}}{r}\right)^{1/2}. \tag{4.12}$$

In other words, the assumption is that the gas inside r_{in} is falling toward the event horizon in a noncircular orbit and exerts no torque on the gas at r_{in}. Such an assumption must be tested with detailed numerical simulations using real viscosity sources. Preliminary investigation of this type suggests that for thin disks with $H/r \ll 1$, this is probably a good approximation. However, thicker disks can be different, and there are indications, from detailed numerical calculations, that a significant amount of emission can originate from inside the ISCO.

Standard thin accretion disks are defined by the preceding properties as well as their inner (r_{in}) and outer (r_{out}) radii. As explained, the ISCO provides a good approximation for r_{in}, and its exact value depends on the spin parameter of the BH, a (see Table 3.1). Thus, for a nonrotating BH, $r_{in} = 6r_g$, whereas for a fast-rotating Kerr BH, with $a = 0.998$, $r_{in} \simeq 1.24r_g$. Regarding the outer radius, r_{out}, this is more difficult to define and is a function of the pressure and the gravity at large distances. Some approximations of r_{out} are given in § 4.1.5. The various disk parameters are illustrated in Figure 4.1.

Thin accretion disks provide a mechanism to transfer angular momentum outward and matter inward. An interesting and important question is the fate of the outward-going angular momentum. The process must result in the expansion of the outer parts of the disks. Some of the outward-going angular momentum may

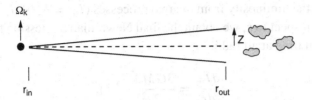

Figure 4.1. Definitions of basic disk parameters. The inner boundary of the disk is assumed to be the location of the ISCO. The outer boundary, r_{out}, is the self-gravity radius of the disk beyond which it breaks into self-gravitating blobs (courtesy of K. Sharon).

be transferred to gas and stars in regions far away from the BH. The inner torus in AGNs, which is discussed in § 7.5, may be influenced by this.

4.1.2. Luminosity emissivity and temperature

The energy released by a standard, optically thick, geometrically thin accretion disk can be obtained by using simple energy conservation arguments assuming a complete conversion of gravitational potential energy to electromagnetic radiation. The local energy release at radius r is determined by the loss of gravitational energy of the inward-going material (which must be positive) and the work done by the torque on the exterior disk (which can be positive or negative). The gravitational luminosity associated with the first process, L_G, is given by the time derivative of the mechanical energy E_G. We derive this energy by considering, again, a circular ring of mass dm at a distance r. For this ring,

$$dE_G = \frac{1}{2}v^2 dm - \frac{GMdm}{r} = -\frac{1}{2}\frac{GMdm}{r}. \qquad (4.13)$$

Using $\dot{M} = dm/dt$ for the inflowing gas, we find

$$L_G = \frac{dE_G}{dt} = -\dot{M}\frac{GM}{2r}. \qquad (4.14)$$

The second contribution to the luminosity is associated with the work done by the local torque on the exterior disk. The energy related to the torque N is

$$dE_N = Nd\theta, \qquad (4.15)$$

and the associated luminosity is

$$L_N = -\frac{dE_N}{dt} = -N\Omega = -\dot{M}\frac{GM}{r}f(r), \qquad (4.16)$$

where, in the last step, we substituted the earlier defined expressions for N (Equation 4.11) and for Ω_K (Equation 4.6).

Combining the luminosity from the two processes ($L_r = L_G + L_N$), and differentiating with respect to r, we obtain the final Newtonian expression for the energy release per unit time at all radii,

$$\frac{dL_r}{dr} = \frac{3GM\dot{M}}{2r^2} f(r).$$ (4.17)

To summarize the Newtonian approximation part, we note that a simple integration of Equation 4.17 over the parts of the disk far from its center gives $L_r \propto GM\dot{M}/r$. The total radiated power, obtained by integrating over the entire disk, in the nonrelativistic limit is

$$L = \frac{1}{2}\frac{GM\dot{M}}{r_{\text{in}}}.$$ (4.18)

This luminosity is exactly half the total available power, meaning that, neglecting relativistic corrections, the gas at r_{in} still retains a kinetic energy that is half the potential energy it has lost.

The previous derivations must be modified close to the BH, where GR effects are significant. The generalization for the case of strong gravitational fields requires several modifications. For example, a more accurate expression for dL_r/dr that includes GR terms contains an expression that looks like $Q/BC^{1/2}$ and replaces $f(r)$. The expressions for B, C, and Q have been accurately calculated and are being used in numerical accretion disk calculations. Additional GR terms are used for the angular momentum, the pressure, and the torque terms. There are also first-order approximations for a modified gravitational potential that improve the accuracy of the calculations. One such idea is to use the innermost allowed radius, r_g, to scale better the change of the gravitational potential at small distances. One such expression is the Paczynsky–Wiita pseudo-Newtonian potential,

$$\Phi(r) = \frac{-GM}{r - r_g},$$ (4.19)

which gives adequate results for Schwarzschild and slowly rotating BHs at small radii. Subsequently, we keep the Newtonian approximation for all expressions involving thin-disk emissivity, luminosity, and temperature.

The relativistic corrections are particularly important in thin accretion disks around fast-spinning BHs. Much of the radiation in such cases is released very close to r_{in}, which modifies the correction factor $f(r)$ in a significant way. Including those terms, one finds that, for a Schwarzchild BH, 50 percent of the energy is emitted between 6 and about 30 r_g, whereas for extreme Kerr BHs, the 50 percent limit is much closer. In particular, in thin AGN accretion disks, much of the ionizing UV radiation is released at radii of order $\sim 10 r_g$.

The expression for dL_r/dr enables us to calculate the emissivity per unit area, $D(r)$, on each of the two surfaces of the disk. Because dL_r is released over an area of $2\pi r dr$, we get

$$D(r) = \frac{1}{2} \frac{1}{2\pi r} \frac{dL_r}{dr} = \frac{3GM\dot{M}}{8\pi r^3} f(r), \qquad (4.20)$$

where the additional factor of $1/2$ takes into account the two surfaces of the disk. Differentiating once more, we can find the radius of maximum emissivity. This radius is at about $1.36r_{in}$.

The next step involves the estimate of the radius-dependent disk temperature in the disk. The simplest estimate is obtained by assuming that the local emission is by a perfect blackbody, which translates to $D(r) = \sigma T(r)^4$. This gives

$$T(r) = \left(\frac{3GM\dot{M}}{8\sigma\pi r^3} f(r) \right)^{1/4}. \qquad (4.21)$$

Expressing the radius in units of r_g and the accretion rate in normalized units of $\dot{M}/\dot{M}_{Edd}$ enables us to obtain a general expression for all thin accretion disks around various mass BHs. The expression for the normalized temperature is

$$T\left(\frac{r}{r_g}\right) \propto M_8^{-1/4} \left[\frac{\dot{M}}{\dot{M}_{Edd}} \right]^{1/4} \left[\frac{r}{r_g} \right]^{-3/4} f(r). \qquad (4.22)$$

To introduce scaling into this equation, we note that for a thin accretion disk around a BH with $M_8 = 1$, the maximum temperature is roughly $T(6r_g) \sim 10^5$ K. Such disks emit most of their energy in the UV part of the spectrum. In a similar way, the maximum blackbody temperature associated with thin accretion disks around stellar-size BHs is about 10^7 K, with most of the energy emitted in the X-ray part of the spectrum. Thus accretion disks around massive BHs are much cooler than disks around stellar-size BHs or neutron stars. A Newtonian approximation to the disk effective temperature is

$$T(r) \simeq 8.6 \times 10^5 \dot{M}_1^{1/4} M_8^{-1/2} f(r)(r/r_g)^{-3/4} \text{ K}, \qquad (4.23)$$

where $\dot{M}_1$ is the accretion rate in units of $M_\odot$/yr. Figure 4.2 shows this approximation for $a = 0$ and $a = 0.9$. It also shows a line corresponding to the case $f(r) = 1$, illustrating the simple $r^{-3/4}$ form far from r_{in}.

4.1.3. Viscosity

The only variable that is not yet fully defined is the radial velocity, $v_r(r)$. This must depend on the friction between adjacent rings, which is determined by the viscosity in the disk. The disk properties that can affect the viscosity are related to the local microphysics, that is, the specific atomic or molecular processes, the local

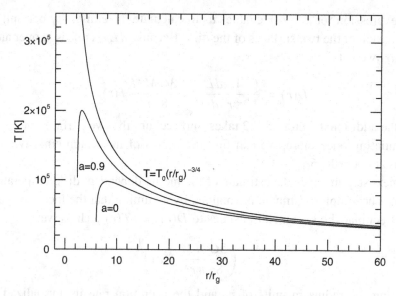

Figure 4.2. Newtonian thin-disk temperatures for stationary ($a = 0$) and fast-rotating ($a = 0.9$) BHs. In this case, $M_{\mathrm{BH}} = 10^8 M_\odot$ and $\dot{m} = 0.1\ M_\odot$/yr. The calculated temperatures are very close to the approximation of Equation 4.23. The top rising curve shows the $a = 0.9$ case with $f(r) = 1$.

or global turbulence, and the magnetic field strength and structure. In what follows, we describe all these by using the kinematic viscosity, ν, which is the viscosity coefficient divided by the density.

A full treatment of the viscosity and hence $\nu_r(r)$ requires a detailed calculation of the viscous torque at all locations. We start by defining the rate of shearing, A,

$$A = r\frac{d\Omega}{dr} = r\frac{d}{dr}\left([GM]^{1/2} r^{-3/2}\right) = -\frac{3}{2}\Omega, \qquad (4.24)$$

and the absolute value of the viscous stress,

$$\nu\rho A = \nu\frac{\Sigma}{H}A = \frac{3}{2}\nu\frac{\Sigma}{H}\Omega. \qquad (4.25)$$

We can now write another expression for the torque using, this time, the viscous force at the location r. The viscous force per unit length is $\nu\Sigma A$, and the torque associated with it is

$$N(r) = (2\pi r)(\nu\Sigma A)r. \qquad (4.26)$$

Substituting A from Equation 4.24 and comparing with the previous torque equation ($N \propto \dot{M}(GMr)^{1/2}$; Equation 4.11) allows us to solve for $\dot{M}$:

$$\dot{M} = -3\pi\nu\Sigma/f(r). \qquad (4.27)$$

The continuity condition (Equation 4.5) is $\dot{M} = -2\pi r \Sigma$, and the comparison of the two equations for $\dot{M}$ can be used to derive an expression for v_r that involves the viscosity v:

$$v_r = \frac{3v}{2rf(r)}. \tag{4.28}$$

This shows that the higher the viscosity, the larger is the radial velocity of the inflowing gas. It also means that for a given accretion rate, a higher viscosity means smaller Σ.

The preceding discussion completes the definition of all important radial-dependent disk parameters. We note that unlike the real physical parameters of the disk, the derived expression for the viscosity is only a recipe that involves no real physics. There have been many attempts to calculate the viscosity in various types of thin accretion disks from first principles. So far, none has resulted in a rigorous, well-justified, and well-understood expression. The physical conditions most likely to affect the viscosity in AGN disks are turbulence and tangled magnetic fields. Much work has been done on magnetic viscosity that results from the so-called magnetorotational instability (MRI). The magnitude and importance of such processes can only be estimated from detailed numerical computations.

Many disk models handle the viscosity by adopting a simple prescription for v that allows us to derive its value from the better understood properties. Perhaps the most useful prescription, and definitely the one most commonly used, is the one proposed by Shakura and Sunyaev in 1973. According to this work, the viscosity depends on a typical length (H in our case) and a typical velocity (the sound speed, v_s) in the disk. The assumption is that

$$v = \alpha' v_s H, \tag{4.29}$$

where the parameter α' includes all the unknowns in the microphysics of the gas. This results in a family of models referred to as α-*disk models* (ADs).

The actual Shakura and Sunyaev prescription is somewhat different and requires a modified parameter:

$$\alpha = \frac{3\alpha'}{2}. \tag{4.30}$$

Using this α, one finds that the viscous stress is given by

$$v\rho A = \alpha P, \tag{4.31}$$

where P is the local pressure. The rationale is that fluid turbulence is a natural process in accretion disks. Such turbulence could be associated with fluctuations in the radial and azimuthal velocity components. The suggestion is that the turbulence velocity obeys $v_{\text{tur}}^2 < v_s^2 = P/\rho$ and that vA is proportional to v_{tur}^2. Thus, $v\rho A < P$.

This idea can be generalized to other situations, for example, a fluctuating magnetic field.

The α prescription has been used in the literature in various ways. In some papers, P is taken to be the gas pressure, and in others, it is the total pressure, $P = P_g + P_{rad}$, where P_{rad} is the radiation pressure. Another variation based on the same principle uses the geometrical mean of the two, $P = \sqrt{P_g P_{rad}}$.

Having made these assumptions, we can now derive a specific expression for the radial velocity in the disk. Starting from the radial velocity equation (Equation 4.28) and the viscosity equation (Equation 4.29), we find

$$v_r f(r) = \frac{3\nu}{2r} = \alpha v_s \frac{H}{r}. \tag{4.32}$$

The value of α depends on the gas density; the larger the density, the smaller is α. The values that are usually used are those found to give good agreement with the observations of accretion disks in dwarf novae. In those cases, $10^{-2} \le \alpha \le 1$. The simplest way to use this prescription is to assume the same α for all parts of the disk, that is, a viscosity that depends on the pressure (or the sound speed) in exactly the same way throughout the disk.

Thin accretion disks are defined by the condition that $H/r \ll 1$ (see justification later). Because $\alpha \le 1$, we conclude that $v_r \ll v_s$, which means that the inward velocity is small compared with the sound speed. This also means that the radial inflow is subsonic.

4.1.4. Important time scales

We can now compare several time scales associated with thin accretion disks. First we write an expression for the *viscous time scale*, t_{vis}, which is the radial drift time through the disk,

$$t_{vis} \simeq \frac{r}{v_r} \simeq \frac{2r^2}{3\nu}. \tag{4.33}$$

Assume $\alpha \sim 0.1$ and $H = 10^{-2}r$; we find

$$\frac{r}{v_r} = \frac{r^2}{\alpha v_s H} \simeq \frac{10^3 r}{v_s}. \tag{4.34}$$

Because $v_s \simeq 17\sqrt{T/10^4}$ km s^{-1},

$$t_{vis} \simeq \frac{200 M_8}{\sqrt{T/10^4}} \frac{r}{r_g} \text{ yr.} \tag{4.35}$$

For thin accretion disks around massive BHs, the viscous time at $r = 100 r_g$ is of order 10^4 yr. This raises an interesting question regarding the origin of AGN variability. The radiation emitted by steady-state disks that are fed from their outer radius can only vary on a viscous time scale, as a result of long-term changes in the accretion rate. However, AGN observations clearly show much shorter time-scale variations: days to weeks for optical–UV variations and hours to days for X-ray variations. Obviously, other mechanisms that affect the disk luminosity must operate in such systems.

The additional time scales that can provide some explanation to the optical–UV continuum variations are as follows:

The sound-crossing time scale:

$$t_s \simeq \frac{0.3 M_8}{\sqrt{T/10^4}} \frac{r}{r_g} \text{ yr} \qquad (4.36)$$

The dynamical time scale:

$$t_{\text{dyn}} \simeq 0.005 M_8 \left[\frac{r}{r_g} \right]^{2/3} \text{ yr} \qquad (4.37)$$

The thermal time scale:

$$t_{\text{th}} = t_{\text{dyn}}/\alpha \qquad (4.38)$$

The light-crossing time scale:

$$t_l \simeq 0.016 M_8 \frac{r}{r_g} \qquad (4.39)$$

The light-crossing time scale is relevant to the case of an irradiated disk, which is discussed later. The other time scales are important in various cases involving instabilities in the disk.

4.1.5. Disk geometry

AGN accretion disks can be thin, slim, or thick, depending on the mass accretion rate. This can be seen by considering the part of the disk that is dominated by the gravitational field of the central BH. Assume g_z is the vertical component of this field, that is,

$$g_z = \frac{GM}{r^2} \sin \theta \simeq \frac{GM}{r^3} z = \Omega_K^2 z. \qquad (4.40)$$

We can write the hydrostatic equation in the following way:

$$\frac{dP}{dz} = -\rho g_z = -\rho \Omega_K^2 z, \qquad (4.41)$$

where

$$P = P_g + P_{\text{rad}} = \frac{\rho k T}{\mu m_p} + \frac{1}{3} a T^4. \qquad (4.42)$$

This allows us to consider various cases of interest. Assume first that the total pressure is determined by the gas pressure, $P = P_g$. In this case, $P = \rho v_s^2$, where v_s is the sound speed. Thus,

$$\frac{dP}{dz} = v_s^2 \frac{d\rho}{dz} = -\rho \Omega_K^2 z, \tag{4.43}$$

with the following simple solution for the vertical dependence of the density:

$$\rho = \rho_0 \exp\left[-\frac{z^2}{2H_0^2}\right], \tag{4.44}$$

where ρ_0 is the mid-plane density of the disk and we have defined a new parameter, a disk scale height, H_0, given by

$$H_0 = \frac{v_s}{\Omega_K}. \tag{4.45}$$

The disk scale height is related to the scale height used earlier by $H_0 \simeq 0.5H$. It can be compared to the radial size of the disk by using the sound and Keplerian velocities,

$$\frac{H_0}{r} = \frac{v_s}{v_K}. \tag{4.46}$$

The accretion disks considered so far are indeed geometrically thin since, under the conditions assumed here, $v_s \ll v_K$. A rough estimate of $H_0(r)$ can be obtained by using $v_s \simeq 17\sqrt{T/10^4}\,\mathrm{km\ s^{-1}}$, $T \simeq 10^5 (r/r_g)^{-3/4}$ K, and $v_K \simeq 1/2c(r/r_g)^{-1/2}$. However, the estimated temperature is related to the assumption that all of the released gravitational energy can be radiated locally as a perfect blackbody that must break down at some point. For example, the disk may not be optically thick to its own radiation, or the density may be too low to radiate away all the locally released energy on the local dynamical time. In such cases, the local temperature increases and the assumption of a thin ($v_s \ll v_K$), gas-pressure-dominated disk no longer holds. Such conditions may occur when the accretion rate through the disk drops below a certain critical value. They can result in radiation-inefficient flows, which are discussed in the following section.

The hydrostatic solution applies only to thin accretion disks that are dominated by gas pressure and only in the parts of the disk where gravity is entirely due to the central BH. However, self-gravity must become important far enough from the center. The vertical component due to self-gravity, g_z^{self}, is given by

$$g_z^{\text{self}} = 2\pi G \Sigma z = 4\pi G \rho H_0 z, \tag{4.47}$$

which should be compared to the vertical component due to the central gravitational field, g_z. We expect the disk to break into small fragments at those radii where $g_z^{\text{self}} > g_z$. This provides a natural outer boundary for the disk, r_{out}. This boundary, together with the inner boundary, r_{in}, and the disk scale height, $H(r)$, completely specifies the thin-disk geometry (Figure 4.1).

Consider now some typical numbers that specify the properties of optically thick, geometrically thin accretion disks around massive BHs. The example is calculated for $M_8 = 1$. In this case, the self-gravity radius is roughly

$$r_{out} \simeq 0.01 \left[\frac{\alpha}{0.03}\right]^{1/2} \text{pc} \simeq 2 \times 10^3 \left[\frac{\alpha}{0.03}\right]^{1/2} r_g, \qquad (4.48)$$

and for $\alpha = 0.03$, the mass of the disk between r_{in} and r_{out} is about $3 \times 10^5 \, M_\odot$, that is, $M(\text{disk})/M \simeq 3 \times 10^{-3}$. The time it takes for such a disk to completely disappear, that is, for all the mass in the disk to be accreted onto the BH in the absence of additional material, is $M(\text{disk})/\dot{M}$. For an accretion efficiency of $\eta = 0.1$, and for $L/L_{Edd} = 0.1$, this is of order 10^6 yr. Such an event (i.e., the accretion of the entire disk's material) can be defined as an *accretion episode*. In the case in question, one needs several hundreds accretion episodes lasting about 10^6 yr each to double the mass of such a BH. Obviously, the real situation is more complex, and the length of a single accretion episode depends on the exact BH mass, the exact accretion rate, and other parameters. However, the actual numbers do not change by much. These results have important implications regarding the growth times and growth modes of supermassive BHs.

If BH growth is indeed the result of a large number of short discrete episodes, due to the noncontinuous mass supply from the galaxy, then we do not expect the same angular momentum direction of the external accreted material. In such cases, the BH spin can increase, decrease, or even completely change its direction over a short time scale. This has important implications to the growth time of BHs, to the mass-to-luminosity conversion efficiency, and to the duty cycle of AGNs. The issue of BH growth is further discussed in Chapter 9.

4.1.6. The emitted spectrum

The expressions for the emissivity and temperature of thin accretion disks can be combined to obtain the locally emitted (single surface) monochromatic luminosity of the disk,

$$dL_\nu = 2\pi r (\pi B_\nu) dr, \qquad (4.49)$$

and the total monochromatic luminosity,

$$L_\nu = \int_{r_{in}}^{r_{out}} dL_\nu = \frac{8\pi^2 h \nu^3}{c^2} \int_{r_{in}}^{r_{out}} \frac{r \, dr}{\exp(h\nu/kT) - 1}, \qquad (4.50)$$

where we multiplied by an additional factor of 2 to account for both surfaces of the disk. The integral can be evaluated by making the substitution $x = h\nu/kT$ and

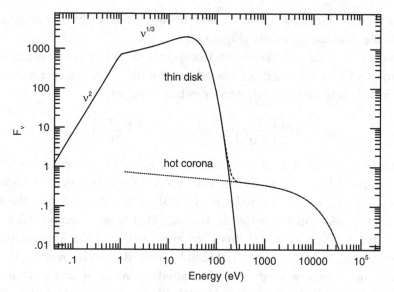

Figure 4.3. A schematic of a combined disk–corona spectrum. The maximum temperature of the geometrically thin, optically thick accretion disk is $T_{\mathrm{max}} = 10^5\,\mathrm{K}$, and its outer boundary temperature is determined by the conditions at the self-gravity radius. The disk is surrounded by an optically thin corona with $T_{\mathrm{cor}} = 10^8\,\mathrm{K}$.

noting that $T \propto \nu/x \propto r^{-3/4}$. Thus $r\,dr \propto \nu^{-8/3} x^{5/3}\,dx$, which allows the integration between r_{in} and r_{out}. If $r_{\mathrm{out}} \gg r_{\mathrm{in}}$, the spectral shape over a certain frequency range is given by

$$L_\nu \propto \dot{M}^{2/3} M^{2/3} \nu^{1/3},\tag{4.51}$$

where we made use of the temperature dependence on the mass and accretion rate (Equation 4.22). This expression is occasionally referred to as the *canonical disk spectrum*.

The simple schematic $\nu^{1/3}$ dependence of L_ν holds only over a limited energy band because of the physical boundaries of the disk. The outer boundary imposes a minimal disk temperature and a typical frequency associated with this temperature, ν_{out}. Below this frequency, the spectrum resembles a blackbody with a frequency dependence of $L_\nu \propto \nu^2$. At the inner boundary, there is an inner temperature and a typical associated inner frequency, ν_{in}. Beyond this frequency, L_ν drops exponentially with a functional dependence corresponding to the maximum disk temperature. A schematic representation of the spectrum of such disks is shown in Figure 4.3.

The part of the SED that shows the $L_\nu \propto \nu^{1/3}$ dependence is important because this is where we can estimate $\dot{M}$ from direct measurements of L_ν. We can write two useful approximations for this part of the spectrum by defining $M_8 = M/10^8 M_\odot$,

$\dot{M}_\odot = \dot{M}/M_\odot$ yr^{-1}, and i as the inclination to the line of sight. The first equation gives the monochromatic luminosity at long wavelengths,

$$L_\nu \simeq 9.4 \times 10^{29} \cos i [M_8 \dot{M}_\odot]^{2/3} \left[\frac{\lambda}{5100 \text{ Å}} \right]^{-1/3} \text{ erg s}^{-1} \text{ Hz}^{-1} \qquad (4.52)$$

and the second the mass accretion rate,

$$\dot{M}_\odot \simeq 2.6 \left[\frac{L_{5100,45}}{\cos i} \right]^{3/2} M_8^{-1}, \qquad (4.53)$$

where $L_{5100,45} = L_{5100}/10^{45}$ erg s^{-1}. For low-redshift AGNs with BH mass of order 10^7–10^8 $M_\odot$, the wavelength of 5100 Å is far enough from the peak of the SED to use this estimate of the accretion rate. For higher-mass BHs, longer wavelengths are preferred.

4.1.7. Viewing angle dependences

So far we have not specified the viewing angle of the disk, which, given the flat geometry, must have a significant effect on the observed luminosity. Accounting for viewing angle dependences requires a combination of a simple geometrical correction factor, $\cos i$, with an additional frequency-dependent parameter, $b(\nu)$, which is the equivalent of the limb-darkening factor in stellar atmospheres. A simple way to express this relation is to assume

$$\frac{L_\nu}{L_\nu(\text{face} - \text{on})} = \frac{1}{1 + b(\nu)} \cos i (1 + b(\nu) \cos i). \qquad (4.54)$$

The exact value of the parameter $b(\nu)$ depends on the properties of the atmosphere and the gravitational potential at the specific location in the disk. The dependence on the optical depth of the atmosphere is similar to what is found in stellar atmospheres. The additional inclination dependence originates from changes in the paths of the emitted photons due to the strong gravitational field. This is most important for the highest-energy photons, which are emitted close to the last stable orbit. Such photons suffer the largest gravitational redshift and are also bent toward the plane of the disk. Thus, the spectrum of the innermost rings is softer in the perpendicular (face-on) direction and harder at large inclination angles. Some of those photons suffer such a large deflection that they hit the disk surface at a different location. This changes the local energy balance and must be taken into account when calculating the disk spectrum. All of this does not amount to a very large change in $b(\nu)$ for the optical–UV photons and, for a large range of conditions, $b \simeq 2$. This is not very different from what is found in a thick electron scattering atmosphere where $b = 2.06$.

Obviously real disks are not simple blackbody emitters, and full radiation transfer calculations must be used to properly account for the changes in effective

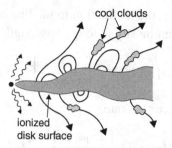

Figure 4.4. Complex magnetic disks may include a nonuniform geometry like the one shown schematically in this diagram. Winds and cool blobs can flow from the surface of the disk along the magnetic field lines. Such gas is exposed to the central radiation field at some height above the disk surface and can be driven in the radial direction by radiation pressure force (courtesy of K. Sharon).

temperature and opacity with viewing angle. This must be taken into account when attempting to derive the total disk luminosity from a single line-of-sight measurement of the SED. These considerations are also important when comparing disk luminosity with other sources of radiation that are more isotropic, for example, the luminosity and the EWs of certain emission lines (§ 7.1.11).

4.2. Real AGN disks

There are good reasons to believe that real AGN disks are quite different from the simple, optically thick, geometrically thin accretion disks described earlier. Strong magnetic fields are likely to affect the physics and the structure of such systems. They provide an important source of viscosity and may drive massive winds from the surface of the disk. Very hot coronae, with temperatures reaching 10^7K and beyond, are likely to exist in such systems, giving rise to strong X-ray radiation that hits the disk surface and changes the local energy balance. Mass accretion does not necessarily originate at the outskirts of the disk and may not be time independent. A schematic of a disk that is affected by such processes is shown in Figure 4.4. Sophisticated numerical methods are being developed with the aim of solving the structure and the spectrum of such systems. Several such models are already providing more realistic solutions for the vertical structure of isolated rings and the behavior of the gas near the ISCO. However, the global disk properties, and the combination of the α-disk with a realistic disk corona (see later), is still missing.

4.2.1. Comptonization and the disk corona

The more realistic calculations of disk atmospheres include several processes that affect the emergent spectrum. In particular, they take into account the temperature

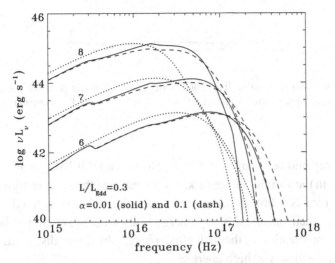

Figure 4.5. Calculated spectra of three accretion disks around BHs with $M_{\rm BH} = 10^6$, 10^7, and $10^8\,M_{\odot}$, all with $L/L_{\rm Edd} = 0.3$. The viewing angle of the disk is 60 degrees. The dotted lines show the much softer simple spectra obtained by assuming a combination of blackbodies without Comptonization (from Hubeny et al., 2001, reproduced by permission of the AAS).

gradient across the disk and the temperature increase in the atmosphere. This increase, combined with the large Compton depth, leads to Comptonization of the softer-emitted photons and a considerable hardening of the emitted spectrum (see § 2.3). Figure 4.5 shows the results of several calculations that include this effect. The diagram shows that the inclusion of Compton scattering results in the extension of the emitted spectrum to considerably higher frequencies compared with the simple local blackbody approximation. The spectral changes depend on all the parameters that affect the disk density and temperature.

A considerable fraction of the energy of many AGNs is emitted in the X-ray band, between $\sim$0.2 and $\sim$100 keV. The large-amplitude, short-time-scale variations observed in this band suggest a size that is smaller than the size of the central accretion disk. The fraction of the bolometric luminosity emitted in this part of the spectrum is luminosity dependent. In low-luminosity AGNs, it can reach 20–30 percent. In the highest-luminosity sources, it is much smaller.

Basic disk theory indicates that the maximum temperature of typical AGN disks cannot exceed few $\times\,10^5$ K. Comptonization in the disk atmosphere can increase the energy of some of the photons by another factor of a few. However, even the highest energy obtained in this way cannot exceed about 100 eV in disks around BHs with $M_{\rm BH} > 10^7\,M_{\odot}$. Moreover, the observed X-ray SED is significantly different from what is predicted for the hottest disks (Figure 4.5). Thus, the origin of the

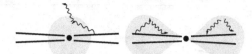

Figure 4.6. Two schematic disk–corona structures showing possible locations of the corona and the scattering geometry of the disk-produced and corona-produced photons.

energetic X-ray photons cannot be the disk itself, and it is likely that an additional hot medium, in the vicinity of the disk, is the source of the more energetic photons. A plausible idea is a hot, dilute gas in which the soft disk-emitted photons are upscattered to their observed X-ray energy. This process should not be confused with the Comptonization in the disk atmosphere, which was discussed earlier and cannot reach the observed high energies.

The conditions for the formation of a hot corona around thin accretion disks are still unknown. One possibility is that the density gradient in the vertical direction is large and that some of the accretion proceeds via the lower-density, outer layers of the disk. This can result in a large energy dissipation and sharp temperature rise. It causes an expansion of the outer layers and the creation of a hot corona. A combined solution for the hydrostatic and radiative transfer equations for the disk–corona structure is complicated and cannot be treated within the simple analytical theory presented here. The simplest assumption involves an optically thin corona with a small scale height. The local temperature in this case is determined by the fraction of the total accretion power released in the corona. Even a very small fraction, as little as 1 percent of the total accretion rate, can result in very high temperatures of $\sim 10^8$ K. Such coronae radiate through free–free processes that dominate the emission at hard X-ray energies. The schematic spectrum of such a corona is shown in Figure 4.3.

An interesting possibility involves a corona that is heated through the combined effect of magnetic fields and the original disk radiation. There are several reasons to assume strong magnetic fields and associated turbulence in such systems. This can provide a disk viscosity as well as an efficient local source of corona heating. Numerical modeling is required to follow the development of such systems.

Regardless of the origin of the corona heating, the presence of such a structure is a source of heating for a cooler disk. A cartoon of several plausible geometries of this type is shown in Figure 4.6.

A large Compton depth corona will result in a significant Comptonization that, unlike scattering in the relatively low temperature disk atmosphere, boosts the photons to much higher energies (because of the $4kT_e$ factor). As explained in § 2.3, the overall boost depends on the Compton parameter y. It is possible to show that the combination of an exponentially decreasing probability of scattering and the exponential gain in photon energy leads to a power-law spectrum with an energy index of

$$\alpha = \sqrt{\left(\frac{9}{4} + \frac{4}{y}\right) - \frac{3}{2}}. \tag{4.55}$$

For $y \sim 1$, the spectral index is $\alpha \simeq 1$, agreeing with observations of many AGNs. However, it is difficult to explain the X-ray properties of the more extreme, highly variable AGNs, such as NLS1s, with this model.

4.2.2. Irradiated disks

External radiation sources can illuminate the disk and change the local energy balance. Important examples are illumination by an X-ray-emitting corona and the irradiation of the outer areas of a flaring disk by its inner part.

The calculation of the spectrum of irradiated disks is complicated because of the unknown geometry and the combination of external sources of energy and locally dissipated energy. Here we discuss only the simplest case, in which an external point source of radiation, L_*, is located along the rotational axis of the BH at a height H_*. We look for a specific expression that relates the illuminating flux per unit area to the location on the disk surface, r. We assume $F_{\text{irr}} \propto r^{-\beta}$ and look for the the value of β that is most appropriate for the geometry of the disk in question. This, plus the assumption of local blackbody emission, is converted to a spectral slope s such that $L_\nu \propto \nu^s$. Note that for a standard optically thick, geometrically thin accretion disk, $s = 1/3$.

Assume that the "typical" disk albedo averaged over wavelengths is a. Neglecting GR effects like light bending, we can write the expression for the irradiating flux as

$$F_{\text{irr}} = \frac{L_*(1-a)}{4\pi r^2} \cos\alpha, \tag{4.56}$$

where α is the angle between the disk normal and the incoming radiation of the external source. If $H_* \gg H(r)$ and $r \gg H(r)$, this is reduced to $F_{\text{irr}} \propto L_* r^{-3}$.

Returning to the local emissivity, we assume thermal blackbody emission and local energy balance, which is dominated by the external source of radiation. This gives $F_{\text{irr}} \propto T^4$ and hence $T(r) \propto L_* r^{-3/4}$ and $\beta = 3$. This radial dependence is identical to the one derived for the standard α-disk (Equation 4.22). Thus, the emergent spectrum of a disk that is dominated by a point source of radiation above its axis of rotation, with $H_* \gg H(r)$, has the same SED as the standard thin accretion disk (Equation 4.50). In particular, some parts of the spectrum are expected to show disk SED with $L_\nu \propto \nu^{1/3}$.

Another interesting case of an irradiated accretion disk is a flaring disk in which the source is located near the center, $H_* \ll H(r)$, and reprocessing is important at the outskirt of the system. This configuration distributes the illuminated flux over a larger area of the disk and results in smaller β. It is easy to show that if $H(r) \propto r^\gamma$, then $\beta = 1 - \gamma$ and $s = (1 - 3\gamma)/(3 - \gamma)$. For $\gamma = 1$ we get $\beta = 2$ and $s = -1$, similar to a typically observed slope in the optical–UV part of the spectrum.

The most important cases of irradiated AGN disks are related to the illumination of optically emitting accretion disks by an external X-ray source such as a disk corona. The presence of a hot corona can significantly alter the disk structure, local level of ionization, local temperature, and, consequently, local emitted spectrum. In particular, hard photons that are created (or scattered) in the corona can either escape or else be absorbed by the cool disk and change the local energy balance. Likewise, soft disk photons can be absorbed by the corona in some parts but freely escape in other parts, depending on the global geometry. This is true for a hot X-ray-emitting corona that covers the entire disk area and also for external X-ray sources that are situated outside, yet close to, the center of the accretion disk.

Figure 4.7 shows the spectrum of a thin disk exposed to an external X-ray source situated close to the BH ("above" and "below" the center of the disk). The diagram shows the illuminating X-ray spectrum as well as the resulting disk emission and reflection. The characteristics of the disk spectrum are very different from those of the nonilluminated disks. The spectrum extends well into the X-ray regime and shows distinct X-ray spectral features.

The nomenclature used here and in later chapters is adopted from the field of X-ray astronomy and requires some explanation. The term *reflection* refers to radiation whose origin is outside the disk. It describes the part of this radiation that is either scattered or else absorbed and reemitted by the gas in the disk. *Disk emission* in this context refers to radiation from the disk material itself. The energy source of this emission is the external radiation field, but the emission frequency depends on the specific atomic processes; that is, the absorption and emission frequencies can be very different. X-ray fluorescence lines belong to this category.

4.3. Slim and thick accretion disks

The situation discussed in § 4.1.5 changes completely when the accretion rate, which determines the disk temperature and hence the local radiation pressure, exceeds a critical value. This results in very high temperatures and hence in $P_{\text{rad}} \gg P_g$ over large parts of the disk. The simple hydrostatic solution for a gas-pressure-dominated structure (Equation 4.44) no longer holds, and the value of H/r can increase significantly. Such disks are no longer geometrically thin. The terms used to describe these systems are *slim* or *thick* accretion disks, depending on their exact geometry. Much of the prior analysis does not hold for slim or thick accretion disks, and calculating their temperature, luminosity, and spectrum requires very different methods.

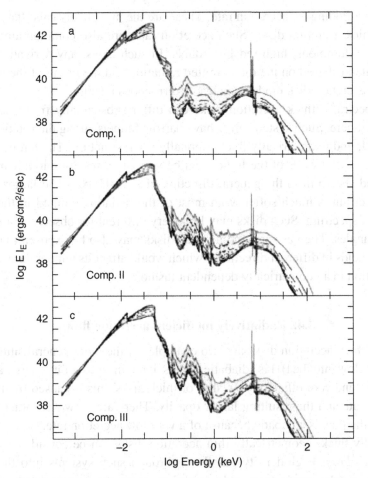

Figure 4.7. Various models illustrating the spectra of X-ray-illuminated accretion disks under different conditions (Rozanska and Madej, 2008; reproduced by permission of John Wiley & Sons Ltd.).

State-of-the-art calculations show that most AGN disks are dominated by radiation pressure. However, for $L/L_{\mathrm{Edd}} \lesssim 0.3$ with $\eta \sim 0.1$, $H/r \le 0.1$, which is small enough for the disk to be considered thin. Larger accretion rates can result in large optical depths and inefficient emission since the accretion (inflow) time scale can be shorter than the time it takes for the radiation to diffuse to the surface of the disk. Photons created in the radiation-pressure-dominated regions of such disks are trapped in the accretion flow and advected to the BH. The result is low radiation efficiency and $\epsilon_r < \eta$. According to some slim-disk models, the relation between total luminosity L and normalized accretion rate $\dot{m} = \dot{M}/\dot{M}_{\mathrm{Edd}}$ is $L = a(1 + b \ln \dot{m}) M_{MH}$ where a and b are constants.

Observations show (Chapter 7) that some low-luminosity, low-redshift AGNs, especially those known as NLS1s, are accreting at rates that are close to, or even exceed, the Eddington accretion rate. These are the most likely candidates to host slim or thick accretion disks. Such accretion rates are also common among very large mass, luminous, high-redshift AGNs. In such cases, any derivation of BH mass, which is based on the total emitted radiation and the fitting of the spectrum by a thin accretion disk model, can lead to erroneous results.

The spectra of thick accretion disks can differ substantially from the spectra considered here. Such systems may have narrow funnels, along the rotational axis of the BH, and geometrically thick, optically thick structure. The temperature of the gas along the walls of the funnel can be very high, resulting in a hard X-ray-dominated spectrum in the general direction of the BH axis. The larger viewing angle spectrum is much softer, with most of the radiation emitted in the optical part of the spectrum. Such disks may look very different for observers at different viewing angles. The gas surrounding such disks may also be exposed to different radiation fields in different directions, which would affect its temperature and level of ionization in a geometrically dependent fashion.

4.4. Radiatively inefficient accretion flows

Slim and thick accretion disks are two examples of the more general situations in which the flow into the BH is adiabatic, that is, it retains most of its energy and does not radiate and cool efficiently. In slim or thick disks, this is caused by the larger accretion rate and the resulting larger opacity. There are, however, other systems in which the flow is adiabatic because of a very low accretion rate.

Optically thick, geometrically thin accretion disks can be considered *cooling-dominated flows*. High-density gas flows through such systems into the center, losing its angular momentum and increasing its kinetic energy due to the local viscosity. The cooling of the high-energy particles, mostly the high-energy electrons, is through various atomic processes that will be discussed in the next chapter. The cooling efficiency in most of these processes is proportional to N_e^2, where N_e is the electron density. For high enough densities, the cooling is very efficient, with typical time scales that are very short compared with the inflow time scale. Accretion disks with very low accretion rates may have much lower densities and hence cooling times that equal or even exceed the inflow time of the gas. In the extreme case, low accretion rates can have very low density. In this case, cooling is much less efficient, and the particles retain the dissipated gravitational energy for a longer time. The temperature in such cases can rise to the virial temperature, and the particles can be advected into the BH without releasing their energy. The basic energy equation that governs such cases can be written, schematically, as

$$H = C + \text{ADV}, \qquad (4.57)$$

where H stands for all the local heating, C stands for all local cooling, and ADV is the advected energy measured in the same units.

The general term used to describe the preceding flows is *advection-dominated accretion flows* (ADAFs). They are predicted to be less luminous for the same accretion rate and to emit a completely different spectrum. A more general term that describes an advection-dominated flow is *radiatively inefficient accretion flow* (RIAF). Coulomb collisions between ions and electrons in low-density gas are slower and less efficient in sharing the total kinetic energy. Since the ions carry most of the energy of the gas, because of their larger mass and because the electrons are more efficient coolants, the RIAFs represent cases in which ions and electrons can have very different temperatures.

To better understand the physical conditions in various RIAFs with low accretion rates, and to estimate the particle density and temperature, we return to several of the basic concepts related to Bondi accretion (§ 3.3.2). In these cases, the accretion rate is given by $\dot{M}_{\text{Bondi}} \simeq 4\pi\lambda v_s \rho r_A^2$, where r_A is the Bondi (or critical) radius, $\lambda \simeq 0.1$ (see Equation 3.22), and v_s is the sound speed. Mass continuity requires a density profile of

$$\rho = \rho_0 \left(\frac{r}{r_A}\right)^{-3/2}, \tag{4.58}$$

where ρ_0 is determined by the conditions at the Bondi radius, r_A, where $v_s \sim v_r$. The original Bondi model involves the assumption that the gas retains its temperature as it falls into the center. However, more realistic scenarios, for example, those involving magnetic fields, lead to the conclusion that the gas must be heated and reach very high temperatures. In the case of nonnegligible magnetic fields, the sum of the kinetic and magnetic energies equals the gas potential energy. For low-density gas, there is very little cooling, and hence the gas temperature close to the BH reaches very large values of order $10^{12} r_g/r$ K. If the density of the gas is low enough and the Coulomb interaction inefficient, this is the ion temperature. The electron temperature is some 2 orders of magnitude lower. Most of this energy is advected into the BH.

We can compare the accretion rates in Bondi (spherical) and disk accretions. Accretion through a disk involves gas with angular momentum and viscosity. The term ADAF is used to describe such a process in the limit of very low density (i.e., advection-dominated RIAFs). This is the case in which the ions and the electrons have thermal energy distributions yet they differ in their temperatures by a factor of ~100. Assuming a virial temperature, the sound speed v_s is roughly the Keplerian velocity. In this case, the radial drift velocity in the disk is roughly $\alpha(GM/r)^{1/2}$, where α is the viscosity parameter with typical values of $0.1 \lesssim \alpha \lesssim 0.01$.

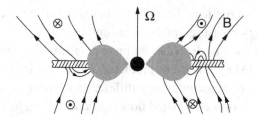

Figure 4.8. Two-temperature disks are similar to the standard optically thick, geometrically thin accretion disks far from the BH. Their very low accretion rate results in low densities and inefficient cooling in the inner parts. The gas in these regions cannot radiate away the released gravitation energy. It heats to very high temperatures, which results in a significant increase in the thickness of the disk. A large fraction of the released energy is advected to the center, and the mass-to-radiation conversion efficiency drops significantly.

This is lower than the inflow velocity in a Bondi spherical flow by a factor α; therefore

$$\dot{M}_{\mathrm{ADAF}} \sim 0.1 \dot{M}_{\mathrm{Bondi}} \tag{4.59}$$

$$L_{\mathrm{acc}} < 0.1 \dot{M}_{\mathrm{Bondi}} c^2. \tag{4.60}$$

Note again that ADAF is only allowed for very low densities and hence very low accretion rates, $\dot{M}/\dot{M}_{\mathrm{Edd}}$ of order 1 percent or less.

ADAFs are characterized by the inefficient conversion of gravitational to radiated electromagnetic energy because much of this energy is advected. Thus, the conversion efficiency factor ϵ_r can attain values that are far below η of the standard disk. This results in low luminosities, yet the mass accretion rate may not necessarily be very low. In principal, such disks can extend to very large radii, up to r_A. ADAFs with small $\dot{M}/\dot{M}_{\mathrm{Edd}}$ may also be associated with a dramatic change in the disk structure. Such a flow can be viewed as a combination of a standard thin accretion disk far from the BH, where all the gravitational energy is radiated away, and a very hot, radiatively inefficient disk closer in, all the way to the ISCO. This is usually referred to as a *two-temperature disk* and is one of the suggested models to explain the spectrum of low-luminosity AGNs (Chapter 6). A schematic view of such systems is shown in Figure 4.8.

Accretion flows with other properties can also result in RIAFs. One of these involves both inflow and outflow of gas and is coined *advection-dominated inflow outflow solution* (ADIOS). A second case involves *convection-dominated accretion flow* (CDAF) and is related to various instabilities in the standard ADAF model. The properties of all these flows are strongly influenced by the properties of the local magnetic field. The treatment of such structures requires sophisticated, high-resolution magnetohydrodynamic (MHD) calculations, some of which are not yet at a stage that enables a meaningful comparison with observations.

The large deviations of the properties of the various RIAFs from those of the well-formulated, well-understood accretion disks suggest a need for a modified accretion efficiency. A general definition should include the case where not all of the released gravitational energy is converted to electromagnetic radiation. For example, we can use ϵ_r, the general radiation efficiency parameter that was introduced earlier, that is,

$$L_{\text{bol}} = \epsilon_r \dot{M} c^2. \tag{4.61}$$

In the case of a high accretion rate through an optically thick, geometrically thin accretion disk, $\epsilon_r = \eta$. For RIAFs, the accretion rate is slower and $\epsilon_r < \eta$, in some cases by a very large factor.

The next step is to modify the definition of the Eddington accretion rate (Equation 3.17), which refers only to the emitted radiation such that

$$\dot{M}_{\text{Edd}} = \frac{L_{\text{Edd}}}{\epsilon_r c^2}. \tag{4.62}$$

The modified normalized Eddington accretion rate is therefore

$$\dot{m} = \frac{\dot{M}}{\dot{M}_{\text{Edd}}} = \frac{\epsilon_r \dot{M} c^2}{L_{\text{Edd}}}. \tag{4.63}$$

Given these definitions, we can search for the critical normalized accretion rate, $\dot{m}_{\text{cr}}$, below which not all of the gravitational potential energy is converted to electromagnetic radiation. This is the accretion rate below which ϵ_r starts to deviate from η. Some RIAF models suggest that this happens at around $\dot{m} = 0.01$, but the exact value is not known.

Putting all the preceding together enables us to characterize the entire accretion process, from fast accretion to the low-accretion-rate RIAFs, by the value of ϵ_r:

$$\epsilon_r = \begin{cases} \eta & \text{if } \dot{m} > \dot{m}_{\text{cr}} \\ \eta \left(\frac{\dot{m}_{\text{cr}}}{\dot{m}}\right)^p & \text{if } \dot{m} \leq \dot{m}_{\text{cr}}, \end{cases} \tag{4.64}$$

where p is a parameter that depends on the details of the RIAF. Suggested values of this parameter range from $p = 1/2$ to $p = 1$. Detailed RIAF calculations show that such systems suffer mass loss through winds and that the value of $\dot{m}$ at the ISCO can be substantially smaller than the value at the transition point, inside which the RIAF dominates the flow. A plausible model may involve $p = 1$ just inside the transition radius and $p = 1/2$ close to the BH horizon.

In the important case of a RIAF involving mass loss through winds, a large fraction of the gravitational energy of the inflowing gas is converted to kinetic energy of the wind, E_K. Such gas can leave the nucleus on a short time scale and, if moving through the galactic ISM, or perhaps through the intergalactic medium in a cluster of galaxies, can deliver much of its kinetic energy to the surrounding gas.

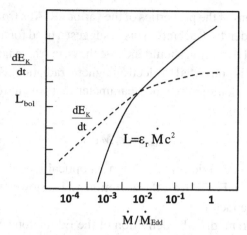

Figure 4.9. Schematic of the two modes of accretion in AGNs. The radiation power, L_{bol}, and the kinetic energy power, dE_K/dt, are plotted against the normalized accretion rate, $\dot{m} = \dot{M}/\dot{M}_{Edd}$. The transition from the radiative (AGN) mode to the kinetic energy (radio) mode is at $\dot{m} = \dot{m}_{cr}$, which, in this example, corresponds to $\dot{m} \approx 0.01$. The mass-to-radiation conversion efficiency ϵ_r (Equation 4.64) approaches at high accretion rates the radiation conversion efficiency η.

The interaction with this gas raises its temperature and, in several cases, enables us to measure the emitted X-ray luminosity and hence to estimate the kinetic energy of the outflow. There are well-documented cases in which dE_K/dt estimated in this way is larger than L_{bol}.

The current AGN literature refers to radiation-dominated processes in high L/L_{Edd} sources as the *radiative mode* or the *AGN mode* (also the *QSO mode*). The ones in which most of the released energy is kinetic are designated *radio mode* because of the observational evidence that such flows are often associated with powerful radio jets. A schematic of this situation is shown in Figure 4.9.

4.5. Further reading

Standard α-disks: The seminal paper on the physics of α-disks is Shakura and Sunyaev (1973). Optically thick, geometrically thin disks are discussed in numerous books and articles. The treatment adopted here is similar to the one discussed in greater detail in Frank et al. (1985). A useful comprehensive review is given in Blaes (2007). State-of-the-art calculations of thin-disk spectra can be found in Blaes et al. (2001), Hubeny et al. (2001), Davis et al. (2007), and references therein. Davis and Laor (2011) describe the method to estimate a BH accretion rate regardless of its spin.

Comptonization and disk corona: See Reynolds and Nowak (2003), and references therein.

X-ray emission from irradiated disks: For the general geometry, see Blaes (2007). For X-ray features, see Reynolds and Nowak (2003), Ross and Fabian (2005), Dovčiak et al. (2004), Rozanska and Madej (2008), Cao (2009), and references therein.

Slim and thick accretion disks: There are relatively few calculations of such disks and their spectra; see, for example, Madau (1988), Kawaguchi et al. (2001), and Wang and Netzer (2003). For super-Eddington accretion, and the connection to NLS1s, see Collin and Kawaguchi (2004), and references therein.

RIAFs: The physics of various advection-dominated flows is discussed in numerous papers, including many that are not related to accretion onto massive BHs. Useful comprehensive descriptions with many additional references are Narayan and McClintock (2008) and Cao (2010).

5

Physical processes in AGN gas and dust

Ionized AGN gas is found at various locations, from very close to the central BH at $r \simeq 1000 r_g$ and out to distances of several kpc from the nucleus. Most of the gas is directly exposed to the strong central source of radiation, making photoionization the dominant heating and ionization process. Dust grains are found at almost all locations beyond a distance of about 1 pc from the BH. Their properties are intimately related to the conditions in the ionized gas. Collisional ionization is another important process, especially in SF regions across the host galaxy. This chapter discusses the details of these processes and their observed signatures. The underlying assumption is that the gas density is always low enough, $N_H < 10^{13}$ cm^{-3}, where N_H is the hydrogen number density, such that the conditions are very far from thermodynamical equilibrium.

5.1. Photoionized gas

Consider gas at a distance r from a point source of radiation with bolometric luminosity L and monochromatic luminosity L_ν. The fractional ionization of ion X of a certain element Z at this location is determined by the ionization rate per particle, due to the local ionizing flux, I_X, and by the recombination rate per particle for this ion, R_X. Assume σ_ν is the frequency-dependent photoionization cross section of ion X from a given atomic level (normally the ground level) whose threshold ionization frequency is ν_X. The total recombination coefficient for the ion in question is $\alpha_X(T)$, and the frequency-dependent optical depth for absorption by this level is

$$\tau_\nu = \int_r N_X \sigma_\nu dr. \tag{5.1}$$

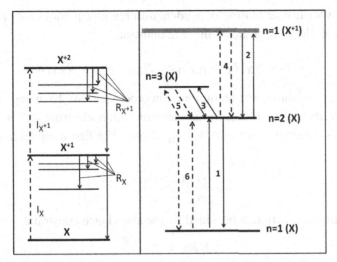

Figure 5.1. (left) A general scheme showing photoionization from the ground states of ions X and X^{+1} (upward dashed lines) and radiative recombination to all levels of these ions (downward solid lines). (right) Various population and depopulation routes of level $n = 2$ in ion X. Solid lines represent direct electronic transitions and dashed line transitions involving emission or absorption of line and continuum photons. (1) Collisional excitation and deexcitation involving the ground level. (2) Collisional ionization and three-body recombination. (3) Collisions to and from level 3. (4) Photoionization and radiative and dielectronic recombination. (5) 2–3 absorption and 3–2 emission. (6) 1–2 absorption and 2–1 emission.

These parameters can be combined to write an equation for the photoionization rate per particle,

$$I_X = \int_{\nu_X}^{\infty} \frac{(L_\nu / h\nu)\sigma_\nu e^{-\tau_\nu}}{4\pi r^2} d\nu, \qquad (5.2)$$

and the recombination rate per particle,

$$R_X = \alpha_X(T)N_e. \qquad (5.3)$$

The total ionization rate of the ion (cm^{-3} s^{-1}) is therefore $N_X I_X$, and the total recombination rate is $N_{X+1}R_X$. The two processes are illustrated, schematically, on the left side of Figure 5.1.

Photoionization, as well as other processes to be considered later, affect the population of levels with $n > 1$. This is illustrated on the right-hand side of Figure 5.1, where direct electronic transitions are shown separate from transitions that involve the emission or absorption of radiation.

We can now write a set of time-dependent differential equations for the fractional ionization of all the ions of the element in question,

$$\frac{dN_X}{dt} = -N_X[I_X + R_{X-1}] + N_{X-1}I_{X-1} + N_{X+1}R_X, \qquad (5.4)$$

where all other ionization and recombination processes have been neglected at this stage. The steady-state solution for an element with n electrons and $n + 1$ ions is given by the solution of a set of $n + 1$ equations. The first n equations are of the type

$$\frac{N_{X+1}}{N_X} = \frac{I_X}{R_X}, \qquad (5.5)$$

and the additional equation is provided by the abundance constraint,

$$\Sigma N_X = N_Z, \qquad (5.6)$$

where N_Z is the total abundance of the element in question. Thus, the level of ionization of a gas in photoionization equilibrium is proportional to I_X/R_X, which, in turn (see Equations 5.2 and 5.3), is proportional to the photon number density divided by the electron density.

The preceding set of equations leads to the definition of two important time scales: the recombination time, $t_{rec} = 1/R_X$, and the ionization time, $t_{ion} = 1/I_X$. The first is the typical time required for ion $X + 1$ to recombine to X, when the radiation source is turned off. The second gives the time that is required for X to get ionized to $X + 1$ following a large increase in the ionizing flux. The recombination time depends on the particle (electron) density, and the photoionization time depends on the photon number density. Thus, following a time-dependent recombination process provides a way to probe the local conditions in the gas, while following a time-dependent ionization process enables us to set limits on the local radiation field.

The simplest example of time-dependent equations of this type is the one describing the ionization and recombination of hydrogen. In this case,

$$\frac{dN_{H^0}}{dt} = -I_{H^0}N_{H^0} \quad \frac{dN_{H^+}}{dt} = -\alpha(H)N_e N_{H^+}, \qquad (5.7)$$

with the trivial solutions,

$$N_{H^0} = N_{H^0}(0)\exp(-I_{H^0}t) \quad N_{H^+} = N_{H^+}(0)\exp(-\alpha(H)N_e t). \qquad (5.8)$$

For $T = 10^4$ K, $\alpha(H) \sim 10^{-13}$ cm^3 s^{-1}. In this case,

$$t_{rec} = \frac{1}{\alpha N_e} \simeq \frac{10^{13}}{N_e} \text{ sec.} \qquad (5.9)$$

The recombination and ionization time scales can be quite different, depending on the radiation field and the gas density. Moreover, the coupling between the various stages of ionization suggests that, under various conditions, the fractional abundance of ion X may change very slightly or remains almost constant, despite large changes in the radiation field. Following the details of all such processes leads to a somewhat different definition of the recombination time that involves positive contributions (recombination from X^{+1}) as well as negative (recombination to X^{-1}) contributions. Such expressions can be found in various textbooks and reviews. Perhaps more important are the global recombination and ionization time scales, which are the times required for all the gas to undergo a significant change in the level of ionization. Both are roughly given by $1/R_X$, with X being the *most abundant* ion of the element in question.

While photoionization from the ground level and radiative recombination followed by cascade to the ground level are the dominant processes in many cases of astrophysical interest, other processes must also be considered. In AGN gas, the most important additional processes are as follows:

Dielectronic recombination: This name refers to recombination through excited states. The process involves one free electron and one bound electron that combine, for a short time, to form a highly excited state above the ionization threshold of the ion. The excited state can decay via *autoionization* (i.e., returning to its original state) or via recombination, which results in a more neutral ion.

Dielectronic recombination is traditionally divided into *low-temperature dielectronic recombination*, due to $\Delta n = 0$ transitions, where n is the energy level in question, and *high-temperature dielectronic recombination*, involving $\Delta n = 1$ transitions. The first type is important for many elements at $T_e = 1 - 3 \times 10^4$ K and the second for a much hotter ($T_e \sim 10^{6-7}$ K) plasma.

Auger ionization: Ionization from inner shells due to absorption of high-energy X-ray photons is important for ions with three or more electrons. The ejection of the inner electron from ion X results in an excited unstable ion $^*X^{+1}$. This unstable ion decays, in most cases, by ejection of one or more additional electrons, which leads to even higher ionization (e.g., ion X^{+2}). In a small fraction of the cases, the relaxation to the ground state of X^{+1} involves the emission of a line photon. In many of the cases, this is due to a $2p$–$1s$ transition. This fraction is called the *fluorescence yield* of the line, and its value depends strongly on the charge of the ion in question. For example, for low-ionization oxygen ions, the fluorescence yield is of order 0.01, while for iron ions, it is typically 0.3.

Auger ionization couples nonadjacent stages of ionization and must be treated accordingly. For the most abundant elements with $Z < 26$, the inner shell ionization involves a doubly ionized atom (two stages above the ion that absorbs the initial high-energy photon, e.g., coupling O^{+1} to O^{+3}).

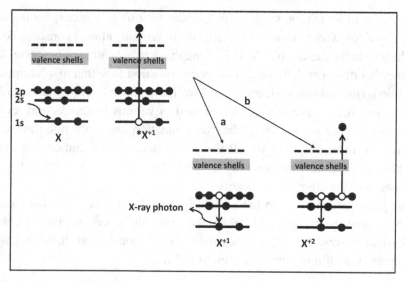

Figure 5.2. Auger ionization of ions with more than 10 electrons. Ion X is photoionized from its $1s$ level by a high-energy photon (top left). This leads to an
excited, unstable ion $^*X^{+1}$ (top right) with a vacancy in its ground level. Stabi-
lization is by one of two routes: (1) a 2–1 transition leading to the emission of
an X-ray fluorescence line photon and (2) ejection of an additional electron from
level 2 resulting in a stable X^{+2} ion.

For iron and heavier elements, there are other routes that can lead to even
more ionization, depending on the number of inner shell electrons. The various routes of the process are illustrated in Figure 5.2.

Heating and ionization by secondary electrons: This process is important in
gas with a very low level of ionization that is exposed to a large flux of high-
energy photons. The energetic electrons that are ejected due to the absorption
of such photons can collide with neutral atoms (mostly hydrogen) and cause
additional ionizations before they are thermalized. They can also collisionally
excite $n > 1$ levels, causing additional line emission. An important example
is the excitation of the hydrogen Lyα transition by such electrons in regions
of almost completely neutral gas, despite the low electron temperatures in
such regions.

Charge exchange: Charge exchange collisions between ions,

$$X + Y^+ \rightleftharpoons X^+ + Y, \tag{5.10}$$

can be very fast, leading, under favorable conditions, to significant changes
in the level of ionization. An important example for photoionized gas at
low temperatures involves neutral hydrogen and singly ionized oxygen. This
reaction is known to be very fast under conditions prevailing near the hydrogen
ionization front. Other important examples are charge exchange of neutral

nitrogen with ionized hydrogen and charge exchange of several low-ionization metal ions with He^+.

Collisional ionization and three-body recombination: These processes can be described as

$$X + e \rightleftharpoons X^+ + e + e. \tag{5.11}$$

They are negligible for most ions for temperatures below few $\times 10^5$ K. Collisional ionization dominates the ionization of the gas in high-temperature conditions such as in starburst regions, near shock fronts, and so on. In AGNs, both processes can become important in a high-density, large-optical-depth gas, where they can dominate the population of the high-n levels of hydrogen and helium.

Compton ionization, heating, and cooling: As shown in § 2.3, photon–electron collisions can heat or cool the electron gas. In particular, the inverse Compton process can be a major source of cooling for a highly ionized gas. High-energy X-ray photons can also knock out valence electrons from bound orbits, contributing in this way to the ionization of the gas.

5.2. Basic spectroscopy

5.2.1. Absorption and emission coefficients

The calculation of the emitted line and continuum spectrum requires the complete solution of the radiative transfer equation (Equation 2.1). In general, there are many lines of many ions that contribute to the emission. There are also several continuum emission and absorption processes that can be coupled to the line radiation field. This requires several more definitions of absorption and emission cross sections and some modifications of the basic quantities defined in Chapter 2.

The treatment of the transitions between levels i and j, with statistical weights g_i and g_j, where $j > i$, is through the use of the Einstein emission and absorption coefficients: the coefficient for spontaneous emission, A_{ji}, the absorption coefficient, B_{ij}, and the stimulated emission coefficient, B_{ji}. The relationships between these coefficients are

$$A_{ji} = \frac{2h\nu^3}{c^2} B_{ji} \tag{5.12}$$

and

$$g_i B_{ij} = g_j B_{ji}, \tag{5.13}$$

where ν is the line frequency. Using these definitions, the isotropic line emission coefficient is

$$\epsilon_\nu = \frac{1}{4\pi} n_j A_{ji} h\nu \Phi_\nu, \tag{5.14}$$

where Φ_ν is the normalized line profile assumed to be identical for both absorption and emission.

The upward-only absorption coefficient, κ_ν, is related to the absorption cross section per particle, σ_ν, by

$$\kappa_\nu = n_i \sigma_\nu = n_i \sigma \Phi_\nu. \tag{5.15}$$

Using the Einstein coefficients, we get

$$\kappa_\nu = \frac{1}{4\pi} \left(n_i B_{ij} - n_j B_{ji} \right) h\nu \Phi_\nu. \tag{5.16}$$

It is clear from this expression that stimulated emission is counted as negative absorption. We can now write an expression for the source function, S_ν, using its standard definition (Equation 2.3):

$$S_\nu = \frac{\epsilon_\nu}{\kappa_\nu} = \frac{A_{ji} n_j}{B_{ij} n_i - B_{ji} n_j} = \frac{2h\nu^3}{c^2} \frac{1}{(n_i g_j / n_j g_i) - 1}. \tag{5.17}$$

It is customary to define the *integrated* (over the line profile) cross section per particle, uncorrected for stimulated emission, using the absorption oscillator strength, f_{ij},

$$\sigma = \frac{\pi e^2}{m_e c} f_{ij} \simeq 0.0265 f_{ij}, \tag{5.18}$$

where m_e is the electron mass. Similarly, the *line center* cross section per particle is

$$\sigma_{lc} = \frac{\sigma}{\sqrt{\pi}}. \tag{5.19}$$

The downward oscillator strength is given by

$$f_{ji} = -\frac{g_i}{g_j} f_{i,j}. \tag{5.20}$$

We can also write a relationship between A_{ji} and f_{ij},

$$f_{ij} = 1.499 \times 10^{-16} \lambda^2 \frac{g_j}{g_i} A_{ji}, \tag{5.21}$$

where λ, the line wavelength, is given in angstrom units.

Manipulation of these expressions gives the following relationship for the incremental *line center* optical depth over a distance dr,

$$d\tau_{ij} = \frac{1.5 \times 10^{-10} \lambda f_{ij}}{v_D}(n_i - n_j g_i/g_j)dr, \tag{5.22}$$

where λ is in angstrom units and v_D is the *Doppler velocity spread parameter* (referred to also as the *line b parameter*), defined by

$$v_D = \left[\frac{2kT_e}{m_i}\right]^{1/2} = 1.29 \times 10^4 \left[\frac{T_e}{A}\right]^{1/2} \text{ cm s}^{-1}, \tag{5.23}$$

where m_i is the mass of the ion and A is its atomic weight. A simple generalization to the case of a turbulent medium with a random velocity field of magnitude v_{turb} is to replace v_D by $\sqrt{v_D^2 + v_{\text{turb}}^2}$. Note again that we use here, and in all other places in the book, the line center cross section and the line center optical depth.

Consider now the energy separation of the levels, E_{ij}, and define J to be the angle and frequency average of the radiation intensity I_ν:

$$J = \frac{1}{4\pi} \int \int I_\nu \Phi_\nu d\nu d\Omega. \tag{5.24}$$

Using this notation, the rate of upward transitions from level i to level j, due to the local radiation field, is $n_i J B_{ij}$, and the rate of downward transitions is $n_j(A_{ji} + J B_{ji})$. These rates enter the statistical equilibrium equations that are described in § 5.3.

5.2.2. *Atomic transitions and selection rules*

This section gives a short summary of the terminology being used to describe atomic transitions. The transition probability A_{ji} between levels i and j is obtained from quantum-mechanical calculations. The transition probability can be high or low, depending on the lifetimes of the levels involved. This in turn depends on various selection rules that are associated with the quantum numbers of the levels.

Quantum numbers

The following quantum numbers are commonly used:

Principal quantum number n: Integer greater than zero.

Spin angular momentum: s for a single electron and S for all electrons; $s = \pm 1/2$ in units of $h/2\pi$. The spin of the electron is $1/2$. The vector sum spin angular momentum, S, takes on integer or half-integer values.

Orbital angular momentum: l for a single electron and L for all electrons. Allowed values are $l = 0, 1, 2, \ldots (n-1)$ in units of $h/2\pi$. $L = 0, 1, 2,$

3, 4, and 5 are designated S, P, D, F, and H, respectively, and similarly for a single electron, s, l, and so on.

Total angular momentum: j is the vector sum of s and l of a single electron. J is the vector sum of S and L. The quantum numbers n, S, L, and J determine the energetic state of the atom. For closed shells, generally, $S = 0$, $L = 0$, and $J = 0$. Thus we are concerned only with electrons outside a closed shell.

Configuration terms and multiplets

LS **and** jj **couplings:** The usual case for light ions is for the individual orbital angular momenta to couple to a certain L and for the individual spins to couple to a certain S. The total angular momentum in this case is $J = L + S$. The number of possible J values is $2L + 1$ if $L < S$ or $2S + 1$ if $L \geq S$. This is called LS (Russell–Saunders) coupling.

In many electron ions, each orbital angular momentum l tends to combine with the individual spin s to form the individual angular momentum j. These then add up to form the total angular momentum J. This kind of interaction is known as jj coupling.

Configuration: This is the set of quantum numbers of an individual electron. There are several possible combinations of LS in each configuration, each one called a *term*. A term can include several levels and can have odd or even *parity*, which refers to its summed electronic l. The designation of a term is written as $^{2S+1}L_J^{\text{parity}}$, for example, $^3P_1$ for even parity and $^3P_1^0$ for odd parity terms with $L = 1$, $S = 1$, and $J = 1$.

Multiplet: All possible transitions between two terms.

Permitted forbidden and semiforbidden transitions

Permitted transitions: Electric dipole transitions with $\Delta n \geq 1$. The typical A-values are 10^8–10^{10} s^{-1} and depend on atomic charge.

Forbidden transitions: All other transitions, mostly magnetic dipole and electric quadrapole transitions. The typical A-values are 10^{-3}–1 s^{-1}. The values are smaller if the transitions are between fine-structure levels inside a term.

Semiforbidden (intercombination) transitions: These are transitions that do not obey the $\Delta S = 0$ selection rule (i.e., they connect configurations with different S). Their typical A-values are 10^1–10^3 s^{-1}.

Selection rules for atomic transitions

LS coupling selection rules for permitted and forbidden transitions are summarized in Table 5.1. The jj coupling selection rules are different and are not listed in the table.

Table 5.1. *Selection rules for LS coupling*

Electric dipole (permitted)	Magnetic dipole (forbidden)	Electric quadrapole (forbidden)
$\Delta J = 0, \pm 1$	$\Delta J = 0, \pm 1$	$\Delta J = 0, \pm 1, \pm 2$
no $0 \rightarrow 0$	no $0 \rightarrow 0$	no $0 \rightarrow 0, 1/2 \rightarrow 1/2, 0 \rightarrow 1$
$\Delta L = 0, \pm 1$	$\Delta L = 0$	$\Delta L = 0, \pm 1, \pm 2$
no $0 \rightarrow 0$		no $0 \rightarrow 0, 0 \rightarrow 1$
$\Delta l = \pm 1$	$\Delta l = 0$	$\Delta l = 0, \pm 2$
	$\Delta n = 0$	
Parity change	no parity change	no parity change
$\Delta S = 0$	$\Delta S = 0$	$\Delta S = 0$

5.2.3. Energy-level diagrams

The range of ionization in AGN gas can be very large. For example, the gas near the illuminated face of a cloud that is situated near the central source of radiation can be highly ionized. Much deeper into the cloud, the material can be almost completely neutral. For oxygen atoms in the gas producing the strongest observed emission lines, the range can include all ionization stages from O I to O VI. Outflowing X-ray ionized gas, and gas in the coronal-line region (Chapter 7), can include even higher stages of ionization, like O VIII and Mg X. Some outflowing systems are suspected to contain almost fully ionized plasma, with Fexxvi and Fexxvii being the dominant iron ions. To illustrate some of the transitions leading to the strongest observed line, we show, in the following figures, the energy-level diagrams of ions with 1–8 electrons. The examples used in the illustration include all the oxygen ions. Also shown is the case of Feii, a more complicated system with 25 electrons.

Single-electron systems: H-like ions. One-electron, "H-like" ions include Hi, Heii, CVi, NVii, OViii, and so on. The energy of the leading $2p$–$1s$ transition in these ions (the Lyα line) is at $E = 10.2 \times Z^2$ eV, where Z is the ionic charge. To a very good approximation, $A_Z(2p - 1s) = A_1(2p - 1s)Z^4$, where $A_1(2p - 1s) = 6.26 \times 10^8$ s^{-1} is the A-value for neutral hydrogen. (The oscillator strength f_{ij} is independent of the ionic charge.)

Figure 5.3 shows a partial Grotrian (energy-level) diagram for H-like ions using, as an example, O VIII. The energies in this and the following diagrams are not plotted to scale and are only meant to illustrate the lowest levels of the ions. The Lyα lines of all abundant H-like ions are observed in AGN spectra. For example, the He II Lyα line at $\lambda = 303$ Å is predicted to be among the strongest lines in the spectrum. However, most of the line photons are produced deep inside optically thick gas, and only a very small fraction of

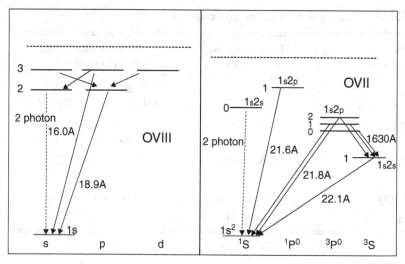

Figure 5.3. Grotrian (energy-level) diagrams for H-like and He-like oxygen show-ing the lower levels and the strongest transitions (energy is not to scale).

these photons can escape. The highly forbidden $2s$–$1s$ transitions in H-like ions give rise to a two-photon continuum. For this transition, $A(2s - 1s) \simeq 8.2Z^6$ s^{-1}. The very small transition probability is not, by itself, a signature of low intensity compared with the Lyα intensity. In fact, in low-density gas, where collisional transitions between $2p$ and $2s$ are slow compared with the radiative $2s$–$1s$ transitions, about 30 to 40 percent of all recombinations followed by cascade lead to the $2s$ level. Thus I(Lyα) $\simeq$ 2I(2-photon). We also note that the $2p$, $3p$, and so on, levels are split into two states with $J = 1/2$ and $J = 3/2$ (not shown in the diagram), which lead to doublet Lyα, Lyβ, and so on, lines. The differences in energy between the two become appreciable only in high-charge ions.

Two-electron systems: He-like ions. Lines of He-like ions, from helium to iron (right-hand side of Figure 5.3), are clearly observed in the spectrum of many AGNs. He I lines at 5876 and 10830 Å are very strong at optical–NIR ener-gies, and the 2–1 transitions of C V, N VI, and O VII and even higher charge ions are common in X-ray spectra. There are three types of 2–1 transitions: permitted or resonance transitions originating from ^1P, forbidden transitions from ^3S, and intercombination transitions from ^3P. The relative intensities of these lines are important density diagnostics because of their very differ-ent transition probabilities. The forbidden line, with the smallest A-value, is more easily suppressed by collisions with fast thermal electrons, while the permitted line is the one least affected by collisions. Intercombination tran-sitions are in between. The range of densities that can be probed using the

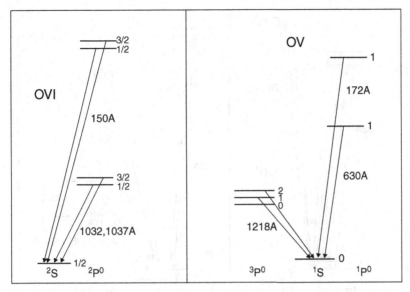

Figure 5.4. Same as Figure 5.3, but for Li-like and Be-like oxygen.

relative intensity of the He-line triplet of oxygen is about 10^{10}–10^{13} cm^{-3} for temperatures in the range 10^5–10^6 K.

The relative intensities of the 2–1 lines depend also on the gas temperature. In a low-temperature photoionized gas, with temperatures that are much below the excitation temperature of the levels in question (for O VII, this means few $\times\ 10^5$ K), the level populations are determined almost exclusively by recombination and cascade. In this case, and assuming densities that are well below the critical densities of the levels, the relative recombination rates determine the relative line intensities. For a much higher temperature gas, with kT_e approaching the energies of the 2s and 2p levels ($T \sim 6 \times 10^6$ K for O VII), collisional excitations from the ground dominate the line intensities.

Three-electron systems: Li-like ions. The Grotrian diagram for O VI is shown in Figure 5.4. Emission lines from the most abundant Li-like ions are clearly seen in AGN spectra, mostly in the optical–UV part. The strongest lines in the BLR and the NLR are the doublet resonance lines due to ^2S–$^2P^0$ transitions, for example, C IV λ1549 and O VI λ1035. The two lines in this doublet have the same A-value and hence f-values that differ by their statistical weight (Equation 5.21). Their relative intensities can be used to infer the optical depths of these transitions. Such observations can only be obtained in low-velocity narrow emission line because of the small wavelength separation between the lines ($\sim$1650 km s^{-1} for the O VI doublet and $\sim$498 km s^{-1} for the C IV doublet).

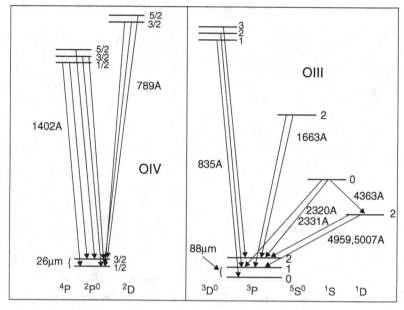

Figure 5.5. Same as Figure 5.3, but for B-like and C-like oxygen.

Four-electron systems: Be-like ions. These lines (right side of Figure 5.4) are also common in AGN gas. The strongest are due to the $^3P^0-^1S$ transitions. Examples are [O v] $\lambda1218$, [N iv] $\lambda1486$, and [C iii] $\lambda1909$. The transitions include both forbidden ($^3P_5-^1S$) and intercombination ($^3P_3-^1S$) lines, and their ratios provide useful density diagnostics. For $T_e \sim 10^4$ K, the critical densities for the intercombination lines are in the range 10^9-10^{11} cm^{-3}, and those for the forbidden lines are 4–5 orders of magnitude smaller.

Five-electron systems: B-like ions. Strong lines of the B-like ions C ii, N iii, and O iv are commonly observed in the UV spectrum of AGNs. The intercombination transition $4P-2P$ can be used to deduce gas densities in the range $10^{10}-10^{12}$ cm^{-3}. The forbidden fine-structure lines in the ground term of B-like ions are observed to be strong in the MIR spectrum of many AGNs. An important example is [O iv] 25.9 μm, which, in many cases, is the strongest line in the 5–40 μm part of the spectrum. Similarly, the 158 μm line of C ii is very strong in the FIR spectrum of AGNs and SF galaxies. The Grotrian diagram for O iv is shown in Figure 5.5.

Six-electron systems: C-like ions. Lines of C-like ions (right-hand side of Figure 5.5) are observed to be strong in low-density gas in HII region, in the NLR of AGNs, in planetary nebulae, and in other environments. Figure 5.5 shows the O iii transitions. The [O iii] $\lambda5007$ line is seen in the spectrum almost all AGNs. The combination with [O iii] $\lambda4363$ can be used to determine temperatures in low-density gas. The semiforbidden line at

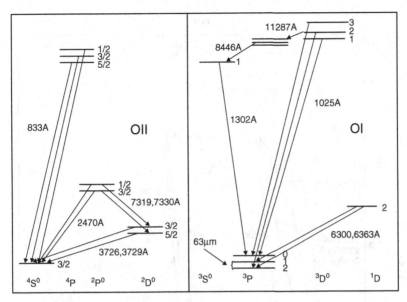

Figure 5.6. Energy-level diagrams for N-like ions and for O I.

1663 Å is a common but weak feature in the spectrum of many BLRs. The fine-structure line at 88 μm provides important diagnostics for gas in AGNs and SF galaxies.

Seven-electron systems: N-like ions. Seven-electron ions, like O II, emit forbidden lines that are useful density diagnostics in low-density gas. Such lines, near 3727 and 7320 Å, are shown in Figure 5.6. The resonance transitions near 833 Å are hardly ever observed because of the high energy (considering the low-ionization ion) and because they are easily absorbed by neutral hydrogen, which is abundant in such an environment.

Eight-electron systems: O-like ions. Figure 5.6 illustrates the energy-level diagram of O I, an important example of an eight-electron system. The strongest low-density lines, at 6300 and 6363 Å, are observed in the spectrum of many AGNs. The permitted $^3D^0-^3P$ lines demonstrate the interesting case of *line fluorescence*. Several of these lines coincide in wavelength, almost exactly, with the hydrogen Lyβ line. Since O I and H I occupy the same region in photoionized gas, this leads to the possibility of absorption of the Lyβ photons by O I followed by the emission of the 11,287 and 8446 Å O I lines (see diagram). These lines, especially the one at 8446 Å, are indeed strong in planetary nebulae, and in AGN gas, confirming this route. Another important example of an eight-electron ion is Ne III. The strongest lines in this case are the forbidden $^1D^0-^3P$ transitions at 3868 and 3967 Å.

Twenty-five-electron systems: The special case of Fe II. To complete the list of examples, we show in Figure 5.7 a small part of the energy-level diagram of Fe II. This ion is very abundant in the low-ionization part of AGN clouds.

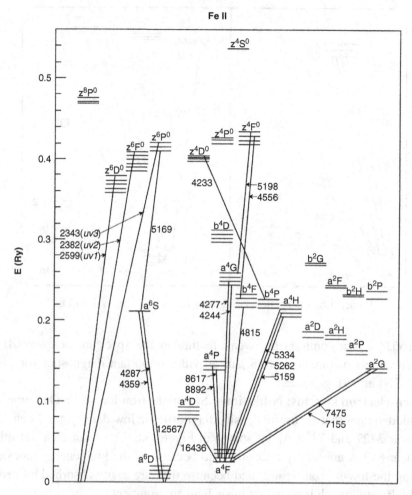

Figure 5.7. Energy-level diagram for Fe II showing some of the multiplets that are important in AGNs (from Sigut and Pradhan, 2003; reproduced by permission of the AAS).

Its many low-energy levels produce a rich spectrum with thousands of emission lines that overlap in wavelength and cover much of the 2000–5500 Å continuum of AGNs.

5.3. Thermal balance

The thermal structure of the gas is determined by the balance between various heating and cooling processes. The common notation in describing these processes is to use H for heating and C for cooling (both in erg cm^{-3} s^{-1}). The heating and cooling of astrophysical plasma can be described by a set of equations that involve *energy* terms rather than *ionization and recombination* terms. The most important heating in AGN gas is due to photoionization (bound–free heating). For ion X, the heating term is given by

$$H_{bf,X} = N_X \int_{\nu_X}^{\infty} \frac{L_\nu \sigma_\nu e^{-\tau_\nu}}{4\pi r^2} d\nu. \qquad (5.25)$$

Note that in this equation, we sum over the absorbed energy, while in the photoionization equation (Equation 5.2), the sum is over the number of ionizing photons. Similarly, an important cooling term is *recombination cooling*, which, for ion X, is given by

$$C_{bf,X} = N_{X+1} N_e \alpha_X(T)[h\nu_{1,\infty} + h < \nu_{\text{free}} >], \qquad (5.26)$$

where $h\nu_{1,\infty}$ is *the minimum* ionization energy from the ground level of the ion and $h < \nu_{\text{free}} >$ is the mean energy of a free recombining electron that is of order kT_e.

The heating and cooling terms are simple manifestations of energy conservation. They are different from standard heating and cooling expressions given in other books, where *the energy of the free electrons* is considered. In that case, the term L_ν in the heating integral should be replaced by $(L_\nu/h\nu)(h\nu - h\nu_X)$, and the recombination cooling equation does not include cascade to the ground level (the $h\nu_{1,\infty}$ term). The two treatments are equivalent, and all other heating and cooling terms can be computed with either approach, provided they are used consistently.

The free–free process is another cooling mechanism, which has already been discussed in § 2.5. It is especially important in fully ionized plasma, where it can be the dominant way of losing energy. The cooling rate for ion X with charge Z was given in Equation 2.40 and is repeated here for completion:

$$C_{ff,X} = 1.42 \times 10^{-27} Z^2 T^{1/2} g_{ff} N_e N_{X+z}. \qquad (5.27)$$

Several cases of astrophysical interest include gas that is not fully ionized. In many of these, the most important cooling term is line cooling due to bound–bound (*bb*) transitions. The computation of this term, denoted C_{bb}, requires a complete solution of the statistical equilibrium (level population) equations.

The most important contributers to the level populations in AGN gas are recombination and cascade, emission and absorption of line photons, and collisional excitation and deexcitation of bound states. Consider first bound–bound collisional transitions between levels i and j, with statistical weights g_i and g_j, where $j > i$. The collisional excitation rate between the levels is

$$q_{ij} = C_{ij} N_e, \qquad (5.28)$$

where

$$C_{ij} = \frac{8.63 \times 10^{-6}}{\sqrt{T}} \frac{\Omega_{ij}}{g_i} e^{-E_{ij}/kT}, \qquad (5.29)$$

where Ω_{ij} is the effective collision strength that is obtained from quantum-mechanical calculations. Similarly, the collisional rate from j to i is

$$q_{ji} = C_{ji} N_e, \qquad (5.30)$$

where

$$C_{ji} = C_{ij} \frac{g_i}{g_j} e^{E_{ij}/kT}. \qquad (5.31)$$

Next we add radiative transitions between levels which depends on the level populations. Solving for the level populations involves a large set of time-dependent statistical equilibrium equations. To illustrate this, we consider a simple two-level system and only three processes: recombination with a rate $\alpha_{\text{eff},2}$, collisional excitation and deexcitation with rates q_{12} and q_{21}, respectively, and line (but no continuum) photon emission and absorption. Consider first a case of very small optical depth in the 2–1 transition. The time-dependent equation for level 2 can be written as

$$\frac{dn_2}{dt} = N_{X+1} N_e \alpha_{\text{eff},2} + n_1 q_{12} - n_2 q_{21} - n_2 A_{21} - J(n_2 B_{21} - n_1 B_{12}), \qquad (5.32)$$

where the frequency and direction averaged intensity, J, was defined in Equation 5.24. The additional constraint that allows the full solution, once N_X and N_{X+1} are known, is

$$n_1 + n_2 = N_X. \qquad (5.33)$$

The level populations are time dependent since the ionic abundances and the electron temperature are time dependent. Additional processes that have been neglected in these equations, and were mentioned earlier as potentially important in AGN gas, are dielectronic recombination, collisional ionization of both levels, three-body recombination, and photoionization from both levels. An additional "continuum pumping" process, which will be explained later, is in fact included through J. Most of these can easily be added to the preceding equations to obtain a more complete and accurate solution.

Having solved for n_1 and n_2, we can now write an expression for the time-dependent line emissivity, ϵ_{21} (erg s^{-1} sec^{-1} cm^{-3}). For optically thin gas, this is given by

$$\epsilon_{21} = n_2 A_{21} h\nu_{21}. \qquad (5.34)$$

The expression contains the assumption that the emitted radiation can *freely escape* the medium. In such a case, a photon emitted in a certain location carries with it the maximum possible energy, $h\nu_{21}$. This results in maximum cooling of the gas. However, this assumption is not always correct. Some lines, in particular, the leading resonance lines of hydrogen, helium, and the more abundant elements, can become optically thick to their own radiation. A line photon emitted in such a transition can be absorbed in a different location in the gas, contributing additional heating at that point and increasing the n_2 level population there. This process is sometimes described as *self-absorption*. For lines with large optical depths, the absorbing location is close to the location of emission. For lines with small optical depths, the absorbing location can be in a different part of the cloud.

A simple way of dealing with this complication is to introduce a new function, the *line escape probability*, which gives the probability of line photons to leave the point where they are created without being absorbed by another 1–2 transition. This escape probability, which is indicated here by β_{12}, can be specified as an average property over the entire cloud (a *mean escape probability*) or as a local quantity (a *local escape probability*). All that is required is to replace A_{21} in Equations 5.32 and 5.34 by an effective A-value $\beta_{21} A_{21}$, that is, not to count those photons that are absorbed on the way out. Thus, the local n_2 level population is increased by a factor of $1/\beta_{12}$ relative to the previous case of negligible line opacity.

The more rigorous treatment is based on the fact that $(1 - \beta_{12})$ is the probability of the photon to be trapped. This gives the following modified definition of J:

$$J = S(1 - \beta_{12}). \tag{5.35}$$

Using the definition of S from Equation 5.17, and the rate equation for level 2 (Equation 5.32), it is easy to show that A_{21} should be replaced by $\beta_{21} A_{21}$, exactly as obtained from the simplistic argument used earlier.

The additional modification required in the escape probability approach is to replace the old definition of the line emission per unit volume by

$$\epsilon_{21} = n_2 \beta_{21} A_{21} h\nu_{21}. \tag{5.36}$$

For a simple two-level system, and in the absence of nonradiative processes, the net result is no change in the local cooling and in the emergent line emission, relative to the case of very small optical depth where $\beta_{12} = 1$. The only change is the increase in the $n = 2$ level population. In reality, this is not the case because transitions to and from other levels, which had not been considered so far, must be included in the level population equations. In addition, collisional suppression of the upper levels of optically thick transitions becomes more important because of the smaller effective A-value, and there are other global effects.

The escape probability formalism provides a simple way to replace the more rigorous, full radiative transfer solution. It enables us to obtain a full local solution for the level population but neglects the radiative coupling of different locations in the gas. Part of its success relies on the proper definition of the escape probability function. Comparison with detailed radiative transfer calculations shows that a reasonable approximation is

$$\beta_{12} = \frac{1}{1 + b\tau_{12}}, \tag{5.37}$$

where τ_{12} is the optical depth between the two levels and b is a constant of order unity.

Extensive theoretical studies show that the escape probability formalism, in which each line is assigned its own escape probability, is an adequate method to treat many resonance transitions, those leading to the ground states of the atoms. There are well-known cases in which the approximation is more problematic. For example, treating the emission of hydrogen lines in AGN gas is problematic since, in such cases, the strong ionizing radiation field can result in clouds of large column densities where hydrogen is almost fully ionized. In such cases, the hydrogen level 2 population can be large and the optical depth in the Balmer lines significant. This, coupled with the extremely large optical depth of the Lyman lines, makes line transfer and photon diffusion in the cloud difficult to treat and reduces the usefulness of the escape probability approach.

Line photons can also be absorbed by the *bf* continuum. For example, in large-column-density AGN clouds, the hydrogen Balmer continuum can attain a significant optical depth. Resonance line photons, like Civ λ1549 or Mg II λ2798, can be absorbed by this continuum. This results in additional ionization of hydrogen and in the attenuation of the absorbed emission lines. The escape probability formalism provides a simple way to take into account the effect of this process. Consider again the two-level atom, a 1–2 line absorption cross section of κ_l and a continuum absorption cross section, at the line frequency, of κ_c. Define

$$X_l = \frac{\kappa_l}{\kappa_l + \kappa_c}, \qquad X_c = \frac{\kappa_c}{\kappa_l + \kappa_c}, \tag{5.38}$$

$$\beta_{lc} = \beta(\tau_l + \tau_c). \tag{5.39}$$

The escape probability formalism amounts to the replacement of the line-only escape probability, $\beta(\tau_l)$, by an effective escape probability,

$$\beta_{\text{eff}} = X_c + X_l\beta_{lc}. \tag{5.40}$$

The emergent line flux, per unit volume, becomes

$$n_2 A_{21}\beta_{lc}h\nu_{12}, \tag{5.41}$$

and the number of line photon absorptions, (i.e., the photoionization of the ion providing the continuum opacity) is

$$n_2 A_{21} X_c(1 - \beta_{lc}). \tag{5.42}$$

Finally, we have to consider the contribution to J due to continuum radiation reaching the location in the gas where we solve for the n_1 and n_2 level population. If the opacity to the illuminated surface at the line frequency ν_{12} is τ_{in}, and the monochromatic continuum luminosity L_ν, the additional contribution to J in the escape probability formalism is

$$\beta(\tau_{in})\frac{L_\nu}{16\pi^2 r^2},\tag{5.43}$$

where r is the distance from the point in the gas to the source of continuum radiation.

We now return to the cooling associated with line emission, C_{bb}. This is simply the sum over all emission lines,

$$C_{bb} = \Sigma_{j>i}\epsilon_{ji}.\tag{5.44}$$

Using this sum, and having computed all other heating and cooling processes, we can write the general energy equation,

$$H = C,\tag{5.45}$$

where

$$H = \Sigma_i H_i\tag{5.46}$$

$$C = \Sigma_i C_i.\tag{5.47}$$

The solution of this equation can be used to obtain the kinetic temperature of the gas. This is relatively simple in steady state, where the level population and ionization fractions are not time dependent. The main complications are due to radiative transfer in optically thick lines and continua that couple the level populations to the gas temperature through collisional excitation and deexcitation processes. Because of the coupling, several iterations are normally required to obtain a full solution of the ionization and thermal structure of the gas, even in a steady state, time independent situation.

5.4. The spectrum of ionized AGN gas

5.4.1. Ionization parameter

The earlier discussion suggests that the level of ionization of photoionized gas depends on I_X/R_X, which is proportional to the ratio of the ionizing photon density to the gas density. Given the conditions of photoionization equilibrium, one can define an *ionization parameter* proportional to this ratio:

$$U = \int_{E_1}^{E_2} \frac{(L_E/E)dE}{4\pi r^2 c N_H}, \tag{5.48}$$

where c, the speed of light, is introduced to make U dimensionless and E_1 and E_2 are the energies defining the limits of integration over the ionizing continuum.

There are various ways to define U such that its value provides a good indication for the level of ionization of the gas. The best is to make sure that E_1 is close to the minimum energy required to ionize the most important (cooling-wise) ions. For gas that is ionized by soft-UV radiation, E_1 should be chosen such that it is related to the UV ionizing field, for example, the energy of the hydrogen Lyman limit at 13.6 eV. An appropriate name in this case is U(hydrogen). Similarly, a suitable choice for gas whose level of ionization is dominated by a strong soft X-ray source is $E_1 = 0.1$ keV and $E_2 = 10$ keV. This is usually assigned the symbol U(X-ray) or U_X. Such gas can be transparent to UV photons, and U(hydrogen) is not a good indicator of its level of ionization. An alternative suitable choice for the case of high-ionization X-ray-dominated gas might be $E_1 = 0.54$ keV, corresponding to the K-shell threshold ionization of oxygen, the most important emitting and absorbing element under such conditions. A suitable name is U_{oxygen}.

There are other, somewhat different definitions of the ionization parameter that are used in the literature. A common one, ξ, which is used in many X-ray studies, is defined by the energy flux, rather than the photon flux, and is given by

$$\xi = \int_{13.6\ \text{eV}}^{13.6\ \text{keV}} \frac{L_E}{r^2 N_H} dE. \tag{5.49}$$

Using the "right" ionization parameter, we normally find that $U = 10^{-1}$–10^{-2} corresponds to gas that produces strong emission lines in the energy range in question. For example, $U_{\text{oxygen}} = 0.1$ results in strong 0.5–3 keV emission lines. A much smaller U results in more neutral gas, with very little X-ray emission, and a much larger U results in very highly ionized gas and insignificant line emission. Table 5.2 gives a list of several ionization parameters that are useful in various situations and are commonly used in AGN studies. The right column of the table allows a simple conversion between the various ionization parameters for the case of a simple power-law continuum with $L_E \propto E^{-1.3}$.

5.4.2. *Line and continuum emission processes*

The dominant continuum emitting processes in low-temperature, photoionized gas are due to bound–free transitions. The clear signature of such processes in the optical–UV band is the bound–free edges of levels $n = 1$ and $n = 2$ of hydrogen and helium. An example is shown in Figure 5.8. In the X-ray band, the strongest

Table 5.2. *Various ionization parameters used in AGN research*

Ionization parameter	E_1	E_2	Conversion factors, $L_E \propto E^{-1.3}$
U_{hydrogen}	13.6 eV	∞	1000
U_{helium}	54.4 eV	∞	165
$U_{\text{X-ray}}$	0.1 keV	10 keV	73.3
U_{oxygen}	0.54 keV	10 keV	8.2
ξ	13.6 eV	13.6 keV	31.1

observed transitions are those leading to the ground levels of H-like and He-like carbon, nitrogen, oxygen, neon, magnesium, and silicon. Free–free emission is also seen mostly at radio frequencies due to low-temperature gas and at hard X-ray energies due to very high temperature gas.

Bound–bound transitions dominate the heating–cooling balance in low-ionization ($U(\text{hydrogen}) \sim 0.01$) photoionized gas. The highest-intensity transitions are due to forbidden lines in low-density gas and to recombination and collisionally ionized permitted lines in high-density gas. For highly ionized X-ray gas ($U_{\text{oxygen}} \sim 0.01$), most cooling is via bound–free transitions because the lowest accessible atomic levels are at energies that are much higher than kT. This gas is dominated by permitted, intercombination, forbidden, and recombination lines. Starburst-heated X-ray gas, which is sometimes found in AGNs, is characterized by higher temperatures and lower densities. The most intense lines in such cases

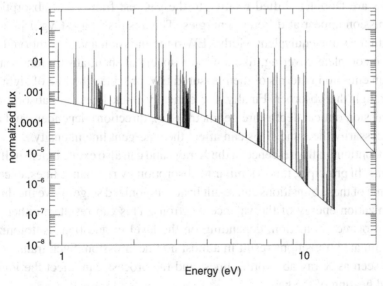

Figure 5.8. The calculated spectrum of optically thin $T_e \sim 10^4$ K $N_H \sim 10^6$ cm^{-3} photoionized gas, showing the strong bound–free recombination edges (in this case, the Lyman, Balmer, and Paschen jumps at 13.6, 3.4, and 1.51 eV, respectively) and many emission lines.

are produced via collisional excitation of the $n = 2$ and $n = 3$ levels of the more abundant elements.

Several additional line and continuum emitting processes become important under favorable conditions. Two of these are as follows:

Continuum fluorescence: This process, sometimes referred to as *photoexcitation* or *continuum pumping*, is the result of populating low-lying levels by absorbing the incident continuum in various resonance transitions. This increases the level population and results in additional line emission. The way to include this process in the level population equations is given by Equation 5.43.

The emission-line flux resulting from continuum fluorescence can be very strong, especially in cases where the line opacity is not very significant. However, the effect of this line emission on the global energy balance is negligible, especially in low-density environments, because most of the emitted radiation can escape the gas. Thus, heating and cooling almost exactly balance each other.

The increase or decrease in the observed line intensity, due to continuum fluorescence, is geometry dependent. For example, in a static spherical atmosphere, or in a static spherical thin shell around a point continuum source, the process can be considered as pure scattering because emission is exactly balanced by absorption. In a moving spherical atmosphere, the emitted photons are Doppler shifted relative to the gas rest frame, and absorption and emission appear at different energies. This results in a P-Cygni line profile with the same equivalent widths (EWs) for emission and absorption. In cases of incomplete covering, like a broken spherical shell, absorption is stronger than emission in terms of the measured EW, provided the line of sight passes through the absorber. Finally, for optically thick lines, the absorption and emission optical depths are geometry (e.g., direction) dependent and are not necessarily the same. This can affect the emergent line intensity.

Continuum fluorescence in the X-ray band can also excite inner shell transitions. In principle, this is similar to absorption by resonance lines, except that some of these transitions can result in an autoionized stage above the threshold ionization energy of the valence electrons. This can result in either ionization or recombination, depending on the level in question. Autoionization-dominated transitions result in almost no line emission. Such transitions can be seen as X-ray absorption lines, and the process can affect the ionization and heating of the gas.

Line fluorescence: As explained earlier, the ejection of inner shell electrons can lead to line emission (*fluorescence lines*). This process is important at X-ray energies where lines of this type, most notably iron Kα and Kβ

lines, in the 6.4–9.0 keV energy range are occasionally observed. The fluorescence efficiency (*fluorescence yield*) depends on Z^3 and is roughly 0.3 for iron. Thus, the most abundant elements, like oxygen, do not produce strong fluorescence lines, despite the order of magnitude difference in abundance between iron and oxygen. This difference can partly be compensated for by the larger incident flux around 0.54 keV (the K-shell threshold for neutral oxygen), compared with the flux near 7.1 keV (the K-shell threshold of neutral iron).

5.4.3. The hydrogen line spectrum

Hydrogen emission lines are unique in several important ways. They can be used to estimate the number of ionizing Lyman continuum photons, they are very good reddening indicators in low-density gas, and their ratios with other emission lines are the basis for computing the metal abundance in the gas. For low-density AGN gas, where the Balmer line opacity is very small, the hydrogen line spectrum is not very different from what is expected in the standard recombination theory, that is, the case where radiative recombination is the only recombination process. Simple radiation theory is occasionally divided into case A and case B recombination. The first is a case in which all hydrogen transitions are optically thin, and the second is a case in which the optical depths in all Lyman lines, and the Lyman continuum, are very large but all other lines are optically thin. Conditions in ionized gas in HII regions, planetary nebulae, and other low-density nebulae that are ionized by a blackbody source of radiation are very similar to the simple case B recombination conditions.

Photoionization models of low-density AGN gas (the gas in the narrow-line region; see Chapter 7) suggest only slight variations from case B conditions. The only significant exception is the Lyα line, which, in some cases involving a small α_{ox} continuum (hard ionizing spectrum) and a low-ionization parameter, can be enhanced, significantly, by collisions from the ground state. A smaller enhancement of Hα can also affect the spectrum of ionized AGN gas. It is useful to summarize the expected Lyα, Hα, and Hβ line intensities under the various conditions: a blackbody radiation source that gives results that are very similar to the case B conditions and a typical AGN source. Table 5.3 gives such a summary.

5.5. Gas composition

There are several standard ways to measure the composition of ionized plasma. The most accurate methods are based on the comparison of metal recombination lines, with known effective recombination coefficients, with the intensities of hydrogen recombination lines. Such methods depend on the ability to measure the

Table 5.3. *Hydrogen line intensities relative to Hβ for a 40,000 K blackbody and various low-density AGN models*

Model	N_H	U(hydrogen)	Lyα/Hβ	Hα/Hβ
40,000K BB	10^3 cm^{-3}	10^{-1}	25	2.67
40,000K BB	10^6 cm^{-3}	10^{-1}	35	2.80
AGN SED	10^3 cm^{-3}	10^{-3}	31	2.71
AGN SED	10^3 cm^{-3}	10^{-2}	28	2.73
AGN SED	10^3 cm^{-3}	10^{-1}	25	2.78
AGN SED	10^6 cm^{-3}	10^{-3}	46	3.20
AGN SED	10^6 cm^{-3}	10^{-2}	43	2.95
AGN SED	10^6 cm^{-3}	10^{-1}	37	2.82

Note: The blackbody case gives results that are very similar to those of case B recombination with $T_e = 10^4$ K. All AGN models are solar composition, constant-density models with a column density of 10^{22} cm^{-2} and a "canonical" AGN SED with $\alpha_{ox} = 1.1$.

recombination lines that are typically very weak. All these metal lines are too weak in AGNs and cannot be use to derive the gas composition.

A less accurate but more practical method is based on the comparison of the intensities of various collisionally excited lines, mostly forbidden lines. These methods were developed for low-density gas, like in planetary nebulae and HII regions. They fail in AGN gas for three reasons. The first, which is most important in high-density AGN gas, is that under such conditions, many of the levels are collisionally suppressed, especially the upper levels of the strong forbidden lines that are used to derive the abundance. The second reason, which is significant in gas with large optical depth in the hydrogen lines, is that under such conditions, the intensities of the optically thick emission lines are subjected to other processes and cannot be used to compare with the metal line intensities. Finally, the methods are calibrated for stellar SEDs that are very close in shape to a pure blackbody. Such radiation fields result in a clear separation of the line-emitting gas into regions where one of the ions in question, for example, O^{+2}, dominates the emission. The typical AGN SED causes an overlap between neighboring zones of emission by neighboring ions and results in a degeneracy with respect to the derived gas composition.

The obvious way to go is to rely on photoionization calculations that are applied to the conditions in AGNs, including the range of density, ionization parameter, and shape of the ionizing continuum. Such methods have been used to calculate the metallicity of low-density AGN gas with a limited degree of success. Obviously, one must look for permitted and/or semiforbidden line pairs that are not too sensitive to the assumed SED, or the ionization parameter, and are reliable metallicity indicators.

Several methods of this type have been developed to circumvent these difficulties. In particular, the N v $\lambda1240$/C iv $\lambda1549$ line ratio (hereinafter the Nv/Civ method) has been suggested as a good N/C abundance indicator in BLR gas. The broad lines in question are strong and easy to measure using spaceborne spectrograph in low-redshift sources or from the ground in high-redshift AGNs. The suggestion is that the abundances of all α elements (C, O, Mg, etc.) relative to hydrogen are increasing together since they are proportional to the the star formation rate. Thus C/H, O/H, and so on, are all proportional to the metallicity (Z) of the gas. Nitrogen is a secondary element, thus N/C or N/O are expected to be proportional to Z. For high enough metallicity $(Z \geq 0.2)$, N/C $\propto Z$ and C/H $\propto Z$, thus N/H $\propto Z^2$. This allows the use of the N v $\lambda1240$/C iv $\lambda1549$ line ratio as a metallicity indicator.

The accuracy of the N v $\lambda1240$/C iv $\lambda1549$ method is not a trivial issue due to the rather different levels of ionization of C^{+3} and N^{+4} (i.e., the two ions do not occupy the same region in the photoionized gas) and the somewhat different excitation energies of the two lines. This question has been investigated, theoretically, by computing numerous photoionization models with various assumptions about the ionization parameter, gas density, and SED. Most models show a good correlation between the computed line ration and assumed N/C in the model. However, there are also notable exceptions.

A more promising method is to search for line ratios that are insensitive to ionization and local temperature conditions, that is, lines from ions with overlapping production regions and similar excitation energies. This helps to eliminate the strong dependence on temperature through the $\exp(-E_{ij}/kT)$ factor. Some examples of such line pairs are [O iii] $\lambda1663$/[C iii] $\lambda1909$, which can be a good O/C indicator, and [N iv] $\lambda1486$/C iv $\lambda1549$, which is a good N/C indicator. This method, which is in principle more accurate, is difficult to implement because most of the lines in question are weak and difficult to measure. Moreover, several of the lines are due to semiforbidden transitions with critical densities that are close to the typical density of the BLR gas. Thus, there are uncertainties due to the unknown gas density which affect the observed line intensities. The NLR gas provides a better environment to apply these methods. The application of the preceding methods to AGN samples, mostly type-I AGNs, is discussed further in § 7.1.

5.6. Clouds and confinement

The basic line-emitting entity in AGNs is normally referred to as a *cloud*. The assumption of clouds, or more generally condensations, is reasonable given the observations of galactic Hɪɪ regions and the interstellar medium, where condensation is indeed observed. Clouds are also required by line intensity considerations, mainly the observation that typical line widths, for example, in the BLR, are similar, within a factor ~2 for low- and high-ionization lines. A configuration in which every ionization stage is associated with a certain location and thus a certain velocity cannot explain the observed line profiles. Conversely, an ensemble of small clouds, each optically thick and each producing many emission lines with a large range of ionizations, is more plausible. This does not rule out stratification and dependence of line width on location; it is only a general picture that allows a large range of conditions at each location.

The idea of well-defined clouds is somewhat problematic because such entities must be either gravitationally bound or else confined by external pressure. Likely candidates for the origin of the self-gravitating clouds are stars in the central star cluster. In particular, winds produced by "bloated stars" have been considered as the origin of material in the broad- and narrow-line regions. It it interesting to note that "typical" NLR clouds are not just theoretical entities since direct observations of nearby AGNs show such condensations containing masses of about 1 $M_\odot$. Much lower mass clouds, containing as little as 10^{-8} $M_\odot$, have been considered in various BLR models. Such clouds must be confined or else be created and destroyed on a sound speed crossing time, which must be very short. For the BLR clouds, this time is of order 1 yr.

The most likely confinement mechanisms for clouds that are not self-gravitating are pressure confinement either by gas pressure (thermal confinement) from a low-density, hot intercloud medium (HIM) or by magnetic pressure. In both cases, the external pressure equals the gas pressure in the cloud. The conditions inside and outside of the clouds are obtained from general considerations involving a two-phase medium, similar to those used in the ISM.

Consider a medium where the fractional ionization of the gas and its kinetic temperature are determined solely by the central radiation field. Suppose there are two gas components at the same location with different temperatures, densities, and levels of ionization. These are noted here as the cold and the hot components. Consider first the ionization parameters of the two:

$$U_{\text{cold}} \propto \frac{L/c}{r^2 N_H(\text{cold})}, \qquad U_{\text{hot}} \propto \frac{L/c}{r^2 N_H(\text{hot})}. \tag{5.50}$$

At the given distance r, the radiation pressure due to the central source is simply $P_{\text{rad}} \propto L/cr^2$ and is independent of the gas density. This means that $U \propto P_{\text{rad}}/N_H$. Dividing the two values of U by the corresponding temperatures, T_{cold} and T_{hot}, and noting that all quantities are calculated at the same location, we find that the condition for pressure equilibrium due to the same gas pressure in the two components leads to the conclusion that

$$\frac{U_{\text{cold}}}{T_{\text{cold}}} = \frac{U_{\text{hot}}}{T_{\text{hot}}} \propto \frac{P_{\text{rad}}}{P_g}. \tag{5.51}$$

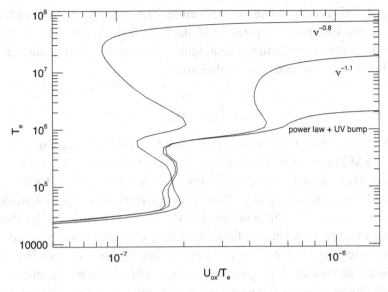

Figure 5.9. Stability curves for solar composition gas exposed to different SEDs: (1) $L_E \propto E^{-0.8}$; (2) $L_E \propto E^{-1.1}$; (3) "typical" AGN SED with a strong blue bump and an X-ray power law. Note the several unstable regions (curve with negative slope) that are less noticeable in the SED containing a UV bump. Such a SED results in lower Compton temperature (T_C) due to the efficient Compton cooling by the radiation in the blue bump.

In other words, P_{rad}/P_g is the same in both components, or $U/T \propto P_g^{-1} \propto (N_H T)^{-1}$. This allows several different temperatures, and hence several values of the density at the same location, provided U/T is the same for all components. Two (or more) such components can survive, side by side, in pressure equilibrium, provided they are stable against thermal perturbations.

The equilibrium situation is illustrated in Figure 5.9, showing the famous "S-curve," which is the photoionization equilibrium curve (sometimes referred to as the *stability curve*) for such gas. Each point on the curve is characterized by thermal equilibrium ($H = C$) at a different temperature. The points outside the curve are regions where either $H > C$ (on the right-hand side of the stability curve) or where $H < C$ (on the left-hand side of the curve). These locations represent pairs of values for U and for T that cannot be found in gas in thermal equilibrium with the composition assumed here.

The various stability curves shown in Figure 5.9 represent several regions of different physical conditions. The upper, almost horizontal branches are stable regions, that is, regions where a small deviation from the equilibrium point results in return to an adjacent point on the curve that is also in equilibrium, that is, $H = C$. For example, increasing the incident flux, and hence U, results in movement to the right of the figure to a point where U/T becomes larger and T is basically unchanged. This is also a point of equilibrium. The temperature on these branches is close to the Compton temperature of the fully ionized gas. This temperature is determined solely by Compton heating and cooling, that is (Equation 2.23; the expression is repeated here for completion),

$$T_C = \frac{h\bar{\nu}}{4k}, \tag{5.52}$$

where $h\bar{\nu}$ is the mean photon energy weighted by the cross section. T_C depends only on the SED and not on the density of the gas. For typical AGNs, $T_C \sim 10^7$ K.

For the SEDs assumed here, all lower branches in Figure 5.9 have positive slopes, and hence they are stable. These branches are typical of photoionized AGN gas with a characteristic temperature of $\sim 10^4$ K. The stability can be illustrated by noting that for all points on these branches, an increase in the incident flux results in increasing U, that is, the point moves into the region where $H > C$. This results in an increase of T (for photoionized gas with such a temperature, increase in flux results in a higher temperature), which will move the point up and back to a new equilibrium point on the same branch, just above the initial point. The new point represents another thermal equilibrium point with a higher ionization parameter.

The intermediate parts of the curves contain one or more regions with negative slopes. Here an increase in flux and in U results in a lower gas temperature. In such regions, a small perturbation to the right (the region where $H > C$) due to additional heating, or to the left ($H < C$) due to reduced heating, will result in further removal from the stability curve. Here there is no nearby $H = C$ solution, and the gas can reach a new thermal equilibrium only in a place with very different values of U and T. This can be on the lower or the upper stable curves where temperatures are very different from their initial values. This is the clear signature of an unstable gas. The curves shown here suggest that for gas exposed to a typical AGN SED, there are values of U/T that correspond to two and even three stable (positive slope) solutions. Thus a two- or three-component medium can be formed in a gas exposed to AGN radiation fields. Such localized components, with well-defined temperatures and densities, can be described as "clouds" or condensations.

An alternative and perhaps a more likely situation is confinement by magnetic pressure. In this case, $B^2/8\pi \geq (N_H + N_e)kT$, and the required magnetic field is of the order of 1 G for the BLR and much smaller for the NLR. Such magnetic fields are likely to occur near the central accretion disk and, perhaps, throughout the emission-line region. The clouds in this case are likely to have a nonspherical shape, perhaps elongated filaments along the magnetic field lines. They are also less prone to disruption by changes in the local density and temperature due to variations in the central source luminosity. An alternative to pressure-confined clouds are condensations or filaments that are constantly produced and destroyed. Such situations are common in the interstellar medium and have also been considered for AGNs.

A somewhat different model for the AGN gas distribution, which attracted much attention, is the locally optimally emitting cloud (LOC) model, which involves a range of densities and column densities at any location. Because of the same distance from the source, and the different densities, the various components in this model are not in pressure equilibrium with each other, and little is known about their formation and stability. The calculated spectral properties are quite appealing since they produce a good fit to the observed optical–UV spectrum of many AGNs. Much like the cloud model, there are several ad hoc assumptions that must be made to specify the various properties of such a medium.

Finally, we consider gas outflow from the central disk, or from other large mass reservoirs, in the form of a continuous wind. Such flows are characterized by continuous changes of density and velocity and are very different from the clouds described earlier. The emission and absorption spectra of such flows are very different from those produced by clouds, mostly because of the different opacity distribution and the locations of the various ionization fronts. The dynamics and the spectra of AGN flows are further discussed in Chapter 7 and § 5.9.

5.7. Photoionization models

5.7.1. Photoionization models for a single cloud

Putting together all that this chapter has discussed thus far is not a simple task. It is normally achieved with the help of complete photoionization codes that include the calculations of all the processes mentioned previously, and more. A model of this type solves, numerically, for the ionization and thermal structure within *one cloud*. Various clouds are then combined to calculate the emitted spectrum. Most present-day photoionization models are time independent since time-dependent processes are more difficult to treat. The transfer of radiation is done, in most cases, with an escape probability formalism, although more sophisticated methods have also been tried. Figure 5.10 shows the results of one such model, which assumes a nonvariable radiation source, a slab geometry, and a constant cloud density. Constant-pressure models have somewhat different structures.

5.7.2. Photoionization models for a system of clouds

Having calculated the ionization and thermal structure, and the steady-state level population of all abundant ions, one can now use these properties to calculate a theoretical emission-line spectrum, which will be compared with observations. Such calculations involve some assumptions about the distribution of clouds and

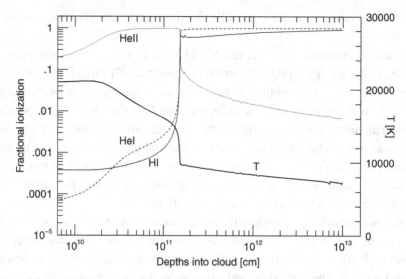

Figure 5.10. The ionization structure of a single cloud exposed to a typical AGN SED. The cloud is assume to have a constant density of $N_H = 10^{10}$ cm^{-3}, a column density of 10^{23} cm^{-2}, and an illuminated face ionization parameter of $U_{\text{hydrogen}} = 0.01$.

their other properties such as density, column density, and shape (i.e., spherical, elongated, etc.). A simple approach is to assume a line-emitting region that is made of numerous small clouds whose densities, dimensions, and other properties depend solely on the distance r. The assumption is that cloud confinement is achieved by hot gas (HIM) or magnetic pressure. The exact mechanism is not important provided the confining pressure is a simple function of the radius,

$$P \propto r^{-s}. \tag{5.53}$$

The contribution of each cloud to the various emission lines is determined by the physical conditions within the cloud and its distance from the center. A simple model of this type assumes a spherical cloud distribution around the central radiation source and radial-only dependence of all cloud properties.

Having defined the external pressure parameter s, we find that the hydrogen number density, $N(r)$ (assumed to be constant within each cloud), is controlled by the equilibrium with the external pressure. Neglecting the small electron temperature variation inside the cloud, we can write

$$N_H(r) \propto r^{-s}. \tag{5.54}$$

Next, we define the cloud column density, N_{col}, by considering spherical clouds of radius $R_c(r)$. We assume that the mass of the individual clouds is conserved, as they move in or out, but it is not necessarily the same for all clouds. Thus,

$R_c^3(r)N(r) = $ const. The cloud mean (over the sphere) column density is thus

$$N_{\text{col}}(r) \propto R_c(r)N(r) \propto r^{-2s/3}, \tag{5.55}$$

and its geometrical cross section is

$$A_c(r) \propto R_c^2(r) \propto r^{2s/3}. \tag{5.56}$$

Finally, the number of clouds per unit volume is defined by a second radial dependence,

$$n_c(r) \propto r^{-p}, \tag{5.57}$$

where p is an additional parameter that is required to fully define the model.

The next step requires the integration over the entire system. This is simplified by assuming only one type of cloud, that is, the same size at the same r for all clouds. The generalization of this scheme can involve a local population of clouds having some size distribution and the same p for all subclasses.

The clouds in question are illuminated by a central source whose ionizing luminosity, $L(t)$, varies in time. Designating $\epsilon_l(r, L)$ as the flux emitted by the cloud in a certain emission line, l, per unit projected surface area (erg s^{-1} cm^{-2}), we get the following relationship for the emission by a single cloud:

$$j_{c,l}(r, L) = A_c(r)\epsilon_l(r, L). \tag{5.58}$$

Assuming the clouds extend from r_{in} to r_{out}, we obtain the cumulative line flux from

$$L_l \propto \int_{r_{\text{in}}}^{r_{\text{out}}} n_c(r)j_{c,l}(r, L)r^2 dr. \tag{5.59}$$

Having determined the location, mass, and density of the emission-line clouds, and having assumed a SED for the ionizing source, we now calculate $\epsilon_l(r, L)$ using a photoionization code. We then follow the preceding formalism to obtain E_l. Because all parameters have been fixed, there is no need to independently specify the ionization parameter, $U(r)$, which, in this model, is given by

$$U(r) \propto r^{s-2}. \tag{5.60}$$

A complete model of this type, with a given gas composition, is specified by the source luminosity and SED, the radial parameter s, the boundaries r_{in} and r_{out}, and the normalization of the density and column density at a fiducial distance. The comparison with observations further requires the normalization of the total line fluxes and hence the integrated number of clouds. This is better defined by using the total covering factor, $C_f(r_{\text{out}})$, which is obtained by integrating

$$dC_f(r) = A_c(r)n_c(r)dr \propto r^{2s/3-p}dr \tag{5.61}$$

between r_{in} and r_{out}. The normalization of $C_f(r_{\mathrm{out}})$ is achieved by comparing the observed and calculated fluxes of the emission lines. In all this we did not consider cloud obscuration and the possibility that radiation emitted by one cloud is absorbed by another. Thus, such models are limited to $C_f(r_{\mathrm{out}})$ smaller than about 0.3.

The variable AGN radiation results in time-dependent local emission-line fluxes. In variable AGNs, the calculated line emission requires integration over time and space. Chapter 7 explains how to take into account such time-dependent variations.

As noted earlier, there are alternatives to the small cloud model. In particular, the LOC model assumes numerous clouds with a range of properties, such as density, column density, and covering factor, at any given location. The main suggestion is that the local ionizing flux results in a large range of ionization parameters at any given location. Lines from high- and low-ionization species are produced, with different efficiencies, at all radii, and the local spectrum reflects the range of physical properties. The $1/r^2$ flux variation across the system is the main reason for the different contributions to the various emission lines at different locations. The general tendency is for lines of higher ionization and higher critical densities to be emitted closer to the center, but there are important exceptions to this behavior. Current versions of the LOC model assume power-law distributions in density and covering factor and constant column densities, but other distributions are likely too. The number of free parameters required to fit the observed spectrum with an LOC model is similar to the number used in the small cloud model.

5.8. Mechanical heating and collisionally ionized plasma

While most of the gas in AGNs is photoionized by the intense, central radiation field, some regions in the host galaxy, particularly those associated with star formation, may be heated and ionized by shocks due to fast stellar winds and supernovae explosions. The resulting collisional ionized plasma can be very different in its thermal and ionization properties from photoionized gas. Its emitted spectrum will carry the signature of higher temperatures, and its analysis can reveal some of these properties.

Besides the starburst activity, the AGN environment itself can also be a source of significant mechanical heating and ionization. For example, small-scale radio jets that are seen in many low-luminosity AGNs can transfer a large amount of mechanical energy into a very small region, producing strong shocks and bright emission knots. High-resolution imaging of such sources in several nearby objects confirms the existence of such regions.

The microphysics of mechanically heated regions is dominated by collisional processes, and typical temperatures can be orders of magnitude larger than those observed in photoionized gas with a similar level of ionization. This section addresses the global energetics involved with such processes and presents the basic tools that are used to analyze the spectra of such regions.

General considerations suggest that mechanical heating cannot be globally important in the type of environment considered here. Consider the mechanical energy produced by a shock of velocity v_{sh}. This is given roughly by $1/2m_{sh}v_{sh}^2$, where m_{sh} is the mass of the flowing material. We can assign an efficiency factor, η_{sh}, to this process, which, in units of mc^2, is $1/2v^2/c^2$. Equivalently, if m_{acc} is the mass of the gas accreted by the BH, then the energy associated with this process is about $\eta_{acc}m_{acc}c^2$, where η_{acc} is the mass-to-energy conversion factor in the accretion process. The velocity of the bulk of the line-emitting gas (the narrow-line region, see Chapter 6) in AGNs is 500–1000 km s^{-1} and, for accretion disks around massive BHs (Chapter 4), $\eta_{acc} \sim 0.1$. The ratio of the energies produced as radiation in the two processes is, approximately,

$$\frac{\eta_{acc}m_{acc}}{\eta_{sh}m_{sh}} \simeq 10^5 \frac{m_{acc}}{m_{sh}}. \tag{5.62}$$

Thus, for shock excitation to be energetically significant, the amount of flowing shocked gas must exceed, by many orders of magnitude, the amount of material accreted onto the BH. Such a large amount of flowing and colliding gas seems to be in contradiction with known properties of AGNs.

Regardless of the preceding, there are clear cases of strong line and continuum emission due to very high temperature gas in numerous AGNs. This emission contributes very little to the total bolometric luminosity of the source. However, it can be an important signature of various processes that connect the general AGN activity to the properties and the evolution of the host galaxy. Some of these processes are described in Chapter 7 and others in Chapter 8.

Consider a region of collisionally ionized plasma in which the temperature is uniform across the zone. It is convenient to express the flux emitted in a given line of a specific ion X as a product of two terms. The first is related to the local physical conditions and is designated here by $g(T_e, AB_X/N_H)$. It includes the collisional excitation and recombination coefficients, the abundance of the ion in question, AB_X, and the electron temperature T_e. The second term describes the global properties of the emission region. It includes the product of the electron and hydrogen number densities, $N_e N_H$, the volume of the emitting region, V, and the volume filling factor, $\epsilon(V)$. This term is noted *emission measure* or *emission integral* and is given by

$$EM = \int_V \epsilon(V)N_e N_H dV. \tag{5.63}$$

Obviously EM is independent of the ion or line in question and gives the global properties of the zone.

Given EM, we can calculate the luminosity of a specific line l:

$$L_l = g(T_e, AB_X/N_H) \times \text{EM}. \tag{5.64}$$

Since g is known for many lines, the relative intensity of two collisional excited lines, with $g(l_1)$ and $g(l_2)$, can be used to derive the density, temperature, and abundances in the formation region of the two. A collisionally ionized region can be divided into subzones, each with its own temperature, density, and emission measure. The conditions in such zones can be found by observing emission lines of different ions and using their known g. Such methods are used to map the density and thermal structure of collisionally ionized plasma.

The intensities of collisionally excited lines depend, sensitively, on temperature. This is useful in determining the temperature since a slight change in T_e can change considerably the relative emitted fluxes. A major problem in such an analysis is the gas filling factor, $\epsilon(V)$, which is not directly observed and must be guessed or obtained from theoretical models or other considerations.

5.9. The motion of ionized gas

5.9.1. *The equation of motion*

AGN gas is exposed to a strong radiation field, which can, under certain conditions, produce large-scale flows. Such flows can be continuous, in which case they are classified as "winds," or else drive clouds or condensations in a ballistic manner (the general term *wind* as used in the literature is not well defined and can include condensations inside a continuous flow).

The general form of the equation of motion for a cloud of mass M_c is

$$a(r) = a_{\text{rad}}(r) - g(r) - \frac{1}{\rho}\frac{dP}{dr} + f_d/M_c, \tag{5.65}$$

where $g(r)$ is the gravitational acceleration, $a_{\text{rad}}(r)$ is the acceleration due to radiation pressure force, and f_d is the drag force. For pure wind flows, we neglect the drag force term and the internal radiation pressure. This gives

$$v\frac{dv}{dr} = a_{\text{rad}}(r) - g(r) - \frac{1}{\rho}\frac{dP_g}{dr}. \tag{5.66}$$

We also need an equation that describes the location-dependent density of the flow. For a wind-type continuous flow, this is obtained from the continuity condition,

$$\dot{M} = 4\pi r^2 \rho v C_f(r) = \text{const.}, \tag{5.67}$$

where $C_f(r)$ is the location-dependent covering factor of the outflowing gas. For ballistic flows, which accelerate clouds, $\dot{M}$ needs to be specified more carefully, given the initial conditions, the number of moving clouds, the filling factor, and the geometry of the flow.

The relative importance of the various terms in the equation of motion depends on the local conditions. Gravity dominates for clouds of low ionization and large column densities, the reason for this being that neutral gas at the part of the cloud far from the illuminated surface is not affected by the radiation pressure force. Radiation pressure force dominates the motion of ionized, low-optical-depth gas near a source with a high accretion rate and large η. This is not the case for fully ionized gas, where only electron scattering contributes to the opacity. The pressure gradient can be the dominant factor when $a_{\mathrm{rad}}(r) \ll g(r)$ and when the volume is filled with highly ionized, high-temperature gas.

5.9.2. Radiation pressure and gas outflows

The equation required for calculating the radiation pressure force at a distance r is obtained by summing over all absorption processes. In this equation, the contribution to a_{rad} due to ion X is

$$a_{\mathrm{rad}}(r, X) = \frac{N_X}{\rho(r)c} \int_{\nu_X}^{\infty} \frac{L_\nu \kappa_\nu e^{-\tau_\nu} d\nu}{4\pi r^2}. \tag{5.68}$$

The term κ_ν gives the total absorption cross section, including all bound–bound, bound–free, free–free, and Compton scattering processes. It is different from the photoionization cross section, σ_ν, used in Equation 5.2, which gives only the bound–free cross section per particle.

The relationship between a_{rad} and the level of ionization can be understood by considering the important case where the largest contribution to the radiation pressure force is due to ionization (i.e., bound–free transitions). To illustrate this, we assume that the mean energy of an ionizing photon is $\overline{h\nu}$ and use the definition of I_X from Equation 5.2. For ionized gas with $N_e \simeq N_H$, this gives

$$a_{\mathrm{rad}}(X) \propto \frac{\overline{h\nu}}{cN_e} N_X I_X. \tag{5.69}$$

In a steady-state case, photoionization is exactly balanced by radiative recombination. This gives

$$N_X I_X = R_X N_{X+1} = \alpha N_e N_{X+1} \propto N_e^2. \tag{5.70}$$

Thus a_{rad} is proportional to N_{X+1}, and for the specific case of hydrogen,

$$a_{\mathrm{rad}} \propto N_e. \tag{5.71}$$

It is customary to introduce the *force multiplier*, $M(r)$, which is the ratio of the total radiation pressure to the radiation pressure due to Compton scattering:

$$M(r) = \frac{a_{\text{rad}}(\text{total}, r)}{a_{\text{rad}}(\text{Compton}, r)},$$ (5.72)

where

$$a_{\text{rad}}(\text{Compton}, r) = \frac{N_e(r)\sigma_T L}{4\pi r^2 \rho(r) c}.$$ (5.73)

$M(r)$ provides a convenient way of expressing the radiation pressure force in the equation of motion. Note that by definition, $M(r) \geq 1$, and that $M(r) = 1$ for fully ionized gas.

For partly neutral gas, the main contributions to $M(r)$ are from bound–bound and bound–free transitions. The former dominates in those cases in which the line optical depths are small. The contribution is much reduced for optically thick lines, except for the illuminated surface of the gas. The Compton and free–free terms are usually small, except for fully ionized plasma. Figure 5.11 shows several examples of the various contributions to the force multiplier in a highly ionized AGN gas.

The gravitational and radiation pressure forces can be directly compared by writing the equation of motion in a slightly different form. This is done by noting that the mass of the central BH is proportional to L_{Edd} and that the radiation pressure force is proportional to L. Denoting $\Gamma = L/L_{\text{Edd}}$ and noting that $L_{\text{Edd}} \propto M_{\text{BH}}$, we obtain

$$v\frac{dv}{dr} \simeq \frac{\sigma_T L}{4\pi r^2 \mu m_H c}[M(r) - 1/\Gamma] - \frac{1}{\rho}\frac{dP_g}{dr},$$ (5.74)

where, again, μ is the average number of nucleons per electron. We can discuss this equation in two, somewhat different ways, related to the cases of continuous (wind) and ballistic (cloud) motion.

For the wind equation of motion, the term $M(r) - 1/\Gamma$ gives the relative importance of gravity and radiation pressure, thus sources of large Γ require smaller force multipliers to drive the wind. For example, a source with $\Gamma = 0.1$, typical of low- and intermediate-luminosity AGNs, requires $M \geq 10$ to drive a wind by radiation pressure force. Such large values of $M(r)$ are normally achieved only in situations where bound–bound transitions are important (see Figure 5.11). Note that the constant multiplying $M(r) - 1/\Gamma$ in Equation 5.74 is defined to within a factor of ~ 1.2 since the force multiplier and the Eddington ratio differ by the value of μ, the mean molecular weight per electron (this is the reason for the use of the $\simeq$ sign in Equation 5.74).

For large-column-density clouds, with a large fraction of neutral gas, the role of radiation pressure force and the resulting changes in cloud orbits are better described by considering the total momentum delivered to the gas by the incident

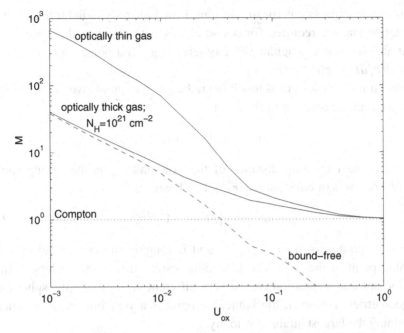

Figure 5.11. The force multiplier calculated for a "standard AGN SED" and solar metallicity over a large range of the ionization parameter U_{oxygen}. The optically thin case includes free–free, bound–free, and bound–bound contribution to M. Bound–bound absorption dominates for all levels of ionization, except for almost completely ionized gas at large values of U_{oxygen}. The large-column-density case is dominated by bound–free absorption since the resonance lines are very optically thick and do not contribute much to M (courtesy of D. Chelouche).

radiation field. The simplest way to describe this is to assign a transmission factor $\alpha(r)$ to a cloud at a distance r such that the luminosity absorbed by the cloud at this location, per unit surface area, is $\alpha(r)L/4\pi r^2$. Consider, for example, a Compton thin, neutral cloud that absorbs all the ionizing radiation (a Compton thin block). In this case, $M(r) \simeq \alpha(r)/(\sigma_T N_{\text{col}})$, where N_{col} is the hydrogen column density. For such a block, $\alpha(r) = L_{\text{ion}}/L$, where L_{ion} is the total ionizing luminosity. In general, $\alpha(r)$ is distance dependent because of the changing column density and level of ionization of the gas and must be calculated for each cloud.

Substituting the relevant numerical factors, we can write

$$a(r) = \frac{L}{r^2}\left[\frac{1.14 \times 10^{-11}\alpha(r)}{N_{23}} - \frac{8.8 \times 10^{-13}}{\Gamma}\right], \tag{5.75}$$

where $N_{23} = N_{\text{col}}/10^{23}$. Radiation pressure is the dominant force when

$$\Gamma \geq 7.7 \times 10^{-2}\frac{N_{23}}{\alpha(r)}. \tag{5.76}$$

The preceding expressions for the limiting Γ include only radial terms. The additional terms that are required for bound clouds with an azimuthal velocity component also include an angular velocity term (e.g., in a polar coordinate system, $a(r) = d^2r/dt^2 - r(d\theta/dt)^2$).

Several important scenarios involving radiation-pressure-driven flows produce a similar asymptotic solution of the form

$$v(r) = v_\infty [1 - r_0/r]^{1/2}, \tag{5.77}$$

where r_0 is the launching distance of the flow and v_∞ is the asymptotic flow velocity. For cases of constant $[M(r) - 1/\Gamma]$, we get

$$v_\infty \simeq v_c [M - 1/\Gamma]^{1/2}, \tag{5.78}$$

where v_c is proportional to $[L/r_0]^{1/2}$ and is roughly the escape velocity at the launching point of the flow. The preceding results are geometry dependent, and the velocities under question can be very different in, for example, spherical and disk geometries. However, the launching point is a very important parameter in determining the largest attained velocity.

AGN outflows are clearly recognized by their UV and X-ray absorption lines. UV absorbing systems indicate a large velocity range, from few $\times 10^2$ to few $\times 10^4$ km s^{-1}. X-ray absorbers are, on the average, similar to slower UV systems, with $v_\infty = $ few $\times 10^2$ km s^{-1}. Some X-ray observations seem to indicate higher velocities, perhaps relativistic. However, these observations are extremely challenging and remain questionable. The observed velocity can be translated into conditions near the base of the flow. As explained, the maximum velocity of AGN flows that are driven by bound–free–dominated radiation pressure force is of the same order as the escape velocity at the base of the flow. The highest-velocity UV-absorbing gas must therefore originate in the vicinity of the central BH and accretion disk. This is also the case for the very fast (still questionable) relativistic X-ray flows. The slower X-ray outflows can originate much further out, perhaps outside the BLR. As explained, bound–bound radiation-dominated flows can reach higher velocities ($v_c \propto \sqrt{M(r)}$).

5.10. Dust and reddening

5.10.1. General dust properties

Dust is a common ingredient in all astrophysical environments containing gas with a high enough metallicity. The obvious exceptions are very highly ionized plasma heated by fast shock waves and regions very close to strong radiation sources where dust grains cannot survive. Thus, dust is likely to be present in various parts of AGNs and, in particular, in their line-emitting regions. Such dust will scatter and absorb some of the ionizing and nonionizing radiation, will reradiate the observed energy at infrared wavelengths, and will modify the observed spectrum in various ways.

Most of the dust properties are well explained by considering a single spherical dust grain, with radius a, situated at a distance r from a radiation field L_ν. The grain is fully exposed to the radiation, and all its emitted radiation can freely escape. The absorption cross section of the grain is Q_ν (the equivalent of σ_ν used earlier for the gas), and its emissivity coefficient, assuming a blackbody source function, is $Q_\nu B_\nu$. The radiation absorbed by the grain is therefore

$$\frac{\pi a^2}{4\pi r^2} \int_0^\infty Q_\nu L_\nu d\nu, \tag{5.79}$$

and the resulting grain emission is

$$4\pi a^2 \int_0^\infty \pi Q_\nu B_\nu(T_g) d\nu, \tag{5.80}$$

where T_g is the grain temperature. Since grain heating is given by Equation 5.79 and grain cooling by Equation 5.80, equating the two can be used to numerically solve for the grain equilibrium temperature, T_g.

For a grain situated inside a gas cloud, the incoming radiation is attenuated by gas and dust inside the cloud. This makes the grain heating less efficient and requires an additional factor of $e^{-\tau_\nu}$ in Equation 5.79. Such a grain is also heated by radiation emitted by other grains and by diffuse gas radiation. A simple (not very accurate) method to account for this effect is to use an escape probability formalism and multiply Equation 5.80 by a term that takes into account the absorption of the radiation emitted by the dust on its way out of the cloud. This makes grain cooling less efficient and roughly accounts for the heating by the internally produced radiation, assuming it is mostly due to dust.

A simple expression for the grain temperature can be derived by noting that most of the absorption takes place in the UV, where the absorption cross section is the largest, and most of the emission is in the IR, because of the relatively low grain temperature. For most known grains, the UV absorption cross section depends slightly on frequency and the IR absorption cross section, and hence emissivity can be described by $Q_\nu \propto \nu^\gamma$, where the index γ depends on the size and the composition of the grain and is in the range 0–2. For example, for large graphite grains, $\gamma \simeq 0$, and for typical silicate grains, $1 \leq \gamma \leq 2$. Such grains radiate a *modified blackbody spectrum* (occasionally referred to as *gray body*),

$$F_\nu(T) \propto \frac{\nu^{3+\gamma}}{\exp(h\nu/kT) - 1}, \tag{5.81}$$

and their grain equilibrium temperature is obtained from the assumption of $L/4\pi r^2 \propto T_g^{4+\gamma}$. We can therefore write

$$T_g^{(4+\gamma)} = A \frac{L_{46}}{r_{pc}^2}, \tag{5.82}$$

where L_{46} is the integrated source luminosity in units of 10^{46} erg s^{-1}. The constant A depends on grain size and composition because both determine the value of γ. Values of A for several cases of interest are given later on.

The properties of ISM dust in the galaxy are normally specified by the grain composition, mostly the fraction of metals in the dust phase (depletion), the particle size distribution, and other specific properties such as grain shape and orientation. Graphitic dust and silicaceous dust are common in many dusty environments and were extensively studied in astronomical objects and in the laboratory. A commonly used grain type (*MRN grains*, after the work of Mathis, Rumpl, and Nordsieck) is made of graphite, enstatite, olivin, silicon cabide, iron, and magnetite, with a size distribution covering the range 0.005–0.25 μm. It beautifully reproduces the observed galactic extinction curve given the assumed optical properties (absorption and scattering cross sections) and a size distribution of the form

$$\frac{dn(a)}{da} = k_d a^{-\alpha}. \tag{5.83}$$

For silicate and graphite grains, the index α is of order 3.5. The minimum grain size is determined by various destruction processes such as photodestruction. The largest size is determined by the growth process of the grains (condensation and sticking). In galactic environments, $a_{min} < 0.01\ \mu$m, but in AGN environments, some of the smallest grains may not survive because of the strong radiation field. The largest size is estimated to be about 0.3 μm.

The constant k_d in Equation 5.83 depends on the dust-to-gas ratio and hence on the depletion of the heavy elements onto dust grains. Studies of ISM dust, whose composition is similar to the solar composition, suggest that the typical depletion is similar to depletions listed in Table 5.4. The optical depths of gas and dust with this depletion and grain size distribution are illustrated in Figure 5.12.

An additional important quantity is the dust sublimation temperature, T_{sub}, which is the maximum temperature attained by the grain before it evaporates. Laboratory and theoretical studies suggest that for graphite grains, $T_{sub} \simeq 1800$ K, while for silicate grains, $T_{sub} \simeq 1400$ K. These numbers can be put together to derive a *sublimation radius*, R_{sub}, which is the minimum radius where a grain of a certain composition can survive the central radiation field without evaporating. In general, smaller grains have larger γ and thus a larger temperature for a given radiation intensity. The result is that larger grains survive at smaller distances, and there is a range of a factor $\sim$3 in sublimation distances for different grains. Averaging

Table 5.4. *Gas composition and metal depletion*

Element	Abundance ($n/n(H)$)	Relative depletion
H	1.0	0.0
He	0.09	0.0
C	2.5×10^{-4}	0.5
N	6.5×10^{-5}	0.4
O	4.6×10^{-4}	0.35
Ne	9.0×10^{-5}	0.0
Al	2.3×10^{-6}	0.98
Mg	3.4×10^{-5}	0.9
Si	3.2×10^{-5}	0.90
S	1.5×10^{-5}	0.1
Ar	1.5×10^{-6}	0.0
Ca	2.1×10^{-6}	0.99
Fe	2.8×10^{-5}	0.97

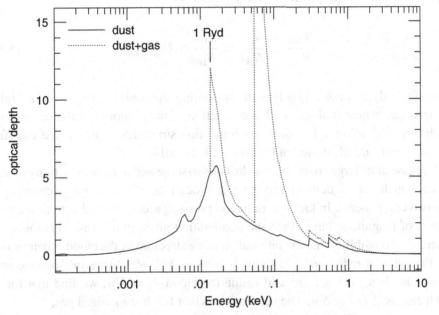

Figure 5.12. Optical depth due to dust (solid line) and the combined dust and gas opacity (dotted line) for a column density $10^{21.5}$ cm^{-2} cloud with galactic dust-to-gas ratio and MRN-type dust. The ionization parameter chosen for this example results in similar gas and dust contribution to the the total opacity over the Lyman continuum wavelength range.

over ISM-type grain sizes and compositions, one obtains the following sublimation distances for graphite ($R_{sub,C}$) and silicate ($R_{sub,Si}$) grains exposed to a typical AGN SED:

$$R_{sub,C} \simeq 0.5 L_{46}^{1/2} \left[\frac{1800}{T_{sub}} \right]^{2.6} \text{ pc} \tag{5.84}$$

$$R_{sub,Si} \simeq 1.3 L_{46}^{1/2} \left[\frac{1500}{T_{sub}} \right]^{2.6} \text{ pc.} \tag{5.85}$$

5.10.2. Dust and ionized gas

The presence of dust can substantially change the level of ionization and the line emissivity in the gas. Such dust competes, effectively, with the ionization of the gas because of its large absorption cross sections at all wavelengths longer than about 0.02 μm. The fraction of Lyman continuum photons absorbed by the dust, relative to those absorbed by the gas, depends on N_{dust}/N_{H^0}. For ionized gas with $N_{H^+} > N_{H^0}$, this is proportional to $U_{hydrogen}$ since

$$\frac{N_{dust}}{N_{H^0}} \propto \frac{N_{gas}}{N_{H^0}} \simeq \frac{N_{H^+}}{N_{H^0}} \propto U_{hydrogen}. \tag{5.86}$$

Therefore, dust is more efficient in attenuating the ionizing radiation in highly ionized gas, where it absorbs a larger fraction of the photons capable of ionizing hydrogen and helium. The different ionization structures of dusty and dust-free environments are illustrated in Figures 5.13 and 5.14.

Because of its large cross section, dust is also subjected to large radiation pressure force, which is then delivered to the gas through the efficient charge coupling of the two components. In fact, the radiation pressure acceleration due to dust can be orders of magnitude larger than the acceleration given to the gas. This raises the interesting possibility that the internal pressure structure of the cloud is determined by the external radiation field through pressure balance. Since just outside such dusty clouds $a_{rad}(r) \propto L/r^2$ and inside the cloud $P_g \propto N_H$, we find that for all such clouds, $L/r^2 \propto N_H$. The conclusion is that for dusty ionized gas,

$$U \propto L/N_H r^2 = \text{const.} \tag{5.87}$$

This fixes both the absolute value of the ionization parameter and the fact that it is location independent. In Chapter 7, we will show that NLR models based on this idea are quite successful in explaining the intensity of many narrow emission lines in AGNs.

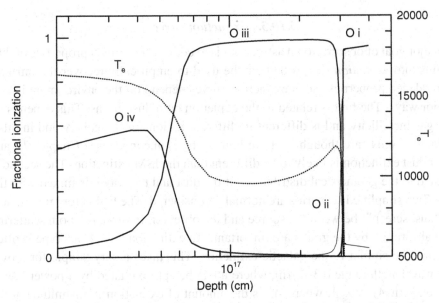

Figure 5.13. The ionization and temperature structure of a constant-density dust-free gas cloud exposed to a typical AGN ionizing continuum. In this case, $N_H = 10^4$ cm^{-3} and $U_{\text{hydrogen}} = 0.03$.

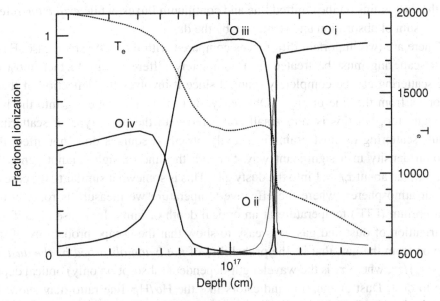

Figure 5.14. Similar to Figure 5.13, except that galactic-type dust with the galactic dust-to-gas ratio is assumed to be present. Note the large reduction in the dimension of the ionized region and the increase in electron temperature mostly because of the depletion of carbon and oxygen, which provide the strongest cooling lines.

5.10.3. Extinction curves

A major goal of observational astronomy is to convert the known properties of dust grains into *extinction curves* that can be used to empirically recover the intrinsic unreddened properties. Such extinction curves depend on the environment in two major ways. The first is related to the depletion onto dust grains. This depends on the gas metallicity and is different in different regions of the galaxy and in other galaxies. AGNs are thought to have high metal abundances (see Chapter 7), and their dust extinction is likely to be different than the ISM extinction. The second is related to the geometrical distribution of the dust and the way it is mixed with the gas. Two simplified scenarios are normally considered. The first is foreground dust (a "dust screen") between the source and the observer. In this case, both scattering and absorption by the grains are important. An extinction law of this type is often taken to be similar to the galactic extinction. This takes a very simple form over the wavelength range 0.3–1 μm, where it can be approximated by a power law in wavelength $A_\lambda \propto \lambda^{-1}$, where A_λ is the amount of extinction in magnitude at the indicated wavelength. Large deviations from this simple approximation are seen both at UV and IR energies.

The second generic situation involves dust that is mixed in with the gas. In this case, the term *attenuation* is more appropriate because it describes the actual effect of the dust grains on the emitted line and continuum fluxes, while *extinction* refers to the sum of absorption and scattering by the dust.

There are two important differences compared with the dust screen case. First, dust scattering must be treated in a completely different way. In fact, most of the scattering can be completely ignored since it involves the deflection of these photons from the line of sight. Obviously, there is some scattering into the line of sight, too, but this is very small compared with the other type of scattering. Thus, scattering by dust grains in heavily obscured sources does not affect the photon density in a significant way. Second, the line of sight cannot penetrate deeper than about $\tau_d = 1$ into the dusty gas. This is somewhat similar to the case of stellar atmospheres, where the effective temperature we measure is roughly the equilibrium (LTE) temperature at an optical depth of unity. In a case of uniform distribution of dust and gas, it is easy to show that the escape probability of the photon from the gas, that is, the probability of *not being absorbed by the dust*, is $(1 - \tau_\lambda)/\tau_\lambda$, where τ_λ is the wavelength-dependent (absorption only) optical depth due to dust. Dust attenuation indicators like the Hα/Hβ line ratio may show, in such cases, dust optical depths of about unity, regardless of the actual optical depth, which may be considerably larger.

The situation regarding dusty HII regions that are ionized and powered by young stars is rather complicated. Observations show that the youngest stars and the line-emitting gas are located deep in the cloud, yet somewhat older stars are less obscured by the dust. The use of the same extinction law for the various components

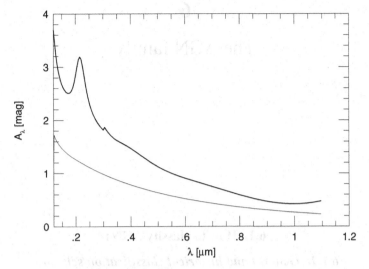

Figure 5.15. Galactic extinction (top line) compared to the Calzetti extinction law (bottom line) assuming the same dust composition. Both curves are normalized to galactic extinction of $A_{5500\,\text{Å}} = 1$ mag.

may lead to erroneous results, and various methods have been developed to apply different attenuation correction factors to the different components. For example, in such systems, the attenuation of the stellar continuum, measured in magnitude, is a factor of 2–3 smaller at 5500 Å compared with the emission-line attenuation. The shape of the attenuation law is different too. For the stellar continua at optical–UV energies, this law is less steep than the galactic extinction law. Approximating these laws by $A_\lambda \propto \lambda^{-s}$, we find that for stars in SF regions, $s \simeq 0.7$, while the galactic law is closer to the case of $s \simeq 1.2$ over the wavelength range 4000–8000 Å. The stellar attenuation law is sometimes described as *gray attenuation*. Two such curves are compared in Figure 5.15.

There have been several attempts to combine the two different laws into one that is appropriate for HII emission lines. One such expression is

$$\frac{\tau_\lambda}{\tau_V} = \frac{A_\lambda}{A_V} = (1 - \mu)\left(\frac{\lambda}{5500}\right)^{s_1} + \mu\left(\frac{\lambda}{5500}\right)^{s_2}, \qquad (5.88)$$

where $\mu \simeq 0.4$ specifies the fractional gray attenuation, $s_1 \simeq 1.3$, and $s_2 \simeq 0.7$. All these considerations apply also to ionized dusty AGN gas.

5.11. Further reading

Comprehensive descriptions of the material discussed in this chapter can be found in several books concerning astrophysical gas and dust, for example, Osterbrock and Ferland (2006) and Emerson (1996). The parts that are more specific to AGNs follow closely Netzer (1990, 2008).

6

The AGN family

6.1. How to classify AGNs

6.1.1. General and historical classification schemes

The classification of AGNs into subgroups is based on the history of research in this area. In particular, the discovery and general understanding of quasars, in the early 1960s, preceded the detailed study of the local members of this group, the Seyfert galaxies. This statement can be challenged by historians because of the seminal paper by Seyfert in 1943, in which he described the spectra of seven members of this group. However, the unusual features of these spectra were neglected for a long time, including the enormous velocities inferred from the widths of the emission lines and the extreme nuclear luminosities. These objects were rediscovered in the mid-1960s and studied in greater detail following the great interest in quasar spectra and redshifts.

The earlier detailed observations of local, low-redshift AGNs provided enough data to define several of the subgroups that are still used today: Seyfert 1 galaxies, Seyfert 2 galaxies, radio galaxies, quasars, blazars, LINERs, and so on. Additional observations of higher-luminosity, higher-redshift objects helped to refine the classification. Many observed properties of such sources have been described in Chapter 1.

Current AGN classification is based on higher-quality observations of a much larger number of sources, on better understanding of the physics of accretion and the line-emitting processes, and on the realization that many of the observed characteristics depend on the luminosity and inclination of the central source. The main subgroups of today are introduced in this book as *type-I radio-quiet AGNs* (in earlier years, Seyfert 1 galaxies and radio-quiet QSOs), *type-I radio-loud AGNs* (in earlier years, BLRGs, radio-loud QSOs, or QSROs), *type-II radio-quiet AGNs* (Seyfert 2s in earlier years), *type-II radio-loud AGNs* (in earlier years, NLRGs), *LINERs*, and *blazars* (BL-Lac objects and OVVs of earlier days). The following are three sets of somewhat different questions that can be posed to subclassify AGNs.

The first group is mostly observational:

What is the "power house" of the source? This type of classification refers
to the nature of the main energy source. Several possible answers are thin
accretion disks around massive BHs, synchrotron and/or inverse Compton
radiation by a nonthermal source, hot free–free emission by a disk corona,
and thermal radiation by hot dust. The answer to this question can help us
isolate two classes of AGNs (blazars and radio-loud AGNs; see later) from
the rest of the population.

What is the observed SED of the source? This is the observational manifestation of the first question. It helps to identify energy bands associated with
the various emission mechanisms, for example, optical–UV for accretion disk
emission, γ-ray for beamed relativistic emission, and so on.

What are the properties of the host galaxy? This is a different kind of question that is related to the idea (see Chapter 8) that the host galaxy properties,
such as mass, gas, and dust content, and the host galaxy evolution determine
the AGN properties. The related question about the gas and dust content in
the center of the galaxy can be used to classify the source into one of several
AGN subgroups.

What is the inclination of the source to the observer's line of sight?
Assuming all AGNs are intrinsically similar, they may still look different
to an external observer. In particular, the source inclination to the line of
sight may cause noticeable observational differences that may affect the
classification.

What is the phase of activity and the amount of gas supply to the center?
This relates to the possibility that the AGN phase of activity, for example,
close to the onset of mass accretion, or its final stage may affect its
appearance.

A more physical approach to classification, which helps to minimize the number
of variables, is based on the properties of the main AGN components: BH, accretion
disk, and nuclear gas and dust. The questions in this category could be as follows:

What is the BH mass? The answer to this question is, in principle, independent
of the BH activity (accretion rate) but must be influenced by BH evolution. For
example, the most massive active BHs are found in earlier epochs (Chapter 9).

What is the spin of the BH? For active BHs, this will determine the mass-to-
energy conversion efficiency (Chapter 3) and can be used to distinguish
different AGNs by several properties such as the mean energy of the emitted
photons, the existence of radio jets, and so on.

What is the accretion rate? This distinguishes AGNs by their total luminosity
and/or kinetic (wind outflow) power.

What is the gas and dust content, and the metallicity, in the nuclear region?
This has observational consequences that can be used to classify AGNs by
means of optical and IR spectroscopy.

What are the properties of the host galaxy? Do different host properties
result in different types of AGNs?

Present-day AGN research makes use of a third list of questions to classify
sources into various categories that were introduced in Chapter 1 and are further
detailed in this chapter. This list of questions is a combination of the first two lists
and is almost entirely observational:

What is the intensity and EW of the observed emission lines? This distinguishes between weak-line objects (blazars and lineless AGNs) and strong-
line objects (all other AGNs).

What is the typical width of the observed emission lines? Do we see only
broad lines (the minority of sources, mostly high-luminosity type-I AGNs),
only narrow lines (many type-II AGNs and LINERs), or a combination of
both (most type-I AGNs)?

What is the level of ionization of the line-emitting gas? This can be used to
distinguish low-ionization AGNs like LINERs from high-ionization AGNs.

How strong is the radio source? This is done relative to the optical–UV and
is used to distinguish radio-loud from radio-quiet AGNs.

How strong is the X-ray source? This relative measure can be used to classify
objects by their bolometric luminosity.

Can we see beamed, nonthermal radiation, and at what energies? This distinguishes radio-loud from radio-quiet sources and blazars from the rest of
the population.

Is there evidence for central obscuration? This is the main way to distinguish
between type-I and type-II AGNs.

What are the variability amplitude and time scale? This distinguishes blazars from other types of AGNs.

This chapter gives a detailed description of the various subgroups of AGNs that
are distinguished by the properties of their supermassive BH, accretion disk, torus,
central jet, and emission and absorption lines. The next chapter explains in detail
the physics of these components.

6.1.2. Diagnostic diagrams

We start by explaining the way to separate galaxies with strong emission lines
into those that are excited mostly by a nuclear, AGN-type source, those that are
excited by early-type stars, and those that show a combination of both. This is

done by emission-line spectroscopy, which provides a meaningful and useful way to probe the physical conditions in the narrow-line-emitting regions of all galaxies. The method is based on several "diagnostic diagrams." It combines various emission-line intensity ratios and compares them with each other. As explained in Chapter 5, the observed emission lines contain a large amount of information about the physical conditions in the line-emitting gas: its level of ionization, the electron temperature, the density and column density, the composition, and the covering factor (the fraction of the sky covered by line-emitting gas, as seen from the central source). Emission-line ratios reflect also the shape of the ionizing SED, hard as in AGNs or soft as in HII regions, where "hard" and "soft" refer to the mean energies of the ionizing photons.

The physical information provided by diagnostic diagrams is contained in various specific line ratios. For example, the I([O III] λ5007)/(Hβ) line ratio provides information about the level of ionization of the gas, the mean energy of the ionizing photons, and the electron temperature. A strong [O III] λ5007 line indicates a relatively high level of ionization (large ionization parameter, U) and large mean ionizing energy, typical of an AGN SED. I(Hβ) is proportional to the number of recombining hydrogen ions and is thus a measure of the total number of ionizing photons absorbed by the gas (Chapter 5). The previous line ratio is, therefore, large in highly ionized AGN gas. A smaller I([O III] λ5007)/I(Hβ) ratio indicates either an AGN continuum with a low U or soft stellar SED with high or low U.

Another example is the I([O I] λ6300)/I(Hα) line ratio. The [O I] λ6300 line is strong in partly neutral gas, which is ionized by an AGN radiation field. The reason is that the X-ray radiation emitted by AGNs can penetrate beyond the H^+ ionization front, into the region where most of the oxygen is neutral. The energy carried by this radiation is large enough to excite O I and make [O I] λ6300 one of the strongest lines in the spectrum. This is especially important in low-ionization AGN gas, such as the one observed in LINERs. Conversely, stellar radiation fields, even those emitted by very massive, early-type stars, cannot penetrate beyond the hydrogen ionization front. The [O I] λ6300 line in such sources is weak and I([O I] λ6300)/I(Hα) is much smaller than in most AGNs. The combination of the preceding two line ratios, plotted one against the other, provides a very useful way of dividing the plane into regions of low and high U and soft and hard SEDs.

A third important diagnostic line ratio, I([N II] λ6584)/I(Hα), provides somewhat similar information to I([O I] λ6300)/I(Hα) and can be used to replace it in those cases in which the latter is too weak to measure. This line ratio provides additional information about the gas metallicity (N/O) in both AGN and stellar environments since nitrogen is a secondary element whose abundance change is different from the abundance of α-elements like oxygen (Chapter 7). Plotting these and other emission-line ratios against each other, for many objects, is a powerful way to

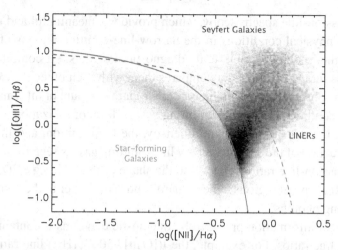

Figure 6.1. The spread of emission-line galaxies from the SDSS on one diagnostic diagram that uses four strong optical emission lines, Hα, Hβ, [O III] λ5007, and [N II] λ6584, to distinguish galaxies that are dominated by ionization from young stars (green points) from those that are ionized by a typical AGN SED (blue points for high-ionization AGNs and red points for low-ionization AGNs). The AGN and SF groups are well separated, but the division between the two AGN groups is less clear. The curves indicate empirical (solid) and theoretical (dashed) dividing lines between AGNs and star-forming galaxies (courtesy of B. Groves).

distinguish low- from high-ionization gas, soft from hard ionizing SED, and low from high metallicity.

Large samples of emission-line objects, like the SDSS and the 2DF samples, with hundreds of thousands of sources, can be classified into various types of sources by diagnostic diagrams. Figure 6.1 shows one such example that has been used to separate star-forming galaxies from AGNs, as well as to distinguish between low-(LINERs) and high-(type-II) ionization AGNs. Diagnostic diagrams are currently the main way to classify type-II AGNs. Type-I classification is more problematic because of the strong broad Hα and Hβ lines.

6.2. Type-I and type-II AGNs

6.2.1. Central obscuration

A basic division that is based on UV, optical, and MIR spectroscopy is to characterize an object by the widths and EW of its permitted emission lines, Hα, Hβ, C IV λ1549, Mg II λ2798, and so on. The implementation of this method depends on source luminosity since very broad emission lines are hard to detect in low-luminosity sources (see later).

Type-I AGNs are those objects with little or no obscuration of the central source of radiation (the central disk or any other source within $\sim 1000r_g$ of the BH) and with very broad permitted lines, more than about $1500(L_{bol}/10^{45}\text{erg s}^{-1})^{0.2}$ km s^{-1}. The dependence on the luminosity stems from a more fundamental dependence on the normalized accretion rate, L/L_{Edd}, which is explained in Chapter 7. Type-II AGNs are those sources with a completely obscured line of sight to the center at UV, optical, and NIR wavelengths and permitted lines with FWHMs that are significantly smaller than the previously mentioned number and are consitent with the velocities of stars in the host galaxy. There are several exceptions to this simplified scheme on emission-line widths. There are type-I sources with low luminosity and permitted lines with FWHM ~ 1000 km s^{-1}. Such objects are coined narrow-line Seyfert 1 galaxies (NLS1s) or, occasionally, narrow-line type-I AGNs (NLAGN1s). They can be identified by their unobscured AGN continuum and by strong FeII and MgII permitted lines that are not observed in type-II sources. There are also type-II sources with low or intermediate L_{bol} and FWHM > 1000 km s^{-1}. Thus, line width by itself cannot be used to distinguish type-I from type-II AGNs.

Most type-I and type-II sources show strong forbidden lines at optical and MIR wavelengths. Notable examples are [O IV] 25.9 μm, [O III] λ5007, [O II] λ3727, [N II] λ6584, and so on. In most type-I sources, the forbidden lines are considerably more narrow than the permitted lines. In type-II sources, the width and other features of the profile are very similar. An additional difference between the groups is the line EW. In high-luminosity type-I AGNs, the forbidden lines are seen against the AGN continuum, and hence their EWs are considerably smaller than in type-II sources, where the lines are seen against the (fainter) stellar continuum. Examples of type-I and type-II spectra were shown in Chapter 1.

The spectral differences between type-I and type-II AGNs can be explained with a simple model that involves central obscuration. The obscurer has axisymmetric structure, such as a thick disk with height and radius of order unity. It is normally referred to as the *central torus*, and its exact structure is a major area of research (Chapter 7). Models involving central tori of different properties are quite successful in explaining many AGN properties, including the NIR–MIR SED and the relative numbers of type-I and type-II sources in the local universe. Such models are normally referred to as *unified models* or *unification schemes* and are discussed in § 6.7.

The first unification models for AGNs were constructed in the mid-1980s following a breakthrough in the understanding of the differences between the two groups. This was the result of various polarization experiments that were able to show the presence of broad emission lines in polarized light in type-II sources. More confirmation of unified models comes from X-ray observations and study of ionization cones in low-redshift AGNs.

6.2.2. *Spectropolarimetry and scattered AGN radiation*

Polarized light is usually associated with some form of asymmetry like a preferred direction in space. Reflection (i.e., scattering) off a highly reflecting surface is one source of polarized light. Such processes are usually inefficient and result in a large loss of radiation. To estimate the scattering probability, we can consider a simple case of dusty clouds that can act as the scattering medium. The probability an emitted photon with wavelength λ to scatter into our line of sight depends on the optical depth of the scatterer along the path of the photon, τ_λ. On its route to the observer, the photons can be absorbed or scattered again, which attenuates the line-of-sight beam by a factor $\exp(-\tau_\lambda)$. The observed luminosity $L_\lambda(\text{obs})$ is related to the source luminosity $L_\lambda(\text{emitted})$ by

$$L_\lambda(\text{obs}) = L_\lambda(\text{emitted})\tau_\lambda \exp(-\tau_\lambda). \tag{6.1}$$

This function has a maximum for $\tau_\lambda \simeq 1$.

Magnetic fields are a common source of polarization because they define a preferred direction in space. For example, high-velocity electrons moving along magnetic field lines result in polarized synchrotron radiation (Chapter 2). Scattering off dust grains that are aligned along magnetic field lines in the ISM is another source of polarized radiation. In this case, the polarization is due to a combination of scattering and absorption by the grains since an aligned grain absorbs light that is polarized along its long axis more efficiently. Thomson or Compton scattering by free electrons is another important source of polarization. Linear polarization is the most common source of polarization in astronomy, but some circular polarization is also seen.

The basic observed properties of linearly polarized light are the percentage polarization P and the angle of polarization θ. P is a measure of the strength of asymmetry and hence the general geometry. θ is a measure of the orientation of the scattering medium, that is, the positions of the source of radiation and the observer relative to the scattering surface. Another common way to define polarized light involves the Stokes parameters Q and U, which will not be described here.

Spectropolarimetry is a way to measure P and θ in narrow-wavelength bins. This is usually a challenging task since many polarized sources have very low percentage polarization, of order 1 percent. This means that the signal-to-noise ratio (S/N) must be greater than 100 to ensure significant measurements. However, the information content can be very large. In AGNs, where the sources of continuum and line emission are detached, polarization measurements allow the separation of line and continuum and provide information on the gas kinematics and its location.

Direct spectropolarimetry of type-I AGNs shows very low levels of polarization, $P < 1$ percent, for both lines and continua in radio-quiet AGNs. The level of

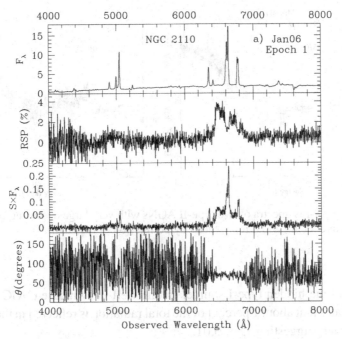

Figure 6.2. Spectropolarimetry of the type-II source NGC 2110 (from Tran, 2010; reproduced by permission of the AAS). The top panel shows the "normal" spectrum of a typical type-II AGN with strong narrow emission lines. The central panel shows the percentage polarization, and the bottom panel is the polarized light obtained only by multiplying the two upper panels by each other at every wavelength. The bottom panel is the angle of polarization. The polarized light shows a typical type-I spectrum with a broad, double-peak Hα line confirming the presence of a continuum source, and broad emission lines, behind the obscuring material.

polarization of the stellar continuum in type-II sources is also very low. However, sensitive spectropolarimetry of a large number of local type-II sources clearly reveals the signature of a hidden type-I source inside or behind the obscuring material. The spectropolarimetry can isolate the scattered light from the more intense foreground light of the stars and provide a clean or "pure" view of the hidden part. One such example is shown in Figure 6.2.

More advanced spectropolarimetry allows us to determine the spatial resolution of the scattering medium and measure its exact location relative to the central BHs. In some cases, the scattering originating in large (>100 pc) regions around the center is most likely due to interstellar dust. In other cases, the scattering medium is located much closer to the center at distances that are compatible with the dimension of the torus (see Chapter 7). This, combined with other properties of the scattered radiation, suggests Compton scattering by free electrons. Detailed modeling of the scattering configuration allows a good estimate of the bolometric

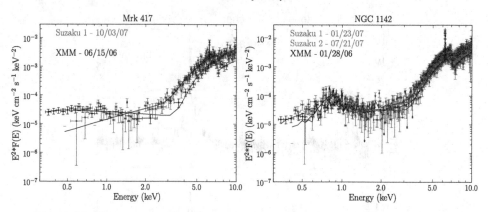

Figure 6.3. X-ray spectra of two type-II AGNs with very large obscuring columns (from Winter et al., 2009; reproduced by permission of the AAS).

luminosity of the hidden type-I source. In one well-known case (NGC 1068), the model indicates that about 1 percent of the total radiation is reflected in the direction of the observer, suggesting $\tau_e \sim 0.01$.

While the fraction of the central continuum radiation that is scattered near the center is of order 1 percent, this is large enough, in high-luminosity AGNs, to affect the observed continuum. This light is directly observed and can dilute the stellar absorption features. This must be taken into account when modeling the stellar continuum to derive stellar and BH mass and star formation rates (Chapter 8).

6.2.3. *X-ray observations and central obscuration*

The X-ray opacity of atomic gas is strongly wavelength dependent. For neutral gas with solar composition, a unit optical depth at 0.3 keV is achieved for hydrogen column density of about $4.5 10^{20}$ cm^{-2}. The corresponding column at 5 keV is about 4.5×10^{23} cm^{-2}. At around a column of $1.5 10^{24}$ cm^{-2}, the gas is Compton thick, which prevents the transmission of almost all the X-ray radiation above about 8 keV. For larger columns, all the X-ray radiation is absorbed.

X-ray observations provide the most efficient way for detecting and measuring the line-of-sight-obscuring column in AGNs. Numerous observations of type-II sources show a wide column density distribution with a peak at around 10^{23} cm^{-2} and a long tail toward very large columns. While X-ray absorption does not depend on the dust content of the gas, much of the obscuring material must be dusty to explain the large opacity at long wavelengths up to 1 μm and perhaps even more (for solar composition dusty gas with galactic gas-to-dust ratio $\tau(5500)$Å $\simeq$ $N_H/1.5 \times 10^{21}$ cm^{-2}). Figure 6.3 shows absorbed X-ray spectra of several type-II AGNs.

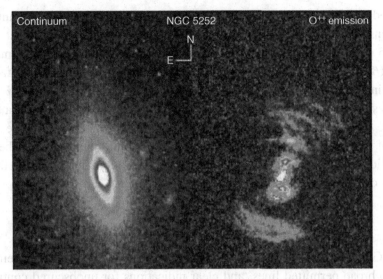

Figure 6.4. Optical images of the type-II AGN NGC 5252. (left) Broadband continuum image. (right) [O III] λ5007 image showing the central double-cone structure of ionized gas (courtesy of C. Tadhunter). (See color plate)

X-ray obscuration is not restricted to type-II AGNs. In fact, most low-luminosity type-I sources show some X-ray absorption along the line of sight with column densities that range between 10^{21} and few $\times 10^{23}$ cm^{-2}. Unlike the neutral obscurer in type-II sources, in these cases, the gas is highly ionized and, most probably, contains no dust grains. This gas is thought to flow out of the source inside or just outside the central opening of the torus. This gas is referred to as the *highly ionized gas* (HIG) or the *warm absorber* and is discussed in more detail in Chapter 7.

6.2.4. Ionization cones

Direct, high-spatial-resolution observations of nearby type-II AGNs in the light of emission lines from highly ionized species show a clear conical or biconical morphology. Such "ionization cones" indicate that the narrow-line gas is not illuminated isotropically by the source of ionizing radiation. The conical shape fits well the opening of the (assumed) central source and is consistent with the fact that the central source itself is not directly observed. An example is shown in Figure 6.4.

The evidence for a strong source of ionizing radiation that is not directly observed is supported by a simple "photon counting" argument. For photoionized gas that is optically thick at the Lyman continuum, the number of Lyα photons emitted by the gas is almost identical to the number of ionizing photons absorbed by the gas (the exact ratio depends on the gas density; see Chapter 5). This can easily be translated to the number of Hα photons, Hβ photons, and so on. For example, in conditions that are typical of the NLR, the number of Hβ photons is expected to be about 15 percent of the number of Lyα photons. Thus, measuring the luminosity of the Hβ line, and making a reasonable assumption about the covering factor of the gas in the cone that can absorb the ionizing radiation (of order 10%), gives a good estimate of the ionizing luminosity of the central source. Given a typical SED, we can estimate the bare continuum luminosity. This is found to be several orders of magnitude more luminous than the upper limit obtained from the direct observations. Thus, the NLR gas receives much more ionizing radiation than we observe, consistent with the assumption of obscuration.

6.2.5. *"Real" type-II AGNs*

Some AGNs that were discovered in large samples show strong narrow emission lines, no broad permitted lines, and clear indications for unobscured continuum. This is confirmed by short-time-scale large-amplitude continuum variations typical of type-I sources and by the lack of significant X-ray absorption. Such AGNs are occasionally referred to as "real" or "true" type-II AGNs.

The other types of "real" type-II sources are obscured AGNs without a hint of broad emission lines. There are two systematic ways to look for such sources. The first is by very deep spectropolarimetry, and the second is by MIR spectroscopy. As explained, spectropolarimetry is a very efficient and sensitive tool to search for scattered broad emission lines. Present-day techniques are sensitive enough to push the limit of such observations to well below 1 percent polarization. Given the distribution of gas and dust around the source, it is hard to imagine that no scattered broad-line photons will be observed by deep observations of nearby type-II sources. Yet such observations indicate that a large fraction of cases, 20 to 50 percent, do not show the signature of scattered radiation. These may well be "real" type-II sources.

While optical line diagnostic diagrams based on narrow-line ratios cannot distinguish "real" type-II from other type-II sources, the narrow MIR lines seem to show somewhat different properties that may distinguish the two groups. Such ratios have been studied by Spitzer (see Figure 7.20), and their interpretation is still under discussion. Moreover, there is a slight change in the shape of the MIR SED (Figure 7.34) that seems to distinguish between the groups. This may indicate that the lack of a BLR may also be associated with a different torus.

There is no simple explanation for the lack of broad emission lines in unobscured AGNs. One possibility is related to the fact that most of the known objects of this type are low-luminosity AGNs (this may be an observational bias since such objects are difficult to identify at high redshift). As will be shown later (Equation 7.22), the minimal width of the broad emission lines, FWHM$_{min}$, depends on $M_{BH}^{1/2} L_{bol}^{-1/4}$ and can be extremely large in low-luminosity sources. It is reasonable to assume also a maximum limit to the line width, $FWHM_{max}$, which may be related to cloud instability, proximity to the central BH (tidal forces), and so on. If this limit is of order 20,000 km s^{-1}, then there are cases with $FWHM_{max} < FWHM_{min}$, that is, no BLR.

6.3. Radio-loud and gamma-ray-loud AGNs

6.3.1. Radio-loud AGNs

FRI and FRII radio sources

As described in Chapter 1, AGNs are divided according the radio loudness parameter R (Equation 1.6) into radio-loud and radio-quiet sources. The optical properties of both classes are very similar, as verified by optical follow-up of radio sources discovered in large-area radio surveys, like the 3C and 4C surveys. This is also confirmed by radio follow-up of optically selected AGNs. In large samples like the SDSS, about 10 percent of all luminous AGNs are radio loud. Many powerful radio sources contain radio cores that are associated, to a very high precision, with the location of the central optical–UV pointlike continuum. Such cores show correlated optical–radio variability, and there are other indications that relate the radio properties to the accretion of gas onto the BH. A brief description of radio-loud AGNs and their SEDs was given in Chapter 1. Here we supply additional information about the classification of such sources and the similarity and differences between radio-loud and radio-quiet AGNs.

A scheme developed by Fanaroff and Riley (FR) classifies radio galaxies according to the extended radio structure and to whether they are edge brightened (FRII sources) or edge darkened (FRI sources). In spatially resolved FRI sources, the radio spots are separated by less than half the overall size of the source, whereas in FRII sources, those points (e.g., the brightest points in the radio lobes) are separated by more than half the total extent of the source. A combined X-ray and radio map of a luminous FRII source is shown in Figure 6.5.

FRI radio sources are much more numerous and less luminous than FRII radio sources. The dividing luminosity between the two groups is close to the break in the radio luminosity function (Chapter 8). The division in luminosity is seen very clearly when plotting radio power versus the optical magnitude of the galaxy. However, the location of the dividing line itself is luminosity dependent such that it is at larger L(5 GHz) in galaxies that are more luminous in the optical and NIR bands. Another division is based on the luminosity and ionization of the narrow emission lines. Low-power radio galaxies with FRI radio structure are invariably weak-line galaxies with typical LINER emission-line ratios.

Most radio galaxies show a synchrotron power-law continuum, $L_\nu \propto \nu^{-\alpha_R}$, extending over a large frequency range with clear indications for synchrotron

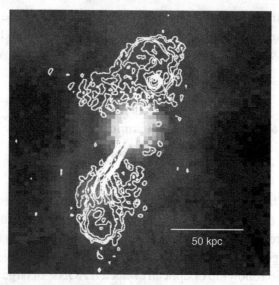

Figure 6.5. The $z = 0.458$ FRII radio galaxy 3C 200 (from Worrall, 2009; reproduced by permission of the Astronomy and Astrophysics review). The blue color is
the smoothed 0.3–5 keV Chandra image, and the contours are the 4.86 GHz VLA
radio map. Nuclear and extended emission are seen at both wavelength bands.
(See color plate)

self-absorption at low frequencies. Core-dominated sources show a "flat" spectrum (flat-spectrum radio-loud AGNs) with $\alpha_R < 0.5$. Lobe-dominated, weak-core
sources show, in many cases, a much steeper spectrum with $\alpha_R > 0.5$ (steep-
spectrum radio-loud AGNs). It seems that much of the difference in the measured
value of α_R is due to the inclination of the central radio jet to the line of sight.

The strong correlation between the radio and optical luminosity of AGNs seems
to extend to other wavelengths. There is a clear signature of nonthermal emission in
the FIR and MIR spectra of many radio-loud AGNs. This may well be the extension
of the lower-frequency radio continuum, which is added to the dust emission in
these wavelength bands. Radio-loud AGNs are also more X-ray bright in a sense
that the X-ray-to-optical luminosity is larger in radio-loud AGNs compared with
radio-quiet AGNs of similar optical luminosity. As argued in the next chapter, this
may well be related to a central X-ray jet.

Radio jets

Another way to subdivide radio-loud AGNs is based on the radio jet properties.
Radio jets are easily found and mapped by modern radio techniques, and their basic
properties, like the degree of collimation and the radio SED, are known for many
sources. Here, again, there is a dichotomy between jets that are strongly collimated
and show indications of relativistic motions and less collimated, less extended jets.
Strongly collimated jets are associated with both FRI and FRII sources, but there are

clear differences between the groups (Chapter 7). Evidence for relativistic motion in such jets comes from radio variability, beaming, and superluminal motion of clumps or blobs very close (a fraction of a pc) to the central BH. The appearance of the jet depends strongly on its viewing angle.

The radio zoo

The long and successful history of radio astronomy, and the many years of study of radio-loud AGNs, resulted in a variety of names that reflect, in many cases, the period of discovery rather than the physical properties of the objects in question. The higher-luminosity radio-loud AGNs were referred to in the past as *quasi-stellar radio objects* (QSROs) to distinguish them from radio-quiet luminous type-I AGNs, which were called *quasi-stellar objects* (QSOs). Here we prefer a division according to radio and/or optical luminosity and avoid the use of QSOs or quasars.

Radio galaxies are classified by their optical properties into broad-line radio galaxies (BLRGs) and narrow-line radio galaxies (NLRGs). BLRGs are more luminous and hence can be observed at higher redshifts. Most of them are FRII radio sources. The optical–UV spectrum of BLRGs is very similar to the spectrum of type-I AGNs with some indications for an additional nonthermal contribution to the optical continuum. Such objects tend to show the broadest emission lines among type-I AGNs, probably due to inclination effects (§ 7.1). There are other subtle differences between BLRGs and radio-quiet type-I AGNs to do with the intensity of the strongest forbidden lines.

The optical emission-line spectrum of NLRGs is less well defined. It covers a broader range of properties, from high-ionization type-II AGNs to low-ionization LINERs. There are clear indications for obscuration in many of them, with direct evidence for hidden broad emission lines in a large fraction of the sources.

The group of NLRGs is occasionally confused with weak-line radio galaxies (WLRGs). Such objects are more numerous than the stronger narrow-line radio galaxies, they are almost invariably FRI radio sources, and they show optical spectra that are almost indistinguishable from LINERs. While the term *weak* is not well defined, a working definition is that radio-loud objects in this class have EW([O III] $\lambda5007$) < 10 Å. Adopting this terminology, we can redefine the term *NLRGs* to include only radio galaxies with narrow, high-ionization lines and EW([O III] $\lambda5007$) > 10 Å.

A small number of radio galaxies are found in dusty spiral galaxies. Such objects suffer large optical extinction and were misclassified in the past into one of several groups, for example, narrow-line X-ray galaxies (NLXGs). Some of these sources show weak broad Hα lines, which led to their optical classification as intermediate-type Seyfert galaxies like Seyfert 1.9.

More physical insight into the differences between the various groups is obtained by plotting the radio-loudness measures against physical properties such as

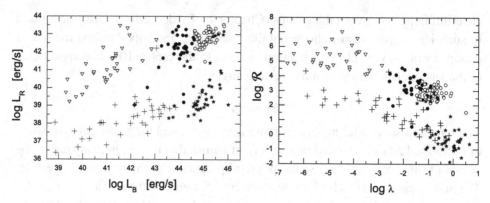

Figure 6.6. (left) Radio luminosity vs. optical (B-band) luminosity for various types of AGNs. (right) The radio loudness parameter R vs. λ (L/L_{Edd}) (from Sikora et al., 2007; reproduced by permission of the AAS).

luminosity and L/L_{Edd}. Some examples are shown in Figure 6.6. Remarkably, at the higher L/L_{Edd} end of the diagram, the more radio-loud objects occupying the upper branch are almost excluded early-type galaxies, while the lower branch of radio-quiet AGNs includes mostly spirals. This raises an interesting possibility that connects the BH spin to the radio loudness of the source. In such a scheme, the most powerful radio jets are those launched by the fastest-spinning BHs with spin parameter a approaching unity. Accepting the idea that large early-type galaxies are the result of a merger between two large disky galaxies (see more in Chapter 9), and that such mergers contribute significantly to the BH spin because they provide a large amount of cold gas that finds its way to the center, seems to explain the connection between the morphology of the host and the power of the radio jet.

6.3.2. Gamma-ray-loud AGNs

Blazars

The group of blazars includes highly variable core-dominated radio-loud sources showing polarization at radio and optical wavelengths. Many blazars are also powerful γ-ray emitters, and some of them show indications of superluminal motion. To be more specific, we define a blazar as an AGN that shows one or more of the following properties:

1. Intense, highly variable high-energy emission in the γ-ray part of the spectrum.
2. Intense, highly variable radio emission associated with a flat radio spectrum and, occasionally, superluminal motion.
3. Radio, X-ray, and/or γ-ray jet with clear indications for relativistic motion.
4. A double-peak SED with a lower-frequency peak at radio-to-X-ray energies and a high-frequency peak at X-ray-to-γ-ray energies (see Figure 1.6).

5. Very weak (small EW) broad and/or narrow emission lines indicative of photoionization by a nonstellar source of radiation on top of a highly variable continuum.

Blazars can be divided into BL Lacertae (BL-Lac) objects (after the first source of this type that showed, for years, no sign of emission lines) and flat-spectrum radio-loud AGNs. The flat radio spectrum blazars are occasionally called flat-spectrum radio quasars (FSRQs) or optically violently variable QSOs (OVVs). BL-Lac objects are often subclassified into low-energy-peaked BL-Lac objects and high-energy-peaked BL-Lac objects.

γ-ray properties of blazars

The understanding of the physical mechanism that drives the blazar phenomenon is strongly coupled to the launch of various advanced X-ray and γ-ray instruments. The launch of the Fermi Gamma-Ray Space Telescope in 2008 revolutionized this field by confirming earlier suggestions that most of the energy in these sources is produced by relativistic jets and by detecting many more blazars. Most of these discoveries are due to observations by the Large Area Telescope (LAT), a wide-field-of-view imaging telescope covering the energy range of $\sim$20 MeV to 300 Gev. As of 2011, there are several hundred Fermi-detected blazars, and the list is growing.

The LAT allows blazar variability to be monitored over a wide range of time scales. It shows that large amplitude variations are very commom in most blazars. The suggestion is that in these sources, most of the nonthermal γ-ray emission arises from relativistic jets that are narrowly beamed and boosted in a specific direction. The jet is launched in the vicinity of the central active BH, and the angle between the line of sight and the axis of the jet is typically a few degrees or less. This explains the superluminal motion often observed in VLBI observations of blazars. As will be shown later, there is good reason to believe that FSRQ blazars are associated with pole-on FRII radio sources and BL-Lac objects with pole-on FRI sources. Because of this, FRI and FRII sources are occasionally referred to as the parent population of blazars.

The jet model raises more questions than answers: how is the jet collimated and confined? What is the composition of the jet close to the launch points and much further out? What are the details of the conversion between the jet's kinetic power and electromagnetic radiation? Some of these issues are addressed in Chapter 7. Simultaneous multiwavelength observations of blazars are perhaps the most important tools for answering these and other questions. Today, such observations can cover a huge energy band, from several centimeters in the radio through MIR, NIR, optical, UV, X-ray, and all the way up to above 100 GeV. Ground-based observatories, like HESS, can extend them to even beyond the LAT energy range.

The fact that most blazars show an SED with two broad peaks (Figure 1.6) is consistent with the jet model. At lower frequencies, from radio to UV and sometimes X-rays, the emission is dominated by synchrotron radiation of high-energy electrons in the jet. The higher-energy part of the SED, from X-rays to γ-rays, is thought to result from inverse Compton emission (IC; see Chapters 2 and 7).

Detailed studies of blazars by LAT reveal real physical differences inside this inhomogeneous group of sources. The lower γ-ray luminosity blazars, classified as BL-Lac objects, have harder γ-ray slope (photon index of about 2) compared with flat-spectrum radio AGNs (photon index of about 2.5). A simple power law is not always a good description of the γ-ray continuum, and in several well-studied cases, the spectrum is better fitted by a broken power law with a steeper, higher-energy part. There are other differences that relate to the galaxy type and morphology. Blazars with strong relativistic jets are usually hosted in elliptical galaxies. However, Fermi found several NLS1s that are also strong γ-ray emitters. These sources are thought to have very large L/L_{Edd} (Chapter 7) and are hosted in spiral galaxies with high star formation rates. All this shows that the subclass of blazars includes objects with very different physical properties that depend on the central energy source, the BH mass and spin, the exact geometry and inclination, and, perhaps, the evolutionary phase of the sources.

6.4. Lineless AGNs

Systematic studies of large AGN samples result in the discovery of a subpopulation of AGNs with extremely weak, sometimes completely undetected emission lines. A typical upper limit on the EW of the emission lines in such sources is 1 Å. The objects show at least one of the four AGN indicators, usually a nonstellar continuum with, occasionally, flux variations. A clear indication for the active BH is an observed point X-ray source in many of the sources. The objects cover a large range in luminosity, from very faint objects in the local universe to very luminous AGNs at high redshift. They are referred to in the literature as *lineless AGNs*, *anemic AGNs*, *dull AGNs*, and other equally original names.

Lineless AGNs differ in their optical continuum properties from blazars. They do not show a power-law continuum; they are mostly radio quiet; their variability, if any, is of very small amplitude; and the typical double-peak SED of blazars is not observed. Figure 6.7 shows a composite spectrum of 15 such sources from the COSMOS survey.

The very luminous lineless AGNs are of special interest and may have a unique role in AGN evolution. These are high-redshift sources with extremely weak broad emission lines that are 1 or 2 orders of magnitude fainter (in term of line EW) compared to other type-I sources. Broad emission lines in AGNs are known to

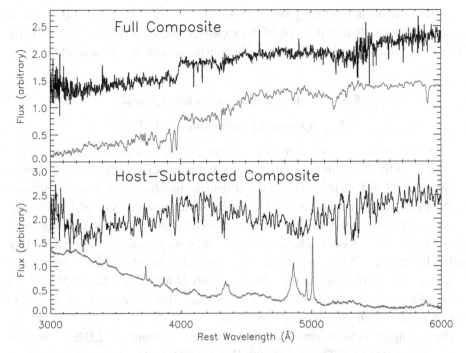

Figure 6.7. The composite spectrum of 15 lineless AGNs with large X-ray-to-optical luminosity (from Trump et al., 2009; reproduced by permission of the AAS). The top panel shows the composite of the 15 sources (top curve) and compares it with the spectrum of a red galaxy. The bottom panel shows the stellar subtracted continuum alongside a composite type-I spectrum.

show a decrease of line EW with continuum luminosity and/or L/L_{Edd} (Chapter 7). The EWs of the very luminous lineless AGNs (in most cases, only upper limits on the EW) are at the very end of these distributions. Nevertheless, extremely large L/L_{Edd} is one possible explanation for the weak broad emission line.

A different type of explanation for the weak emission lines is related to the properties of the central accretion disk in such objects. This applies to very low as well as very high luminosity sources, but for very different reasons. As explained in Chapter 4, a very low accretion rate through the central disk can result in heating of the central part and the onset of radiation-inefficient advection-dominated accretion flow (RIAF) with inefficient conversion of gravitational energy to electromagnetic radiation. Such systems can lack much or all of the (otherwise strong) UV ionizing radiation. This has been proposed as a possible explanation for the very low luminosity of lineless AGNs such as the ones shown in Figure 6.7. Regarding the high-luminosity sources, here the Lyman continuum radiation by the disk depends on the BH mass and accretion rate and can be extremely weak in disks around very massive BHs (Chapter 4). Such systems are likely to show very luminous continua but no line emission.

Lineless AGNs are not fully understood. They seem to be at the tail of one or more well-known distributions of AGN properties, and their relative number in the AGN population (of order 1%) is entirely consistent with this general idea.

6.5. Low-luminosity AGNs and LINERs

6.5.1. Spectral classification of LINERs

Low-ionization nuclear emission-line regions (LINERs) were already mentioned in § 6.1.2. In the local universe, they are found in about one-third of all galaxies brighter than $B = 15.5$ mag. This is larger than the number of local high-ionization AGNs by a factor of 10 or more. Local high-ionization AGNs and LINERs are present in galaxies with similar bulge luminosities and sizes, neutral hydrogen gas (H I) contents, optical colors, and stellar masses. Given a certain galaxy type and stellar mass, LINERs are usually the lowest-luminosity AGNs, with nuclear luminosity that can be smaller than the luminosity of high-ionization AGNs by 1–5 orders of magnitude. An alternative, perhaps more physical name for this class of objects is *low-luminosity AGNs* (LLAGNs).

The strongest optical emission lines in the spectrum of LINERs include [O III] $\lambda5007$, [O II] $\lambda3727$, [O I] $\lambda6300$, [N II] $\lambda6584$, and hydrogen Balmer lines. All these lines are prominent also in high-ionization AGNs, but in LINERS, their relative intensities indicate a lower mean ionization state. For example, the [O III] $\lambda5007/H\beta$ line ratio in LINERs is 3–5 times smaller than in high-ionization type-II AGNs. Line diagnostic diagrams are efficient tools to separate LINERs from high-ionization AGNs. One such example is shown in Figure 6.1, and more discussion and detailed photoionization models are given in § 7.2.

The exact shape of a LINER's SED is still an open issue. In some sources, it is well represented by the SED shown in Figure 1.1. Such an SED has a clear deficit at UV wavelengths compared with the spectrum of high-ionization AGNs. However, some LINERs show strong UV continua and, occasionally, UV continuum variations, and it is not entirely clear what fraction of the population they represent. This is related to the issue of RIAFs discussed in Chapter 4 and the relationship between the mass accretion rate onto the BH and the emitted radiation.

Pointlike X-ray sources have been observed in a large number of LINERs. These nuclear hard X-ray sources are more luminous than expected for a normal population of X-ray binaries and must be related to the central source. Many LINERs also contain compact nuclear radio sources similar to those seen in radio-loud high-ionization AGNs but with lower luminosity comparable to WLRGs (Figure 6.6). The UV-to-X-ray luminosity ratio in LINERs is, again, not very well known. In LINERs with strong UV continua, α_{ox} is smaller than in low-redshift, high-ionization AGNs, consistent with the general trend discussed in Chapter 1 between α_{ox} and L_{bol}. However, α_{ox} is not known for most LINERs because of the difficulty in measuring the UV continuum.

Like other AGNs, LINERs can be classified into type-I (broad emission lines) and type-II (only narrow lines) sources. The broad lines, when observed, are seen almost exclusively in Hα and hardly ever in Hβ. This is most likely due to the weakness of the broad wings of the Balmer lines that are difficult to observe against a strong stellar continuum. As discussed earlier (§ 6.2.5), some, perhaps many, LINERs may belong to the category of real type-II AGNs – those AGNs with no BLR. The phenomenon is expected to be more common among low-luminosity sources and hence to be seen in LINERs. Because of all this, the classification of LINERs is ambiguous, and the relative number of type-I and type-II objects of this class is uncertain even at very low redshift.

6.5.2. Accretion-driven LINERs and imposters

LINER-looking spectra can be produced by various mechanisms that are not associated with BH activity. Shock-excited gas, with typical electron temperatures of $\sim 5 \times 10^4$ K, can result in LINER-looking emission-line ratios. Such gas is found near the boundaries of nucleur radio jets, in the outer regions of interacting systems like ULIRGs, and in other astronomical environments associated with various kinds of winds. A spectroscopic way to distinguish shock-ionized from photoionized gas is to determine the electron temperature of the line-producing gas. For a given level of ionization, the photoionized gas temperature is lower than the temperature of the shock-excited gas since in the former, ionization is mostly by high-energy photons. This process depends on the SED and the ionization parameter and not on the gas temperature. Conversely, ionization in shock-excited gas is due to collisions with fast particles, which require higher gas temperature. For example, the typical electron temperature required to doubly ionize oxygen is about $\sim 5 \times 10^4$ K, while the electron temperature of photoionized gas where O^{+2} is the most abundant oxygen ion is about 10^4 K. Since the gas temperature in such an environement can be measured directly from relative line ratios, for example, the two O III lines at 4363 and 5007 Å, such measurements provide the key for the understanding of the nature of the ionizing source. In most known LINERs, photoionization by a nuclear radiation source seems to be the main source of ionization and excitation of the gas.

Photoionization by evolved stellar populations like post-AGB (pAGB) stars also results in LINER-looking emission-line ratios. This has resulted in much confusion between this population, which is common in early-type galaxies, and nuclear-ionized LINERs. The line-to-continuum ratios in pAGB spectra are very different from those observed in LINERs that are associated with active BHs. In particular, the EWs of Hα and other emission lines in pAGB stars are smaller than in

AGN-type LINERs. A practical way to separate such galaxies from nuclear LINERs is to remove from the sample all sources with EW(Hα) < 3 Å. Taking all this into account, we find that photoionization by low-luminosity AGNs is likely responsible for many, perhaps most, LINERS.

6.5.3. The infrared properties of LINERs

IR studies are useful since they provide the best way to eliminate the effect of dust and reddening on the observed spectrum. However, such studies when applied to LINERs only increase the confusion. Spitzer-based observations of a few dozen LINERs, mostly in luminous host galaxies, indicate two populations that are distinguished by their MIR SEDs, the luminosities of their PAH features (Chapter 8), and the MIR fine-structure lines. The two groups are IR-faint LINERs, with emissions arising mostly in compact nuclear regions, and IR-luminous LINERs, which often show spatially extended, non-AGN emissions. IR-luminous LINERs have MIR SEDs typical of starburst galaxies, while the MIR SEDs of IR-faint LINERs are considerably bluer. Highly excited [O IV] 25.9 μm emission lines are detected in both populations, indicative of AGN photoionization. In fact, some of the more luminous IR LINERs also show the presence of the [Ne V] 14.3 μm IR line that cannot be excited in star-forming regions. The high-ionization lines are accompanied, in many cases, by strong nuclear X-ray sources. Obviously reddening is playing an important role in increasing the confusion.

In some well-documented cases, the IR luminosity of LINERs is very high and approaches the luminosity of ultraluminous IR galaxies (ULIRGs, merging systems with extremely high star formation rates and L(IR) > $10^{45.6}$ erg s^{-1}; see Chapter 8. In some of the sources, the strong high-ionization lines are probably highly obscured, which leads to their erroneous classification as LINERs. Other powerful FIR sources, including some ULIRGs, show LINER characteristics because part of the gas in these strongly interacting systems is shock excited and is not related to BH activity.

An interesting discovery by Herschel is that a large fraction of LINERs at $z \sim 0.3$ have large L(FIR) that, if translated in a simplistic way to star formation rate (SFR), correspond to about 10 $M_\odot$ yr^{-1}. Such SFRs are very different from those found in LINER host galaxies at very small redshift, where the stellar population in the host is old and L(FIR) is much lower. This raises the possibility that some fraction of LINER host galaxies may be forming stars at a high rate, while their nuclear gas, exposed to the active BH, shows a typical LINER spectrum. Another possibility, which is supported by different observations, is that the host galaxies of many LINERS are poststarburst systems with SFR, which is considerably smaller than it was several Gyr ago.

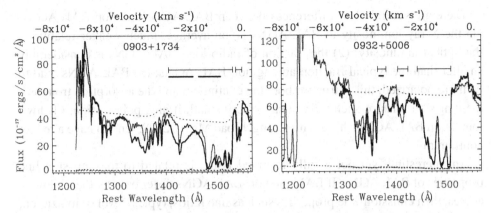

Figure 6.8. Spectra of BAL AGNs showing broad absorption troughs in the (rest) UV part of the spectrum (from Capellupo et al., 2011; reproduced by permission of John Wiley & Sons Ltd.). Velocities relative to the peak emission of the C IV λ1549 line are indicated on the top. Each panel shows two observations of the same source separated by a few months to a few years. In some cases, the spectra are identical, e.g., 1232 + 1325. In others (e.g., 1303 + 3048), there is a very big difference indicating either a change in ionization of the absorbing gas or a motion across the line of sight resulting in a change in column density. The long horizontal lines indicate the boundary of the trough. The thicker, shorter bars indicate the wavelength range where variability is evident.

6.6. Broad-absorption-line AGNs

Broad-absorption-line (BAL) AGNs (also *broad-absorption-line QSOs*, or BAL QSOs) are type-I AGNs with rest wavelength UV spectrum showing deep, blue-shifted absorption features, or troughs, associated in most cases with strong resonance lines of C IV λ1549, Si IV λ1397, N V λ1240, O VI λ1035, and Lyα. Such sources can be referred to as high-ionization BAL AGNs. There are also low-ionization BAL AGNs that show strong, broad Mg II λ2798 troughs and occasionally Fe II lines. About 10 to 20 percent of all high-luminosity type-I AGNs with $L_{bol} > 3 \times 10^{45}$ erg s^{-1} can be classified into one of these categories. The outflow velocities associated with the BAL phenomenon are very large, up to 0.2c and even larger. The widths of the absorption troughs are also very large, typically 10,000 km s^{-1}. Several such spectra are shown in Figure 6.8.

One way to explain the BAL phenomenon is to assume that all high-luminosity AGNs contain high-outflow-velocity systems. The outflow is observed in only a small fraction of objects because of a different viewing angle. If this is correct, it means that the fraction of BAL AGNs in the population, ∼10 percent, is also the typical covering factor of the central source by a large line optical depth outflowing material.

There are several clear differences between BAL AGNs and non-BAL AGNs: (1) the X-ray luminosity in BAL AGNs is, on average, much weaker relative to the optical luminosity; (2) the fraction of radio-loud BAL AGNs is considerably smaller than the typical fraction among non-BAL AGNs; (3) BAL AGNs tend to be more polarized; this is true for both the continuum and the absorption troughs – the emission lines of such objects show very small, if any, polarization; (4) low-ionization BAL AGNs, those with strong, broad Mg II $\lambda2798$, tend to have a redder continuum.

An interesting, yet not a fully resolved issue, is related to the emission-line properties of BAL AGNs. If BAL and non-BAL AGNs differ only by their viewing angles, then emission-line properties such as line ratios, typical level of ionization, line width, and equivalent width should be roughly the same in all sources. This is definitely not the case, at least for some of the lines. These and other known differences between the groups may be related to different amounts of obscuration of lines and continua in the two classes. We come back to this issue in Chapter 7.

6.7. AGN unification

A major observational and theoretical challenge is to construct a general picture that connects the various subgroups of AGNs. Such a scheme starts from the central supermassive accreting BH and the central accretion disk and adds other ingredients, related to the gas and dust in the system and the radio and γ-ray jet. All this goes under the name of *AGN unification schemes*.

The simplest addition of a torus-like obscurer to the two central components, BH and accretion disk, already explains most of the observed differences between radio-quiet type-I and type-II AGNs. Such a structure introduces a viewing angle parameter that determines what AGN components will be seen from a given line of sight. This can account for the different observed properties of MIR, NIR, optical, and UV emission lines in type-I and type-II sources; the luminosity and variability of the optical–UV continuum; the different amount of obscuration of the central X-ray source; and the shape of the MIR continuum. However, it is not as successful in the limits of very low and very high luminosity, where the presence of such an obscurer is questionable. This is the case for various types of lineless AGNs, for the least luminous LINERs, and for the most luminous AGNs at high redshift. A highly simplified scheme of this type is illustrated in Figure 6.9.

Adding radio-loud and γ-ray-loud objects to this scheme requires an additional component and several adjustments. The additional ingredient is a relativistic jet emanating from the vicinity of the BH. The basic properties of such jets include confinement and collimation, radio and γ-ray variability, superluminal motion, and nonthermal emission by relativistic particles. The complications arise from the fact that FRI and FRII jets are rather different in their physical properties (Chapter 7)

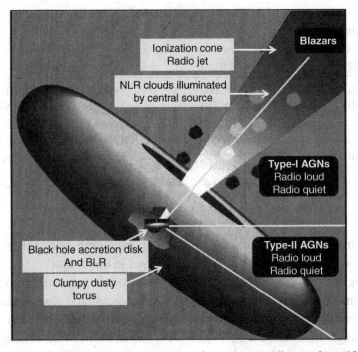

Figure 6.9. A side view of AGNs showing the main ingredients of a unification scheme that does not include LINERs and WLRGs. Strong radio and/or γ-ray jets are present in some 10 percent of all powerful AGNs. Blazars are those objects where the line of sight is along the central jet direction. Radio-loud type-I AGNs are those objects where the line of sight is at some angle to the jet direction, and radio-loud type-II AGNs are those objects where the observer's line of sight to the source is blocked by a large, dusty torus. The cone of ionizing radiation illuminates some, but not all, gas clouds around the source. Radio-quiet type-I sources are situated in the same sector as radio-loud type-I sources, but in this case, the central radio source is much weaker. The same is true for radio-quiet type-II sources. The torus is quite thick, at least in low- to intermediate-luminosity sources, such that the type-II sector subtends a larger solid angle than the type-I and blazar sectors together.

and cannot be considered as a single component in this scheme. It is reasonable to assume that in both cases, the direction of the jet at its launch point is along the spin direction of the BH and perpendicular to the plane of the accretion disk. This results in inclination-angle dependence of the jet luminosity and variability. In a *generalized unification scheme*, blazars are those FRI and FRII objects seen at a very small angle to the inner jet direction. Other radio-loud AGNs are seen at a somewhat larger angle, up to about 50 degrees (i.e., radio boosting is still important). BLRGs are seen at somewhat larger angles, perhaps up to about 60 or 70 degrees (the exact inclination seems to be luminosity dependent; see Chapter 9), and NLRGs are the geometrical equivalent of radio-quiet type-II AGNs. Figure 6.9

Table 6.1. *Main groups of AGNs arranged according to luminosity, obscuration, and radio power*

Nuclear source	Radio quiet	Radio loud
Thin/thick disk	Type-I AGNs	Type-I AGNs (FRII, BLRGs)
Thin/thick disk	Type-II AGNs	Type-II AGNs (FRII, NLRGs)
RIAF	Type-I LINERs	Type I LINERs (FRI, WLRGs)
RIAF	Type-II LINERs	Type-II LINERs (FRI, WLRGs)
Thin/thick disk		Flat radio-spectrum blazars (FRII)
RIAF		BL-Lac objects (FRI)

shows the main ingredients of such a four-component unification scheme and the "location" of the various types of AGNs discussed in this chapter.

Adopting a three-dimensional classification based on nuclear (disk) luminosity, radio luminosity of flat-spectrum FRII radio sources, and obscuration, we can include all radio-loud and radio-quiet high-ionization AGNs discussed in this chapter in the single unification table (Table 6.1). The addition of the missing groups, LINERs, WLRGs, and BL-Lac objects, requires an additional refinement by allowing two types of disks and two types of jets. The suggestion is that FRII AGNs are due to the combination of high-efficiency (large L/L_{Edd}) accretion disks and a jet whose source of inverse Compton photons is the emission from the BLR and the accretion disk. The low-ionization FRI AGNs are due to a combination of advection-dominated disks (RIAFs) and internally produced (synchrotron) photons that undergo inverse Compton scattering in the jet. More discussion of these processes is given in Chapter 7.

The classification into type-I and type-II AGNs provided, over the years, an incentive to subdivide the range into refined spectral groups. An obvious way is to use the relative intensity of the broad and narrow components of a certain line, for example, Hα, as a way to identify a more specific "location" along a continuous sequence. This can start from no narrow component to the line and end with a profile dominated by the narrow component. Such intermediate types have been named *Type-1.5 AGNs*, *Type-1.8 AGNs*, and so on.

Unfortunately, the separation of permitted AGN emission lines into two distinct components is dubious and, occasionally, meaningless because the combined profile is very smooth. Strong narrow forbidden lines, like [O III] $\lambda5007$, can provide a good template for the narrow part of the profile, but the practical separation is still problematic. There is evidence that the relative intensities of the broad and narrow components is luminosity dependent such that the narrow lines are stronger in lower-luminosity AGNs. This enables a more meaningful separation in lower-luminosity sources but very ambiguous results at the high-luminosity end. The "spectral types" that are still being used are Type-1.5 AGNs (narrow component which is about 20%–30% of the total line flux), Type-1.8 AGNs (the narrow component contributes about 50%–70% of the total line flux), and Type-1.9 AGNs (extremely weak broad component seen in Hα but not in Hβ). Type-1.8 and Type-1.9 AGNs are occasionally associated with large line-of-sight reddening and may simply be the result of the more heavily obscured BLR. This is indicated by the Hα/Hβ narrow-line ratio and, in several well-documented cases, a clear dust lane across the host galaxy.

6.8. Further reading

Diagnostic diagrams: General aspects are discussed in Veilleux and Osterbrock (1987). A general review with emphasis on SDSS data as well as references to photoionization modeling is provided in Groves et al. (2006). See also Kewley et al. (2006) for more detailed information on emission-line ratios. Refinements, extension to IR lines, and further modeling are discussed in Netzer (2009), Tommasin et al. (2010), and references therein.

NLR structure and ionization cones: See references in Chapter 7.

Radio-loud AGNs: This is a well-established field that is covered in detail in the books by Robson (1996) and Peterson (1997). For the radio properties of local fainter sources, see Best et al. (2005). For more recent observations and classification, see Sikora et al. (2007, 2008).

X-ray and γ-ray jets and blazars: See Worrall (2009) on the X-ray aspect and a general recent review by Urry (2011).

LINERs: A recent comprehensive review is Ho (2008). For UV continuum, see Maoz (2007). For "fake" AGNs, see Cid Fernandes et al. (2011). IR properties of LINERs are discussed in Tommasin et al. (2012).

BAL AGNs: See references in Chapter 7.

Unification schemes: Comprehensive papers with references to the older schemes are Tadhunter (2008), Ho (2008), and Urry et al. (2002). The IR and radio viewpoint is described in Haas et al. (2005). Classification by emission-line properties is reviewed by Véron-Cetty and Véron (2000). For unification of radio galaxies, see Barthel and van Bemmel (2003).

7

Main components of AGNs

The physical properties of the gas in AGNs depend on its location, density, column density, and composition. These determine the ionization parameter, the relative importance of gravity and radiation pressure force, the dust content, the gas velocity, and more. In this chapter, we consider several possible locations around the central BH and the gas and dust properties in each.

7.1. The broad-line region

Consider large-column-density ($\sim 10^{23}$ cm^{-2}), high-density ($\sim 10^{10}$ cm^{-3}) clouds situated around an accreting BH with $L/L_{\mathrm{Edd}} \sim 0.1$, at a location where $L/4\pi r^2 \simeq 10^9$ erg s^{-1} cm^{-2}. For a very luminous AGN, this is at about 0.1–1 pc from the BH. Assume also that the clouds can survive over many dynamical times either because they are confined or because they are the extensions of large self-gravitating bodies such as stars. For large enough column density clouds, the system is bound because gravity completely dominates over radiation pressure force. The typical Keplerian velocity at this location is ~ 3000 km s^{-1}, which will be reflected in the widths of the emitted lines. We also assume a global (4π) covering factor of order 0.1. This means that we can neglect the effect of the radiation emitted by one cloud on its neighbors.

The above physical properties result in $U_{\mathrm{hydrogen}} \sim 10^{-2}$. This means that only the illuminated surfaces of the clouds are highly ionized. The most abundant ions in the ionized parts are He II–III, O IV–VI, C III–IV, and so on. The strongest predicted emission lines are, therefore, Hα, Lyα, C IV λ1549, and O VI λ1035. The density is high enough to suppress all optical forbidden lines but not all the semiforbidden lines. Strong predicted lines of this type are [C III] λ1909 and [O III] λ1663. A big part of a large-column-density cloud in this location must be partly neutral since only X-ray photons can penetrate beyond a column of $\sim 10^{22}$ cm^{-2}. This part will produce strong lines of H I, Mg II, and Fe II. The observed EWs of the strongest lines in this region depend on the emissivity and covering factor. For the conditions assumed here, they are of order 10–100 Å for the strong emission lines. Absorption lines are predicted to be extremely weak because of the small covering factor. A region with these observed properties and spectrum would justify the name *broad-emission-line region* (BELR) or, in short, *broad-line-region* (BLR).

Can BLR clouds be confined? Magnetic confinement is a likely possibility since the required magnetic field is small, about 1 G. Confinement by HIM is more problematic for various reasons. The required HIM density can be estimated from the known density and temperature of the BLR gas and from the requirement of pressure equilibrium. For the gas in the clouds, $N_e T \sim 10^{14}$ cm^{-3} K. If the ambient medium (HIM) gas is at its Compton temperature of few $\times 10^7$ K, its density is $N_{\rm HIM} \sim 10^7$ cm^{-3}. The dimension of the BLR in the most luminous AGNs is of order 1 pc (see later). This gives a total Compton depth of more than unity for the HIM. The observational consequences would be smearing of the central source variations and broad Compton scattering wings for the broad emission lines. This is not observed in many highly variable AGNs. There are, however, relativistic processes that can raise the HIM temperature beyond its Compton temperature. Such a relativistic HIM may still provide the required confinement.

How many high-density clouds are required to produce the observed broad-line profiles? Given the gas velocity, and the fact that individual confined clouds are likely to emit lines with a typical width of ~ 10 km s^{-1} (the sound speed in the gas), we can estimate the number of clouds required to produce the symmetrical smooth profiles observed in many cases. The number is very large, of order 10^{6-8}. This raises serious questions about the formation and destruction of the clouds and possible collisions between them. A possible way out is to invoke internal turbulent motion that exceeds the sound speed by an order of magnitude or more. This results in much broader single-cloud profiles and can alleviate some of the difficulties. A special geometry and ordered motion of the clouds can help too, for example, a flat, rotating disklike configuration. All these ideas need to be justified observationally as well as theoretically.

7.1.1. The broad-line spectrum

The general considerations outlined in Chapter 5 suggest that photoionization is the dominant physical process in the BLR gas. Indeed, various models constructed for this region are quite successful in explaining the general properties of the broad-line spectrum. Such models have been discussed in detail in § 5.7. The ionization parameter, the gas density, and their distribution are the important factors that determine the BLR spectrum.

Several line ratios are good ionization parameter indicators in the BLR. An important example is the [C III] $\lambda1909$/C IV $\lambda1549$ ratio that decreases monotonously with U_{hydrogen}, except in extreme conditions where the density is too high (greater than few $\times\ 10^{10}$ cm^{-3}) to produce a strong [C III] $\lambda1909$ line. This line ratio was the one used in the early era of AGN research to estimate U_{hydrogen} and is still the one that represents best the typical conditions in the BLR.

The hydrogen line spectrum and isotropic line emission

The relative intensities of the hydrogen emission lines in low-density photoionized gas are well understood within the framework of the simple recombination theory, in particular, the case B recombination. These issues were discussed in Chapter 5. However, the column density in the BLR is large enough to make many of the Balmer and Paschen lines optically thick, which complicates the calculations significantly. For example, the calculated optical depths in typical BLR clouds are $\tau(\text{Ly}\alpha) \sim 10^8$ and $\tau(\text{H}\alpha) \sim 10^4$. The high density compared to galactic HII regions and planetary nebulae adds an additional complication by making collisional excitation and deexcitation (which are not included in the case B calculations) very important. The escape probability formalism introduced earlier provides a simple and limited way to handle these issues, and more sophisticated methods, involving better treatment of the line transfer, have not been fully incorporated into present-day BLR models.

A well-known problem in AGN research, the Lyα/Hβ problem, is most probably related to this issue. Observations show that the observed I(Lyα)/I(Hβ) ratio (about 5–10) is considerably smaller than the one predicted by the simple case B recombination theory (about 35 for gas with $T_e = 10^4$ K and $N_e > 10^5$ cm^{-3}). Photoionization models that use the escape probability formalism go only halfway toward solving the problem. The related issues of the Balmer decrement and the Paschen line ratios are also related to line transfer. In these cases, the escape probability approach gives results that are in better agreement with the observations.

A related issue is the radiation pattern of optically thick emission lines like the hydrogen Lyman lines, most of the Balmer lines, and the resonance lines of the most abundance elements. The simplest case is Lyα, in which strong anisotropy is predicted because a big part of an optically thick gas clump contains neutral hydrogen. Thus, the hydrogen Lyman lines are more likely to escape from the illuminated face of such a clump. This is not necessarily the case for lines like C IV $\lambda1549$ because the transition between C^{+2} and C^{+3} inside the clump is sharp (the exact shape depends on the ionizing SED) and the emitted line photons can escape from both sides of the clump. The hydrogen Balmer lines represent an intermediate case between the two. Unfortunately, radiative transfer calculations that are based on a local escape probability method are limited in their ability to properly model this effect. Thus, anisotropic line emission is expected for some lines, but the exact magnitude of the effect is unknown and depends on the cloud geometry.

Covering factor and optically thin BLR gas

So far we have said nothing about the determination of the covering factor, C_f. This factor is well defined for optically thick gas. It is the fraction of the central source required to be covered by optically thick clouds to explain the total emission by the gas. Photoionization models of the BLR suggest that most of the observed emission is due to gas that is optically thick in the Lyman continuum and for which $C_f = 0.1$–0.2. The covering factor is ill defined for optically thin gas. In such a case, it can be interpreted in a simplistic geometrical way, which can reach unity, or it can be computed by using some effective optical depth.

It is interesting to note that integration over the Lyman continuum, which involves some guesses about its shape, and assuming the preceding covering factor, indicates that the number of Lyman continuum photons is within a factor 2 of the observed number of Lyα photons, that is, close to the prediction of the simple case B recombination theory. Using the Balmer lines instead of Lyα, and assuming simple case B conditions, suggests a much larger covering factor, which disagrees with the observations of several metal lines. Thus, the Lyα/Hβ problem is likely to be due to underestimating the Balmer line intensities rather than overestimating the Lyα intensity.

Some gas in the BLR can be optically thin to the Lyman continuum radiation. This gas emits fewer hydrogen line photons per unit covering factor. However, its geometrical covering factor may approach unity and, combined with a high level of ionization, it can contribute significantly to emission lines, such as C IV $\lambda1549$ and O VI $\lambda1035$, that originate from regions that are very close to the central BH. Such material can have a significant column density and line emission. Assuming a bound cloud system, the profiles of these lines are going to be very broad. In addition, their intensity will be less sensitive to the changing continuum variations, which are discussed in the following sections.

Broad-line reddening

The difficulty in explaining the hydrogen line spectrum may be related to reddening outside the BLR (there is no dust in the BLR gas itself). A galactic-type dust, with relatively small column, can result in a significant reduction of the observed Lyα flux and in the intensity of other broad UV emission lines relative to the intensities of the Balmer lines. Depending on the dust location, such a differential change in intensity may also affect the AGN continuum and result in underestimation of the bolometric luminosity.

There are various claims for reddening of the AGN continuum that are difficult to confirm because of the large scatter in the intrinsic shape of the SED. In particular, there is no signature in the spectrum of type-I AGNs of the prominent 2175 Å galactic dust feature caused by graphite grains. This in itself is not surprising since the very different environment near AGNs may result in different dust properties, for example, different types of grains and/or different grain size distributions. Several models involving such grains have been suggested in an attempt to explain the optical–UV SED and its deviation from a simple disk SED. Perhaps the most sensitive upper limit on the amount of neutral dusty gas along the line of sight to the center is based on the search for Mg II λ2798 and other low-ionization absorption lines expected from the gas associated with such dust.

The best indicators of broad-line reddening are several line ratios in the optical and UV parts of the spectrum. One such ratio, I(O I λ1302)/I(O I λ8446), involves O I lines that are the result of the fluorescence process between Lyβ and O I λ1026 (Chapter 5). The theoretical ratio is relatively easy to compute, but the observations of the weak line at 1302 Å are challenging and the results are uncertain, with some indications for reddening. Another useful ratio is I(He II λ1640)/I(He II λ4686). Photoionization models predict this ratio to be in the range 7–12. Several high-quality observations indicate smaller ratios, consistent with reddening. Here, again, the observations are challenging and the uncertainties quite large.

Fe II emission lines

While the physics of the high-ionization, high-excitation broad emission lines is relatively simple, and the calculated intensities are in reasonable agreement with the observations, this is not the case for some lines of low-ionization species, especially lines of Fe^+. Such lines are observed in almost all type-I AGNs, in the visible, UV, and IR parts of the spectrum. In most objects, they are very strong, with integrated line intensities that exceed the Hβ line intensity. In fact, in some sources, the total Fe II line intensities exceed the combined intensity of all other emission lines in the spectrum. Evidently, the part of the gas containing singly ionized iron is very extended, and the mechanism to pump energy into these transitions is very efficient.

Theoretical calculations of the broad Fe II line spectrum, assuming photoionization by the central source, are in disagreement with most observations. This can be understood by considering the rich emission-line spectrum (Figure 7.1), the lack of essential atomic data, and the various routes that are available to pump energy into these transitions. The fitting of the spectrum of the extreme Fe II emitter shown in Figure 7.1 is a good example of this failure. The largest disagreement between model and observations, regarding individual features, is at UV energies, around 2200–2600 Å, a region that contains hundreds of Fe II lines. There is also a general difficulty in reproducing the relative intensity of the three major Fe II bands: 2200–2600 Å, 2800–3400 Å, and 4000–5300 Å. The relative

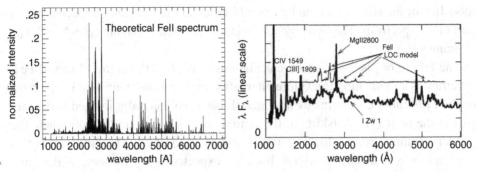

Figure 7.1. (left) A theoretical Fe II model for a typical BLR cloud. The lines are shown with their thermal widths to illustrate the very rich spectrum (courtesy of G. Ferland). (right) Typical calculated LOC Fe II model compared with observations of I Zw 1, an extreme Fe II emitter. The theoretical spectrum is raised by one tick mark to allow easier comparison with observation (adapted from Baldwin et al., 2004). Note the general disagreement, in particular, the two strong UV features that are not seen in the real spectrum.

line intensities across the optical band, between 4000 and 5300 Å, are better reproduced.

A major difficulty in fitting the Fe II spectrum is the very large range of properties across the AGN population. This is illustrated in Figure 7.2, which shows a

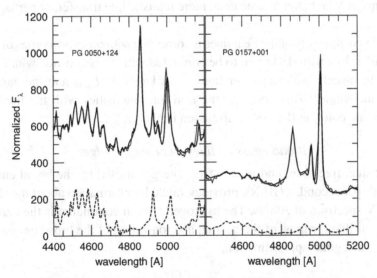

Figure 7.2. Examples of Fe II line deblending in the Hβ region of two type-I AGNs (courtesy of B. Trakhtenbrot). The object on the left is an extreme Fe II emitter, and the one on the right is more representative of the population. The diagram shows the observed spectra, the fitted Fe II models (dashed lines), and the final fits (thick solid lines), which include also the underlying continuum, the [OIII] lines, and the He II λ4686 line (not shown in the plots as individual components).

modeling of the Hβ line region by a combination of the Fe II lines expected in this part of the spectrum. The two objects are later classified as having type A and type B profiles (§ 7.1.4).

The failure of photoionization calculations to explain the entire observed Fe II spectrum is a motivation to explore alternative explanations. One idea is that most of the Fe II emission originates in material that is collisionally ionized and is not part of the BLR. The outskirts of the central accretion disk, in the part where the gas temperature drops below $\sim 10^4$ K, is one such location. The difficulty in this explanation is that other emission lines are expected from the same region, and their intensity relative to the Fe II lines is in conflict with observations. Another suggestion is that much of the observed flux is due to *continuum fluorescence*, a process that was described in Chapter 5 and can pump continuum photons into Fe II absorption lines. The subsequent emission of this radiation, in a different direction, can lead to apparent enhancement of the observed Fe II lines without affecting much the energy balance inside the cloud. This, again, can be calculated and is not in good agreement with the observations. Finally, we note that there are many Fe II transitions with overlapping wavelengths, which can change the line intensity ratio through line fluorescence (Chapter 5). Moreover, some Fe II lines overlap in wavelength with the strong Lyα line. This, again, does not solve the problem, although it can lead to better agreement with observations. It seems that we are missing some basic physics in the modeling of the complex Fe II ion. The missing ingredients may be better atomic data, more realistic line transfer, or perhaps other factors.

The relative intensity of the Fe II lines is correlated with other properties of AGNs. In particular, I(Fe II)/I(Hβ) seem to be correlated with emission-line widths, being stronger in objects with narrower lines, with larger L/L_{Edd} and, perhaps, with the iron metallicity. This connects the iron line intensities with the normalized accretion rate onto the BH and is discussed further in § 7.1.3.

Broad emission lines and energy budget

The total-line, free–free, and bound–free, energy emitted by the broad emission-line clouds per second, $L(\text{BLR})$, provides valuable information about the shape of the far UV spectrum of AGNs. The hydrogen Lyman continuum is the part of the SED responsible for most of the heating and ionization of the broad emission-line gas. Thus, to a good approximation,

$$L(\text{BLR}) = C_f \int_{13.6\,\text{eV}}^{\infty} L_\nu d\nu, \tag{7.1}$$

where C_f is the covering factor of the optically thick gas in the BLR. We can use this equation, together with the assumption that the number of Lyman continuum photons equals the product of C_f and the number of emitted Lyα photons, to write a simple expression for the mean energy of an ionizing photon, $< h\nu >$:

$$< h\nu > = \frac{L(\text{BLR})}{\text{number of Ly}\alpha \text{ photons}}. \tag{7.2}$$

$L(\text{BLR})$ is not easy to obtain because some of the lines, and the Paschen continuum, are emitted in the IR part of the spectrum. There are also relatively strong emission lines at $\lambda < 1000$ Å that are difficult to measure because of various absorption features. Even more challenging is the measure of the so-called small blue bump, which is a blend of strong Fe II lines, and the Balmer continuum emission, which gives an impression of a pseudo-continuum over the 2000–4000 Å band.

Several attempts to measure $< h\nu >$ by this method resulted in a high mean ionizing energy of about 50–100 eV, in disagreement with the simple extrapolation of the observed $E < 13.6$ eV power-law continuum. Thus, the total energy emitted by the BLR gas seems to exceed the energy absorbed by the clouds. This discrepancy has been referred to as the *energy budget problem.*

There were various attempts to explain the energy deficit, including the assumption that the SED observed by us is different from the one "seen" by the BLR gas (e.g., reddening outside the BLR or anisotropic line and/or continuum emission), wrong assumptions about the Lyα emissivity, and a large contribution to the gas heating from the high-energy X-ray continuum. There are also related methods that provide independent estimates of the shape of the (unobserved) Lyman continuum. For example, the intensity of the He II $\lambda1640$ recombination line is directly proportional to the number of photons in the He$^+$ Lyman continuum, at $E > 54.4$ eV. Thus, I(Lyα)/L(He II $\lambda1640$) is a measure of the relative number of $13.6 < E < 54.4$ eV and $E > 54.4$ eV photons and hence the continuum shape. This seems to be consistent with the result obtained by the energy budget method. These unsolved issues are directly related to the properties and inclination of the central accretion disk.

7.1.2. Gas metallicity in the BLR

The standard methods to determine the gas metallicity, Z (measured relative to solar metallicity), and the more specific ways to apply these methods to the BLR gas are explained in § 5.5. The application of the NV/CIV method to a large AGN sample indicates a large metallicity range, $1 \leq Z \leq 5$, with several examples of $Z > 10$. There are interesting suggestions that connect the high measured metallicity to other properties of AGNs. One is that the gas metallicity (more specifically, the N V $\lambda1240$/C IV $\lambda1549$ line ratio) depends on the source luminosity. Other possibilities are a dependence on the black hole mass and/or the normalized accretion rate, L/L_{Edd} (which, as explained, depends also on the spin of the BH). Some of

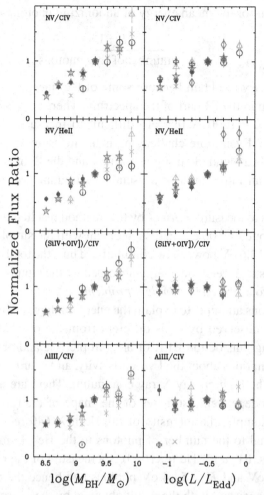

Figure 7.3. Correlations of various line ratios in type-I SDSS AGNs with BH mass and L/L_{Edd}. All line ratios that are metallicity indicators are changing with M_{BH}, but this is not the case for their dependence on L/L_{Edd} (courtesy of K. Matsuoka and T. Nagao).

the better established correlations are those obtained for a large number of SDSS AGNs. Such correlations are shown in Figure 7.3.

The dependence of the various line ratios on L/L_{Edd} is puzzling. The N v $\lambda1240$/C iv $\lambda1549$ ratio is, indeed, proportional to L/L_{Edd}, but this is not the case for other ratios that may depend on metallicity, such as the combination of OIV and SiIV lines near 1400 Å relative to the C iv $\lambda1549$ line. In this respect, the dependence on M_{BH} is more robust. Conversely, evidence from studies of NLS1s, which are known to have large L/L_{Edd}, point in the opposite direction. These objects have relatively low M_{BH}, well below the minimal mass shown in the diagram. However, their N v $\lambda1240$/C iv $\lambda1549$ is large, close to the top of the values shown here. Their inclusion in the M_{BH} diagram would completely destroy the correlation, but they will smoothly join the N v λ1240/C iv λ1549 versus L/L_{Edd} correlation shown on the right-hand side of Figure 7.3.

The various properties of the Fe ii lines discussed earlier, in particular, the correlation of I(Feii)/I(Hβ) with L/L_{Edd}, suggest yet another relationship. As shown from the N v λ1240/C iv λ1549 correlation, L/L_{Edd} is probably related to the gas metallicity. Thus, a larger I(Fe ii)/I(Hβ) ratio may be related to the general metallicity and hence to iron overabundance. This point is still unclear due to the large scatter in I(Fe ii)/I(Hβ), the complexity of the Fe ii spectrum, and the failure of photoionization models to accurately reproduce the Fe ii spectrum of type-I AGNs.

Iron enrichment, if correct, raises some interesting issues regarding AGN evolution. For example, strong Fe ii lines are seen in very high redshift AGNs, at $z \sim 6$. This suggests very fast enrichment of the gas by supernovae explosions. In fact, the time scale is too fast to be consistent with iron enrichment due to type-Ia supernovae. In addition, there are indications for a correlation of high star formation rate (SFR) and L/L_{Edd} in NLS1s. This suggests a direct route between iron abundance in AGNs and SFR in the host galaxy.

All the abundances investigated so far are based on the observations of broad emission lines and are hence related to the gas metallicity in the BLR. This is not necessarily connected to the metallicity in the host. Enriching the gas in the center of the galaxy requires high SFR only in a limited volume of space with a time scale that is considerably shorter than the enrichment time of the entire galaxy. Moreover, if SFR in the center, BLR gas metallicity, and L/L_{Edd} go hand in hand, it is reasonable to assume that there are several metallicity cycles in the history of every AGN, each one corresponding to an episode of high L/L_{Edd}. In other words, unlike the gas metallicity in the galaxy, the BLR metallicity can increase or decrease with time.

7.1.3. Broad-emission-line profiles

Line profile measurements

Emission-line widths, or more generally, emission-line profiles, can differ substantially from source to source. However, there are some general trends that are apparent in most type-I sources. First, lines of more ionized species tend to be broader. The He ii λ1640 and the O vi λ1035 lines are broader than C iv λ1549 and [C iii] λ1909, and all these lines are broader than the Mg ii λ2798 line. One such example was shown in Figure 1.4. This is consistent with a stratified, gravitationally bound BLR where the mean ionization parameter is higher at small distances. The issue becomes trickier when trying to compare metal lines with hydrogen lines since H$^+$ is abundant all over the BLR and since the peak emissivities of the Lyman and Balmer lines occur in different parts of the BLR.

Emission-line profiles are, in general, the best indicators of the gas kinematics in the BLR and the NLR. Following the procedure outlined in Chapter 5, or the one illustrated in § 7.1.10, should lead to the full, coupled solution of the gas distribution, motion, and emission across the BLR. Such solutions are not available, despite of the big efforts in this direction. Aparently, line fitting by simple kinematical models is degenerate with respect to several possible model parameters, and the link to line emissivity is not tight. In fact, attempts to fit two or three Gaussian components, or a combination of a Gaussian and a Lorentzian, to several line profiles give equally good results and similar FWHM. Power-law profiles have also been tried and resulted in satisfactory fits. All these functional forms are not linked, in any direct way, to the ordered motion of the BLR gas. In fact, all observed profiles, except for the very special case of double-peak profiles, which are discussed separately, are consistent with randomly oriented orbits in a gravitationally bound cloud system.

Double-peak emission lines

A small number of well-observed sources can be identified by their unique line profile properties. One such group shows double-peak (or double-horn) line profiles that are clearly distinguished from the rest of the population. Most of these objects are radio-loud AGNs that are seen at a relatively large inclination angle to the direction of the radio jet (steep-spectrum radio sources). Several examples are shown in Figure 7.4.

The fraction of broad, double-peak emission-line AGNs is very small, much below the fraction of radio-loud AGNs. However, some emissivity patterns across a flat inclined disk can result in a more smooth profile, where the individual component widths are larger than their separation, resulting in a less conspicuous double-horn profile. Such objects can be more numerous.

The more successful attempts to fit the double-peak emission-line profiles are based on a completely different model of the line-emitting region where the main source of ionization and heating of the gas is outside the disk. Currently there are no complete models of this type that successfully explain all emission-line intensities and profiles and the emission-line variability pattern expected in this structure (see later).

Type-A and type-B line profiles

Accurate line profile fitting requires high-quality observations. These are not available for a large number of the type-I sources observed in the large spectroscopic samples. However, there are several hundred, or perhaps even thousands, of sources with very high quality observations that allow a more detailed study of this issue. Such studies show that the emission-line profiles of sources with different BH masses and accretion rates can differ substantially. This leads to several

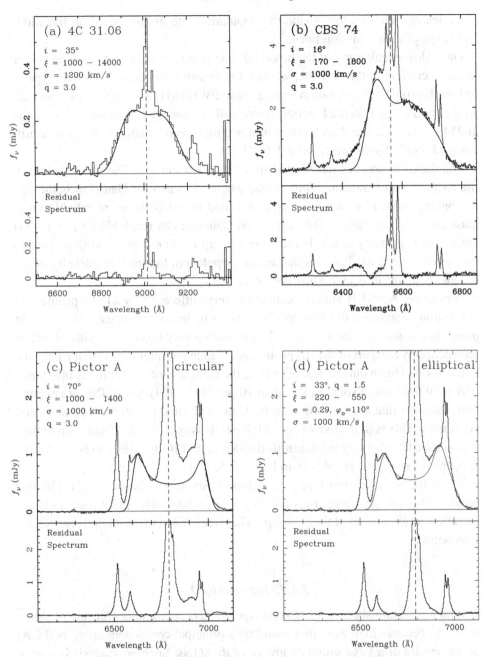

Figure 7.4. Example of double-peak Hα lines (from Eracleous and Halpern, 2003; reproduced by permission of the AAS). The solid lines represent attempts to fit the profiles by emission from the surface of an irradiated disk where q is the Hα emissivity across the disk ($q = 3$ is the case where the line emissivity follows exactly the irradiated energy; see Chapter 4). The inclination angle to the line of sight (i) and the limits of the emission region on the disk in units of r_g (ξ) are also marked.

classification schemes that divide the population into groups based on the width
and shape of their emission lines.

One scheme makes use of the broad Hβ line profile but can be extended to other
lines. According to this scheme, we can distinguish between sources with type-A
or type-B profiles. Population A sources have FWHM(Hβ) < 4000 km s^{-1} and are
well fit by a single Lorentz function. They tend to show lower continuum variations
and little change in the shape of their profile during such variations. At low redshift,
many of those objects are classified as NLS1s.

Population B sources have FWHM(Hβ) > 4000 km s^{-1}. The fit of their profile requires two Gaussian components, where the narrower Gaussian is the one
responding more to continuum variations (and hence is the more reliable virial
mass estimator; see later). The very broad Gaussian can reach FWHM of 10,000
km s^{-1} and, in many cases, is redshifted by up to 2000 km s^{-1} with respect to
the systemic velocity. It seems that in this population, the relative intensity of the
broader component is increasing with L_{bol}.

Population A and B sources can show very different C IV $\lambda1549$ profiles. In
population B sources, the line profile is more or less symmetrical. However, in
population A sources, the C IV $\lambda1549$ line shows a very broad blue wing, which is
unmatched by any part of the Hβ profile. A possible interpretation is that part of the
line is formed in an outflowing gas where the back side of the source is obscured.

A small fraction of sources, less than 10 percent of all type-I AGNs, are population A sources that show strong lines of Al III $\lambda1860$ and [Si III] $\lambda1892$ and several
Fe III lines. This is indicative of very high gas density, 10^{12} cm^{-3} and even higher,
as confirmed by the very weak semiforbidden line of [C III] $\lambda1909$ in such objects.
A couple of examples are shown in Figure 7.5.

While the division into type-A and type-B sources is based on the Hβ line
profile, other physical properties are clearly associated with the different classes
and allow a subdivision of these groups. This is related to the so-called eigenvector
1 sequence.

7.1.4. Eigenvector 1

Several well-established correlations in type-I AGNs appear under the name "eigenvector 1" because they were first found by a principal component analysis (PCA)
of the spectra of a large group of low-redshift AGNs. Such an analysis is a good
way to isolate common properties, in this case, spectroscopic properties, that are
related to each other and are not easy to find by other methods. The PCA shows
that one end of the first principle component (PC1 or also eigenvector 1) sequence,
at low redshift, is occupied by type-A objects. These objects share several correlated properties: narrower than usual broad emission lines, stronger than usual
(in terms of EW) optical Fe II lines, larger than usual Fe II/Hβ, smaller than usual

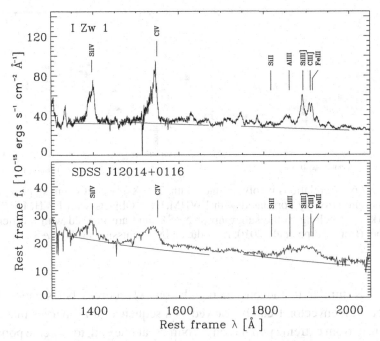

Figure 7.5. The spectrum of type-I AGNs that shows indications of very high density gas in the BLR. The UV part of the spectrum shows strong lines of ALIII and FeIII at around 1900 Å that are seldom observed in the majority of broad-line sources. The [C III] λ1909 line is very weak, and the C IV λ1549 lines show strong blue wings, suggesting fast outflow motion. Many such objects, like I Zw 1, shown on the top, have narrower than usual broad emission lines. Others, like SDSS-J12014+0116, have more typical line widths (courtesy of A. Negrete).

[O III] λ5007/Hβ, and a steeper than usual soft X-ray (0.1–2 keV) continuum. The term *usual* used here refers to the mean of the population regarding a property. Most such objects have low to intermediate bolometric luminosity with M_{BH} in the range $10^{6.5}$–$10^{7.5}\,M_{\odot}$. Population B objects occupy the other end of the sequence and show broader lines, weaker Fe II lines, and so on. Studies at high redshifts suggest the existence of a similar sequence among sources with much higher luminosity and BH mass, including BAL AGNs.

The difference between sources with weak and strong Fe II lines was already illustrated in Figure 7.2, which presented two broad-line AGNs, one with extremely strong Fe II lines and narrow Hβ and one more typical of the majority of the population. As shown in Figure 7.6, the line widths are also correlated with the SED shape and with L/L_{Edd}. In fact, we can order objects along two axes, one of increasing I(Fe II)/I(Hβ) and one of increasing line widths, in a way that agrees with the properties related to eigenvector 1. For example, we can subdivide population A sources into several classes, A1, A2, A3, ..., where I(Fe II)/I(Hβ) increases along the sequence. In this scheme, the mean value of I(Fe II)/I(Hβ) in population

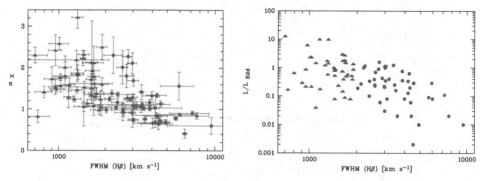

Figure 7.6. Correlations involving line width. The X-ray continuum slope α_x and $L/L_{\rm Edd}$ are strongly correlated with FWHM(Hβ). Objects with FWHM(Hβ) < 2000 km s^{-1} belong to the subgroup of NLS1s and are marked with a different symbol (from Grupe et al., 2010; reproduced by permission of the AAS).

B sources is similar to the one in population A1 sources but smaller than in A2 sources. Eigenvector 1 can be viewed as a sequence of properties that change continuously from extreme population A sources, at one end, to extreme population B sources, at the other end. In particular, there is no specific meaning to the border line of FWHM(Hβ) = 2000 km s^{-1}, introduced in the 1980s to distinguish NLS1s from BLS1s. This is illustrated in Figure 7.6.

The physical origin of the sequence and PC1 is only partly understood. It is most likely related to the normalized accretion rate, $L/L_{\rm Edd}$, where population A sources are at the high $L/L_{\rm Edd}$ end. It may also be related to the gas density and composition, although the correlation between $L/L_{\rm Edd}$ and metallicity is not understood. It is reasonable to assume that under the conditions of fast accretion onto a low-mass BH (large $L/L_{\rm Edd}$), the central disk will have unusual properties that are different from those of the thin α-disk. This can be related to the fast variations and the unusual X-ray slope. It may also explain the relatively weak lines (EW-wise) in extreme type-A AGNs due to the larger dimensions and small covering factor in such cases (see later) and the presence of a disk wind that may drive the C IV λ1549 emitting gas. Conversely, the suggestion of an extremely broad redshift component in population B objects is not known to be related to any known property or physical mechanism in the BLR.

It is interesting to note that the PCA suggests that eigenvector 2, PC2, may be related to $L_{\rm bol}$. It is also important to note that the classification of a specific source along the PC1 or PC2 axis may depends on geometrical issues such as the inclination of a flat BLR to the line of sight.

The various observations that connect the emission-line intensities and the continuum luminosity enable us to test the basic parameters of the BLR model. This involves BH mass and $L/L_{\rm Edd}$ and requires information about the BLR size. This topic, and the related BH mass determination, are discussed later. We return to the issue of line and continuum correlations in § 7.1.11.

7.1.5. Bolometric corrections

The total luminosity of radiatively efficient accretion flows, like accretion through a geometrically thin, optically thick disk, depends on the total accretion rate and the efficiency factor η. Unfortunately, there are no simple ways to determine L_{bol} directly from the observations because the far UV continuum, where most of the energy is emitted, is unaccessible to direct observations. In low-redshift objects, the reason is galactic absorption, and in high-redshift objects, it is absorption by intergalactic neutral gas. This creates a need to define *bolometric correction factors*, *BCs*, that can be used to convert a single-band measurement of L into an approximate L_{bol}.

Bolometric correction factors are likely to differ from one object to the next because they depend on the accretion rate, the accretion efficiency, and perhaps other factors. There are also practical difficulties due to geometrical factors like the disk inclination to the line of sight. The bolometric factors listed later have been obtained from studies of big AGN samples. They represent population means and disk inclination means and are being used to estimate L_{bol} in high-ionization type-I AGNs. More factors that are useful in type-II AGNs are given in § 7.2. All luminosities are reddening corrected and are given in erg/sec.

Optical correction factor, BC_{5100}**:** It is customary to use the optical continuum measured at 5100 Å, λL_λ at 5100 Å (sometimes referred to as L_{5100}). A reasonable approximation in this case is

$$BC_{5100} = 53 - \log(L_{5100}). \tag{7.3}$$

UV correction factor, BC_{1400}**:** To a reasonable approximation,

$$BC_{1400} = 0.5 BC_{5100}. \tag{7.4}$$

L_X **(2–10 keV) correction factor,** BC_{2-10}**:** Here we use a two-step procedure based on α_{ox}. To a good approximation,

$$\log L_{5100} = 1.4 \log L_X - 16.8. \tag{7.5}$$

We then use Equation 7.3 to obtain L_{bol}.

12 μm correction factors, $BC_{12\,\mu\text{m}}$**:** To a good approximation,

$$BC_{12\,\mu\text{m}} = 0.8 BC_{5100}. \tag{7.6}$$

Since the 2–20 μm SED is basically flat in λL_λ, correction factors for other wavelengths in this range are similar.

BC$_{H\alpha}$: The EW of the broad Hα line relative to the intrinsic AGN continuum is roughly constant. This suggests that

$$BC_{H\alpha} \simeq 130. \tag{7.7}$$

Other emission lines can be scaled in a similar way, for example, BC$_{H\beta} \simeq 450$.

7.1.6. Emission-line variability and reverberation mapping

Given enough time, all type-I AGNs show some optical–UV continuum luminosity variations. Given longer time, such a statement can also be made about the broad emission lines. One such example is shown in Chapter 1.

Line and continuum variations are strongly correlated. The general pattern is for an increase (decrease) in the luminosity of almost all broad emission lines following an increase (decrease) of the continuum luminosity. This produces the strongest, clearest evidence that the source of ionization and heating of the broad emission-line gas is the central ionizing continuum. The time lag between continuum and line variation provides a simple estimate of the size of the BLR. Several big campaigns to follow such variations, over months and years, are now providing invaluable information about the gas distribution, the ionization distribution, and the kinematics of the gas. The method that is used to derive this information is *reverberation mapping* (RM).

The basic principle of RM is similar to that which underlies Doppler-weather mapping. The idea is that the time-delayed, Doppler-shifted response of a system to a known input signal can be used to infer the structure and kinematics of the responding system. In the case of the BLR, the input signal is generated by the AGN continuum source, which, given the expected dimension of the accretion disk, is considered to be 10–100 gravitational radii in diameter. We are also making several other simplifying assumptions:

1. The light-travel times across the BLR are in the range of days to months.
2. The time scales for response of individual clouds to changes in the ionizing flux are given by the recombination and ionization times. Given the BLR typical density, this is of order 1 hour or less, that is, much shorter than the typical time scale for continuum variations.
3. The BLR structure and kinematics are constant over the duration of the RM experiment.
4. There is a known relationship between the observed UV or optical continuum and the ionizing continuum that is driving the emission-line variations.

A successful RM experiment is expected to provide two observables: a continuum light curve, $L_c(t)$, and an emission-line light curve, $L_l(t)$. The line light curve is a measure of the integrated line luminosity as a function of time. Since this can be measured in many individual points in the profile, that is, many line-of-sight velocities, we can make use of the two-dimensional line light curve $L_l(v, t)$. The aim is to relate $L_l(t)$ and $L_c(t)$ in a way that will allow mapping of the gas distribution and motion across the region. The standard methods to achieve this are based on the *transfer function*.

7.1.7. The transfer function

The basic equation that relates the line and continuum light curves is

$$L_l(v, t) = \int_{-\infty}^{\infty} \Psi(v, \tau) L_c(t - \tau) d\tau, \tag{7.8}$$

where $\Psi(v, \tau)$ is the transfer function, which depends on the BLR geometry, kinematics, and the physics of the gas that reprocesses the continuum radiation. A unique solution to an integral equation of this form requires a large amount of high-quality data. In practice, high-quality, well-sampled spectra of faint objects are not easy to obtain, and a somewhat less ambitious approach is to focus on the *one-dimensional transfer function*, $\Psi(\tau)$, which describes the response of the *integrated* line luminosity, that is,

$$L_l(t) = \int_{-\infty}^{\infty} \Psi(\tau) L_c(t - \tau) d\tau. \tag{7.9}$$

Thus, $L_l(t)$ is the convolution of $L_c(t)$ with $\Psi(t)$.

As can be seen, $\Psi(t)$ in appropriate units equals the line light curve, $L(t)$, that would result from a δ-function continuum light curve at $t = 0$. For gas that is distributed in a thin shell of radius r, the transfer function is a boxcar-shaped pulse lasting from $t = 0$ until $t = 2r/c$. The rise at $t = 0$ is due to line-of-sight gas that appears to respond instantly (within the ionization time) to the continuum pulse. The line on the opposite side of the shell responds $2r/c$ s later, and there is no further response beyond that. The constant value of $\Psi(t)$ results from the time delay of a ring at a polar angle θ (relative to the direction of the observer) whose time delay is

$$t = \frac{r(1 - \cos\theta)}{c}. \tag{7.10}$$

The surface area of the ring is $2\pi r^2 \sin\theta d\theta$, and the normalized flux, relative to the entire shell area, is $1/2 \sin\theta d\theta$. Differentiating Equation 7.10 and dividing by dt, we get $\Psi(t) = 2r/c$, which is indeed constant in time. In a similar way, the transfer function of a circular ring inclined at an angle i to the line of sight is nonzero between times $r(1 - \sin i)/c$ and $r(1 + \sin i)$, with its center at r/c.

The transfer function of a thick shell is obtained by integrating over many thin shells and weighting the contribution at each radius according to the local

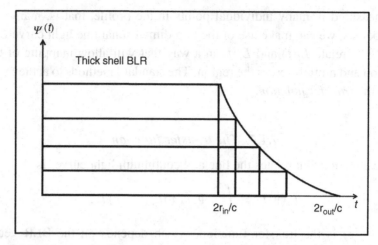

Figure 7.7. The transfer function of a thick spherical shell is made of a combination of the transfer functions of several thin shells. The function is constant up to time $2r_{in}/c$ and declines to zero at $2r_{out}/c$.

emissivity. Such a case is shown in Figure 7.7. It demonstrates the nonzero response between times $t = 0$ and $t = 2r_{in}/c$, followed by a continuous decline until $t = 2r_{out}/c$. Here again, the shape of the declining part depends on the gas emissivity at each radius.

As an example, consider optically thick gas where the energy absorbed at each radius is proportional to the covering factor of the shell at this radius and where the integrated covering factor is much smaller than 1. In this case, $L(t)$ is the total emission of *all lines*, and the weighting factor at each radius is the covering factor, that is,

$$\Psi(t) = \int \frac{dL_l(r)}{r} \propto \int \frac{dC_f(r)}{r}, \tag{7.11}$$

where $dC_f(r)$ is the differential covering factor defined in Equation 5.61.

In principle, $\Psi(t)$ can be obtained from the observed data by applying the convolution theorem,

$$\tilde{\Psi}(\omega) = \frac{\tilde{L}_l(\omega)}{\tilde{L}_c(\omega)}, \tag{7.12}$$

where $\sim$ denotes the Fourier transform. In practice, this is not a trivial task because $L_c(t)$ and $L_l(t)$ are not evenly sampled and the measurement errors are significant. In such cases, Fourier methods may become problematic, and a proper solution to $\Psi(t)$ is difficult to obtain. Quite often, the actual data are so sparse that all we can determine is the cross-correlation function of the line and continuum light curve, CCF(t),

$$\text{CCF}(\tau) = \int_{-\infty}^{\infty} L_l(t)L_c(t - \tau)dt. \qquad (7.13)$$

The CCF is directly related to Equation 7.9. This can be seen by convolving that equation with $L_c(t)$ and by noting that the convolution of $L_c(t)$ with itself is the continuum autocorrelation function, ACF(t):

$$\text{ACF}(\tau) = \int_{-\infty}^{\infty} L_c(t)L_c(t - \tau)dt. \qquad (7.14)$$

Thus,

$$\text{CCF}(\tau) = \int \Psi(\tau')\text{ACF}(\tau - \tau')d\tau'. \qquad (7.15)$$

Cross-correlation of $L_c(t)$ and $L_l(t)$ thus indicates the typical time scale for emission-line response that is often referred to as the emission-line *lag*.

In reality, we have discrete data sets of n observations, each measured to produce two light curves for a certain line and a certain continuum band, each with n points. The CCF is a series of n correlation coefficients, r_i, obtained by shifting the light curves relative to each other n times. For unevenly sampled light curves, this must involve some interpolation. The most important quantity is the peak of the CCF (or its centroid), which is the time corresponding to the maximum r. It is important to note that the centroid of the CCF equals the centroid of the transfer function only when the light curves extend to infinity. Thus, uneven sampling and the limited durations of the light curves are the main limiting factors in obtaining an accurate value for the peak of the CCF.

Equations 7.8 and 7.9 are *linear* equations, which seems to introduce another hidden assumption not always supported by the known physics of the BLR (see preceding discussion about the use of the total BLR emission). However, in practice, the transfer equation is solved by replacing $L(t)$ and $C(t)$ with their difference from the mean values, for example, $\Delta L(t) = L(t) - \langle L \rangle$, which removes the effects of nonvariable components and is equivalent to a first-order expansion of the transfer equation. Thus, mild nonlinearity does not pose a real problem.

Returning to the two-dimensional (velocity and space) transfer function, we consider a simple example of a BLR where the clouds are on randomly inclined, circular Keplerian orbits. Consider first clouds in orbits at inclination $i = 90°$, that is, with the line of sight in the orbital plane. Positions along the orbital path are specified by the polar coordinates r and θ, as defined in Figure 7.8a. Each position on the orbit projects to a unique position in velocity–time delay space, as shown in Figure 7.8b. An isotropically emitted continuum outburst will be followed by an emission-line response that is time delayed by the additional path length to the

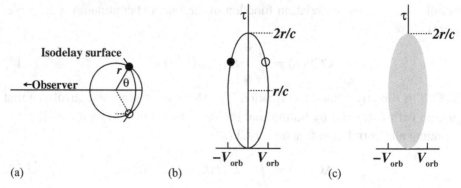

Figure 7.8. (a) An example of BLR clouds that are distributed along a circular orbit centered on the central continuum source at inclination $i = 90°$, with the clouds orbiting counterclockwise. Emission-line clouds respond to a continuum outburst with time delay $\tau = (1 + \cos\theta)r/c$, which, compared with the photons from the central source that travel directly to the observer, is the additional path length (i.e., τc) this signal must travel to the distant observer to the left, as shown by the dotted line. At the time delay shown, two clouds are responding, the upper one approaching the observer and the lower one receding. (b) The points on the circular orbit in (a) project to an ellipse in the velocity–time delay plane. The locations of the two clouds in (a) are shown. (c) For circular orbits at inclinations less than 90°, the axes of both of the ellipses are decreased by a factor $\sin i$, and the center remains at $v = 0$, $\tau = r/c$. Thus, for a random distribution of inclinations, the response of the BLR occurs over the full range of radial velocities and time delays limited by the $i = 90°$ case.

observer that must be traversed by (1) the ionizing photons that travel outward from the central source and are intercepted by BLR clouds and (2) the resulting emission-line photons that are emitted in the direction of the observer (such a time-delayed path is shown as a dotted line in Figure 7.8a). At some time delay τ, the emission-line response recorded by the observer will be due to all clouds that lie on a surface of constant time delay (an *isodelay surface*) given by the length of the dotted path in Figure 7.8a:

$$\tau = (1 + \cos\theta)r/c, \tag{7.16}$$

which is the equation of a parabola in polar coordinates. The intersection of the isodelay surface and the cloud orbit identifies the clouds that are responding at time delay τ. If both clouds shown in Figure 7.8a are moving counterclockwise at orbital speed $V_{orb} = (GM/r)^{1/2}$, where M is the mass of the central source, their observed Doppler-shifted velocities are $v = \pm V_{orb} \sin\theta$, and thus the locations of the two clouds project to the two different points in velocity–time delay space, as shown in Figure 7.8b.

The entire circular orbit is seen to project to an ellipse in the velocity–time delay diagram; the zero time-delay point represents the BLR clouds that lie along the line of sight (at $\theta = 180°$) and line-of-sight velocity ($v = 0$), and the largest line-of-sight velocities are measured at $\theta = \pm 90°$, where $\tau = r/c$. The range of time delays extends up to $2r/c$, corresponding to the response from the far side of the BLR (i.e., $\theta = 0°$). If we consider identical orbits at lower inclinations, it is easy to see that the range of time delays decreases from $[0, 2r/c]$ to $[(1 - \sin i)r/c]$, $[(1 + \sin i)r/c]$, and line-of-sight velocities similarly decrease by a factor of $\sin i$. The projection of such an orbit into velocity–time delay space is thus an ellipse that has the same center (at $v = 0$, $\tau = r/c$) and ellipticity but axes that are smaller by a factor of $\sin i$. For $i = 0°$, the ellipse contracts to a single point at time delay r/c and line-of-sight velocity $v = 0$. For a system of clouds in circular orbits of radius r and random inclinations, the ellipse shown in Figure 7.8b becomes completely filled in, as shown in Figure 7.8c. This is the two-dimensional transfer function for a system of clouds in circular Keplerian orbits of radius r and random inclinations.

It is straightforward to consider more complex models, especially if the response can be assumed to be approximately linear, in which case, transfer functions for complex BLR geometries and continuum anisotropies can be constructed by addition of properly weighted simple transfer functions. For example, partial continuum anisotropy can be modeled as a sum of isotropic and anisotropic geometries. Similarly, the response of a disk can be modeled by a summation of transfer functions for circular Keplerian orbits of various r and fixed i. The transfer function for a thick spherical shell (as in Figures 7.8c and 7.9) can be constructed by adding together transfer functions for thin spherical shells of varying radius; note that as r increases, the velocity–time delay ellipses become taller and narrower, as the major axis is proportional to r and the minor velocity axis decreases like $r^{-1/2}$.

Figure 7.9 shows an example of a thick-shell model that is an extension of the thin-shell model in Figure 7.8 and uses the same continuum beaming parameters. We chose this particular model for two reasons: first, both the one-dimensional transfer function and the variable part of the line profile are double peaked. This makes the important point that such structures are *not* unambiguous signatures of rotating disks or biconical flows.

7.1.8. Reverberation mapping results

The observational goal of reverberation mapping experiments is to use the light curves $C(t)$ and $L(v, t)$ to solve for the transfer function and then use this to test directly various models of the BLR. What makes the reverberation problem different is that the sampling of $C(t)$ and $L(v, t)$ is nearly always irregular, limited in both temporal resolution and duration, and the data are often noisy and plagued by systematic errors. These limitations have led to development of specialized methodologies for time series analysis. The obvious method of Fourier inversion

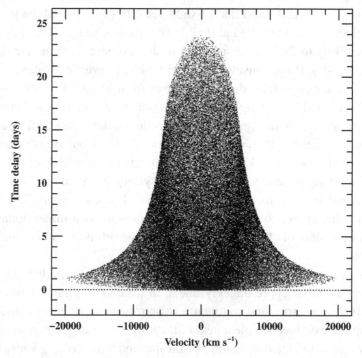

Figure 7.9. A two-dimensional (response as a function of both line-of-sight velocity and time delay) transfer function of a thick shell of clouds moving in randomly oriented circular cloud orbits.

performs poorly on account of the limitations listed. Better methods that were developed to improve this situation go under the names of the *maximum entropy* method, the *SOLA method*, and *regularized linear inversion*.

As of 2011, there are only a handful of cases in which the full two-dimensional transfer function has been recovered. There are about 40 cases of low-redshift sources where a reliable lag has been measured between the Hα and/or Hβ lines and the optical continuum, usually at 5100 Å. There are also about 10 cases showing a measurable lag between C IV λ1549 and the UV continuum at around 1400 Å. A handful of campaigns had enough information to study the CC of several optical and UV continua and emission lines in the same source where clear time lags are detected. One such example, showing the light curves for the UV continuum bands and emission lines of NGC 7469, is shown in Figure 7.10, along with the various CCFs.

The optical RM sample most commonly used until now (2011) includes some 40 AGNs with a range of almost 4 orders of magnitude in L_{5100}. This sample, as it was in 2005, is shown in Figure 7.11, together with various linear fits of the type

$$R_{\mathrm{BLR}} = C_{\mathrm{BLR}} L^{\alpha(\mathrm{BLR})}. \tag{7.17}$$

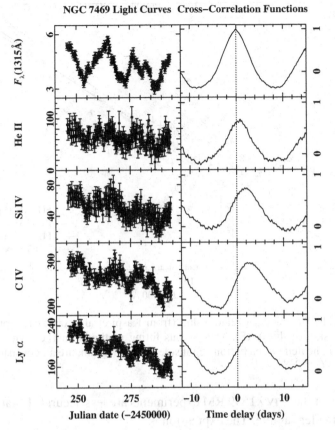

Figure 7.10. (left) The light curves of NGC 7469 obtained with IUE during an intensive AGN Watch monitoring campaign during summer 1996. (right) CCFs obtained by cross-correlating the light curve immediately to the left with the 1315 Å light curve at the top of the left column. The panel at the top of the right column thus shows the 1315 Å continuum autocorrelation function (ACF).

In all fits, $C_{\mathrm{BLR}} \simeq 120$ light days for $L_{5100} = 10^{45}$ erg s^{-1}. The slope of the correlation is $\alpha(\mathrm{BLR}) = 0.6 \pm 0.1$. Major difficulties in determining the slope more accurately are the relative large scatter at the high-L end and the large host-galaxy contribution at the low-L end. In fact, it is not at all clear that a single slope fits the correlation over such a large luminosity range.

Given all this, we find the following estimate for the emissivity-weighted radius of the Hβ line:

$$R_{\mathrm{BLR}}(H\beta) \simeq 0.12 L_{46}^{0.6 \pm 0.1} \ \mathrm{pc}, \qquad (7.18)$$

where L_{46} is the bolometric luminosity in units of 10^{46} erg s^{-1} as derived by using BC$_{5100}$.

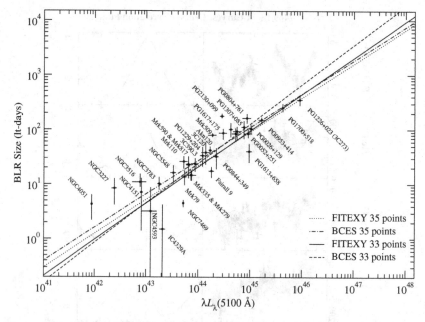

Figure 7.11. The L–R_{BLR} relationship (from Kaspi et al., 2005, 61; reproduced by permission of the AAS). The various fitting methods, marked with straight lines and indicated on the bottom right side of the diagram, are all consistent with $R_{\mathrm{BLR}} \propto (\lambda L_\lambda)^{0.6 \pm 0.1}$.

The results of the C IV $\lambda 1549$ RM experiments are less accurate because they are based on a smaller sample. The expression is

$$R_{\mathrm{BLR}}(\text{C IV } \lambda 1549) \simeq 0.04 L_{46}^{0.6 \pm 0.1} \text{ pc}, \qquad (7.19)$$

where the constant was derived by using the best-fit line to the various lags and the 1400 Å bolometric correction factor, BC_{1400}.

The comparison of the various RM experiments indicates that the emissivity-weighted radius of Hβ is about three times larger than the corresponding radius of C IV $\lambda 1549$. This is an important constraint on photoionization models of the BLR. The radii do not stand for a specific location but rather represent means over a large volume of space. Comparison with theoretical models suggests that for both lines, the main emission volume extends between $0.5 R_{\mathrm{BLR}}$ and $2 R_{\mathrm{BLR}}$.

It is interesting to compare the normalized (by r_g) BLR size,

$$\frac{R_{\mathrm{BLR}}(\text{C IV } \lambda 1549)}{r_g} \simeq 10^4 \frac{L_{46}^{0.6}}{M_8}, \qquad (7.20)$$

with the self-gravity radius of the accretion disk given by Equation 4.48. The comparison shows a tendency for larger BH mass sources to have smaller *normalized* R_{BLR}. In some cases, this radius is similar to or even smaller than the self-gravity radius of the disk. There are interesting consequences to the intensity and profile of the C IV $\lambda 1549$ line since a large part of the BLR emitting this line on the far side of the source can be obscured by the disk. An outflowing, C IV $\lambda 1549$-emitting BLR gas would be seen, in such cases, as an asymmetric line with a strong blue wing.

We can also use the preceding expressions to estimate the total BLR mass, assuming it is made of a collection of individual optically thick clouds. Assume a global covering factor of C_f and a mean column density of N_{col} cm^{-2}; we get

$$M_{BLR} \simeq 23 \left(\frac{C_f}{0.1}\right) \left(\frac{N_{col}}{10^{23} \text{cm}^{-2}}\right) L_{46}^{1.2} M_{\odot}, \qquad (7.21)$$

where the estimate refers to the mass inside the zone emitting roughly half of the $H\beta$ line. The estimate is a lower limit on the mass since it takes into account only isolated clouds that are optically thick and hence efficient line emitters. It is easy to imagine a situation in which much of the gas in the BLR is not emitting efficiently because it is optically thin and hence is not taken into account in this census. For example, in the LOC model, much of the gas at every location is optically thin and contributes very little to the observed line intensities, yet it contributes a lot to the total mass. In such a model, a much larger fraction of the total volume is filled with gas, and the BLR mass can exceed the preceding estimate by a large factor, several orders of magnitude in some cases.

We can also derive the Keplerian velocity at R_{BLR}, $V_{BLR} = (GM_{BH}/R_{BLR})^{1/2}$. For $H\beta$, this is given by

$$V_{BLR} \simeq 1700 M_8^{1/2} L_{46}^{-1/4} \text{ km s}^{-1}, \qquad (7.22)$$

where M_8 is the BH mass in units of $10^8 M_{\odot}$. An interesting consequence is that the line widths in small L/L_{Edd} sources can be extremely large. For example, the combination of $M_8 = 1$ and $L/L_{Edd} = 10^{-4}$, typical of many LINERs (§ 7.2), gives $V_{BLR}(H\beta) \simeq 15,000$ km s^{-1}, larger than observed in 99 percent of type-I AGNs. As shown in § 6.2.5, this may be related to the class of "real" type-II AGNs. It also has important consequences for the ability to detect type-I LINERs (see later).

7.1.9. What determines the BLR size?

Scaling relationships and BLR gas distribution

While much progress has been made in determining, experimentally, the gas distribution in the BLR, a physical explanation for this distribution, and even its basic characters (clouds? LOC type distribution? wind?) is still lacking. The more successful photoionization models use ad hoc assumptions about the distribution and use it to check for consistency with the observations. As explained in Chapter 5, some of the assumptions are plausible, because they relate to realistic magnetic field distributions or other physical processes. However, completely different distributions that are equally plausible are not observed in AGNs.

Perhaps the most important unknown in photoionization models of the BLR is its outer boundary. It determines the actual value of R_{BLR}, the observed gas velocities, and the emission-line ratios. Such properties cannot be inferred from the photoionization calculations, and an outer boundary must be put in "by hand" to explain the observations.

Dust as the outer boundary of the BLR

Reverberation mappings of many BLRs shows that the mean R_{BLR} as determined by the emissivity of the Hβ line (Equation 7.18) is about a factor of 2–3 smaller than the dust sublimation radius (Equations 5.84 and 5.85). This suggests that the existence of dust grains beyond the sublimation radius may be the natural boundary of the dust-free BLR. Dust is such a common component in all known HII regions that it is possible that almost the entire volume of the nucleus, from just outside the BLR to the outer edge of the NLR, is filled with dusty gas.

There is clear observational evidence for dust outside the BLR from two types of independent observations. Monitoring of several nearby AGNs, in the K- and V-bands, clearly shows lag in the variations in the K-band relative to the ones in the V-band. While the V-band emission is due to the central source, the K-band emission is interpreted as radiation by hot dust in the innermost part of the dusty structure around the BH (the dusty torus; see § 7.5). The source of heating is the variable central continuum, and the time lag can be used to estimate the sublimation radius. The comparison of the two relationships is shown in Figure 7.33, and the actual numbers are in good agreement with Equation 5.84.

Another indication comes from observations of dust emission between the broad and the narrow emission-line regions. There are several nearby sources where spatially resolved IR imaging detects such emission all the way down to a few parsecs from the center. The clearest signature is emission in the 10–20 μm range, which cannot be due to stars and is most likely due to the strong silicate dust emission features. Such a search can also be conducted in high-redshift sources since the silicate features can be used to obtain the dust temperature and thus, from Equation 5.82, the distance to the center in sources whose optical–UV radiation is known. Studies of this type are described in § 7.5. They reveal the existence of "warm" ($T_g \sim 200$ K) dust in many high-luminosity AGNs. The typical distance in this case is 100–200 times R_{sub}, which is way outside the BLR and coincides with the innermost part of the narrow-line region.

7.1.10. Black hole mass determination

Determining the black hole mass is a major motivation for the various AGN monitoring campaigns. Besides R_{BLR}, which is measured in such experiments, one needs information about the gas velocity, which is hidden in the observed broad-line profiles. The common assumption is that the system is virialized and individual clouds are moving in their own Keplerian orbits not necessarily with the same inclination and eccentricity. This leads to the expectation that the mean cloud velocity, for example, the one determined from the FWHM of the lines, is proportional to $r_{\rm BLR}^{1/2}$. In a handful of sources, where several emission lines could be used to determine individual $R_{\rm BLR}$ and line widths, this is indeed the case. However, the numbers are too small to generalize this to the entire population. Other velocity measures that are occasionally used are the second moment of the velocity profile (see later) and the FWHM of the variable component of the line.

Assuming bound Keplerian orbits, we can write the following expression for $M_{\rm BH}$:

$$M_{\rm BH} = f(R_{\rm BLR})\frac{R_{\rm BLR}v_l^2}{G} \text{ gr,} \tag{7.23}$$

where v_l is some measure of the velocity obtained from the line profile, for example, the FWHM, and $f(R_{\rm BLR})$ is a geometrical-dynamical factor that depends on the distributions and inclination of the orbits to the line of sight. If v_l is the mean Keplerian velocity, v_K, then $f(R_{\rm BLR}) = 1$. Obviously, v_K is not known, and the entire procedure is designed to estimate $f(R_{\rm BLR})$ from measuring various line profiles in a consistent way.

We can get an estimate of the expected value of $f(R_{\rm BLR})$ by considering several simple cases. The first is a case of complete random orientation of the orbits and isotropic line emission. In this case, the line profile is a Gaussian, and v_l can easily be obtained from the observations by noting that the three-dimensional velocity v is related to v_l by $v = \sqrt{3}v_l$. This means that $v_K^2 = 3v_l^2$ or $f(R_{\rm BLR}) = 3$. For small variations from this ideal case, we can estimate v_l from the FWHM of the line, $v_l = 0.5$FWHM. In this case, $f(R_{\rm BLR})v_l^2 = (3/4)$FWHM^2. Another possibility is to define v_l to be the second moment of the velocity,

$$v_l(\lambda)^2 = <\lambda>^2 - \lambda_0^2 = \frac{\int \lambda^2 P(\lambda)d\lambda}{\int P(\lambda)d\lambda}, \tag{7.24}$$

where $P(\lambda)$ is the monochromatic flux in the line and λ_0 is the weighted line center (the first moment of the velocity),

$$\lambda_0 = \frac{\lambda P(\lambda)d\lambda}{\int P(\lambda)d\lambda}. \tag{7.25}$$

We can compare this v_l with the FWHM of the line in several simple cases: for a Gaussian profile FWHM/$v_l = 2\sqrt{(2\ln 2)} = 2.35$, thus using v_l requires $f(R_{\rm BLR})$ which is 5.52 times larger than the one required when using the FWHM. For a rectangular profile FWHM/$v_l = 2\sqrt{3} = 3.46$. The largest deviation between the two velocity estimates is in cases where the line wings extend to large velocities.

A good example is a Lorentzian profile, which gives a very large $v_l(\lambda)^2$ that can overestimate the derived BH mass.

Assume now a flat system of rotating clouds in the shape of a thick disk with $H/R < 1$ and an inclination i to the line of sight. This can be due to a significant component of turbulent velocity, v_{tur}, in the disk ($H/R \simeq v_{\text{tur}}/v_K$) or some other ordered motion perpendicular to the plane of the disk. In this case,

$$\frac{v_l}{v_K} \simeq \left[\left(\frac{H}{R} \right)^2 + \sin^2 i \right]^{1/2}. \tag{7.26}$$

The mass correction factor in this case is $f(R_{\text{BLR}}) = [(H/R)^2 + \sin^2 i]^{-1}$.

In reality, there is not enough information to determine $f(R_{\text{BLR}})$ from the line profile of individual objects, and we must use another, completely independent way to find its mean value in a sample of objects where R_{BLR} has been measured. Fortunately, the M_{BH}–σ_* method, which is used to determine M_{BH} in bulge-dominated nearby galaxies and is described in Chapter 8, provides such a way. The method is based on a well-established correlation between M_{BH} and the width of stellar absorption lines (σ_*) in the bulge of the galaxy. As of 2011, the values of σ_* can be measured in about 30 low redshift type-I AGNs where successful RM campaigns provide also reliable $R_{\text{BLR}}(H\beta)$. The comparison of Equation 7.23 with Equation 8.2 is then used to obtain an empirical estimate of $f(R_{\text{BLR}})$. For the $H\beta$ line, and for $v_l = \text{FWHM}$, we get $f(R_{\text{BLR}}) \simeq 1.1$.

The use of $f(R_{\text{BLR}})$ to estimate M_{BH} in sources where σ_* is not known is based on the assumption that the value is typical of all AGNs, of low and high luminosity, of small and large R_{BLR}, and of all inclinations to the line of sight. This is a major source of uncertainty and scatter in the entire M_{BH} determination process.

Having found $f(R_{\text{BLR}})$, we can now write more specific expressions to calculate M_{BH} for several continuum bands and several emission lines. An expression that is based on the 5100 Å continuum luminosity and the $H\beta$ line is

$$M_{\text{BH}}(H\beta) = 1.05 \times 10^8 \left[\frac{L_{5100}}{10^{46} \text{ ergs s}^{-1}} \right]^{0.65} \left[\frac{\text{FWHM}(H\beta)}{10^3 \text{ km s}^{-1}} \right]^2 M_\odot. \tag{7.27}$$

Since large spectroscopic surveys like the SDSS result in numerous objects where both the broad $H\beta$ and Mg II $\lambda 2798$ lines are observed, we can calibrate one against the other and derive a similar expression based on the 3000 Å continuum and the Mg II $\lambda 2798$ line width,

$$M_{\text{BH}}(\text{Mg II} \lambda 2800) = 8.9 \times 10^7 \left[\frac{L_{3000}}{10^{46} \text{ ergs s}^{-1}} \right]^{0.58} \left[\frac{\text{FWHM}(\text{Mg II} \lambda 2798)}{10^3 \text{ km s}^{-1}} \right]^2 M_\odot \tag{7.28}$$

(note that FWHM(Mg II $\lambda 2798$) refers to only one of the components in the doublet).

Equations like the ones shown here can be used to obtain *single-epoch mass estimates*, of M_{BH} from with a single continuum point to obtain L and a single line width to obtain FWHM. They have been used to measure BH masses for thousands of sources with an estimated uncertainty of about factor 2–3, when using either Hβ or Mg II as the line in question.

The Mg II-based method applied to ground-based spectroscopic samples can be used to determine M_{BH} up to a redshift of about 2. Many thousands of AGNs in this range have been measured, spectroscopically, in SDSS, 2QZ, and other large samples. Beyond this redshift, the line is shifted into the NIR, and different types of observations, limited to a small number of sources, must be used.

In principle, a similar C IV λ1549-based method can be used at higher redshifts since the line is very strong and can be seen to $z = 4$ and beyond. Moreover, there are several successful RM estimates of the lag of this line with respect to the neighboring 1400 Å continuum. This gives good R_{BLR} estimates to be used in the mass measurements. However, careful comparisons of M_{BH}(Hβ) and M_{BH}(Mg II λ2798) with the single-epoch mass estimate for M_{BH}(C IV λ1549) show large scatter and big disagreements. To understand the problem, we must look again at the basic premise of the virial mass determination method. According to the method, we expect all emissions that come from the same region as Hβ to have similar line widths. A good example is Mg II λ2798, which is predicted by photoionization models to come from gas with similar conditions to the Hβ emitting gas. Indeed, a comparison of the Hβ and Mg II λ2798 FWHMs, as shown in Figure 7.12a, reveals a tight 1:1 correlation over most of the line width range. This trend breaks down at the high-velocity limit, suggesting that some other processes are taking place.

Returning to C IV λ1549, we note that since R_{BLR}(Hβ) > R_{BLR}(C IV λ1549) (Equations 7.18 and 7.19), in a virialized system, FWHM(Hβ) < FWHM (C IV λ1549). This is certainly not the case in Figure 7.12, which shows a large number of sources that have both measured FWHM(Hβ) and FWHM(C IV λ1549). In about a third of the sources, the Hβ line exhibits the broader profile, which indicates that the gas in at least one of the two line-emitting regions is moving with significant non-Keplerian velocities or that the kinematics of the two regions are not correlated. As alluded to (see Chapter 8), there is an independent method to measure M_{BH} in the local universe, based on the velocity dispersion of the stars in the galactic bulge (the $M-\sigma^*$ method). A comparison of BH mass measurements obtained in this way with the ones obtained from the RM-based method for the Hβ line shows very good agreement. Given this, it seems that the mass discrepancy is due to FWHM(C IV λ1549), which likely reflects some nonvirialized gas motion.

The discrepancies between the various mass estimates bring back the issue of radiation pressure force and its affect on the cloud motion discussed in Chapter 5. In that chapter, the conditions under which gravity dominates the motion of the gas

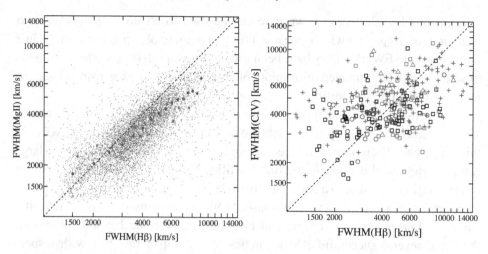

Figure 7.12. (left) FWHM(Hβ) versus FWHM(Mg II λ2798) for intermediate-redshift AGN from the 2SLAQ and SDSS samples (the large dots are means in narrow velocity bins). (right) FWHM(Hβ) versus FWHM(C IV λ1549) for a sample of high-redshift AGNs (adapted from Trakhtenbrot and Netzer, 2012). FWHM(Hβ) is measured from H- and K-band spectroscopy and FWHM(C IV λ1549) from SDSS spectra. Note the complete lack of correlation between the two line widths in this case, suggesting that the simple virial assumption fails in at least some part of those BLRs.

were explained in terms of the column density and level of ionization of the gas (since we are dealing with BLR clouds, dust is not considered as an opacity source). The expression given there (Equation 5.75) shows that for some of the BLR gas, especially in cases of large L/L_{Edd}, this force may affect the cloud motion and hence the observed line widths.

As explained in Chapter 5, the level of ionization of BLR clouds is likely to be location dependent, which means that $a(r)$ and $\alpha(r)$ in Equation 5.75 depend on a complicated function of the distance through the column density of the clouds, N_{23}. Gravity, conversely, depends only on r^{-2}, thus the ratio of the two is a location-dependent quantity. In such cases, even gravitational bound clouds do not move in closed Keplerian orbits. Examples of time-dependent orbits of such clouds are shown in Figure 7.13. Here the cloud mass is conserved and its density is proportional to r^{-s} with $s = 1.2$ (see the series of equations in § 5.7). This results in $N_{23} \propto r^{-4/5}$.

Considering a more realistic configuration of numerous clouds, we can examine the resulting line profiles and compare them with the observations. Examples are shown in Figure 7.14. The application of such models to a realistic BLR model is shown in Figure 7.15. Here the Hβ and C IV λ1549 line profiles and intensities are calculated in two zones, and the resulting combined profile is similar to many observations. Apparently, the motion of BLR clouds in some objects, especially

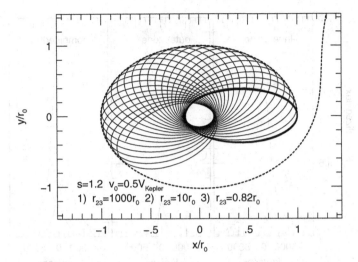

Figure 7.13. Three different orbits of BLR clouds under the combined influence of gravity and radiation pressure force (adapted from Netzer and Marziani 2010). Case 1, with the thick line, refers to a very large-column-density cloud. Here radiation pressure force is negligible and the cloud moves in an elliptical Keplerian orbit. Case 2, with a thin line, corresponds to much smaller column density, which results in bound orbits that are not closed. In case 3 (dotted line), the column is too small to allow a bound orbit, and the cloud is ejected from the system.

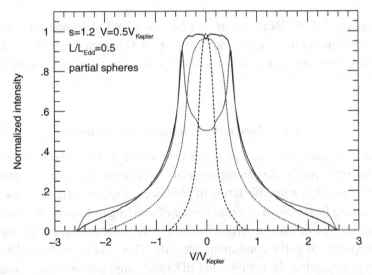

Figure 7.14. Various profiles of BLR clouds moving under the influence of both gravity and radiation pressure force. The narrowest profile (dashed line) represents a sphere where clouds occupy only the section between 0.3 and +0.3 rad relative to the midplane (which is perpendicular to the line of sight). The other cases are for wider coverage, with clouds between 0.9 and +0.9 rad (dotted line) and 1.5 to +1.5 rad (solid line). The double-peak profile illustrates the case of two polar caps where the clouds occupy a sphere whose midsection, between 1.2 and +1.2 rad, has been removed.

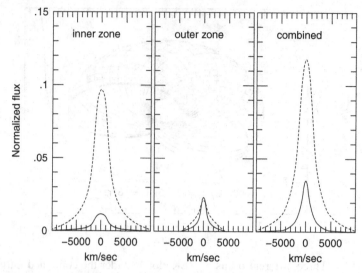

Figure 7.15. Hβ (solid line) and C IV λ1549 (dotted line) profiles from the Netzer and Marziani (2010) model of orbiting clouds. The two profiles in each of the zones are shown separately, and the combined (observed) profile is shown in the right panel. The relative intensity and the relative FWHM of the two lines are similar to those observed in many (but not all) type-I AGNs.

in the inner part of the BLR, where most of the C IV λ1549 line flux is emitted, deviates considerably from Keplerian-type motion. One idea the profile is that much of the C IV λ1549 emission is due to an optically thin wind emanating from the disk.

7.1.11. Line and continuum correlations

Some of the line and continuum properties are strongly correlated with each other. An obvious example is the strong correlation between the line and continuum luminosities, which is a manifestation of photoionization of optically thick gas of a certain covering factor by an external source of ionizing radiation. In this situation, a certain emission line takes a certain fraction in the gas cooling, and its intensity is directly proportional to the continuum intensity. There are subtleties related to the exact level of ionization, the continuum SED, reddening, gas kinematics, inclination to the line of sight, and more. However, the basic photoionization assumption is clearly demonstrated by such line–continuum properties.

Other line–continuum correlations have emerged over the years, some better observed and/or better explained than others. The first is a correlation between Hβ line width, X-ray continuum slope, and L/L_{Edd}. This correlation was shown earlier in Figure 7.6. Other correlations are described later.

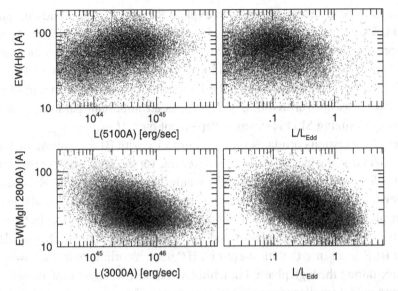

Figure 7.16. Baldwin relationships for Hβ and Mg II λ2798 for type-I SDSS AGNs. The EWs of the two lines are shown as functions of continuum luminosity at a wavelength near the line, and L/L_{Edd}. There is no correlation of EW(Hβ) with both properties. However, there is a strong correlation of EW(Mg II λ2798) in both cases. The Mg II λ2798 diagrams are further divided into two mass groups: objects with $M_{BH} \sim 2 \times 10^9 M_\odot$ in red and objects with $M_{BH} \sim 6 \times 10^7 M_\odot$ in blue. The two groups occupy different regions in the diagrams, and the correlations are significant in both. (See color plate)

The Baldwin relationship

This relationship, discovered by J. Baldwin in 1977, is a correlation of the EW of several emission lines (here "line") with the continuum luminosity measured at a wavelength close to the line wavelength (here λL_λ),

$$\log EW(\text{line}) = a \log \lambda L_\lambda + b, \tag{7.29}$$

where a and b are determined by applying a simple linear regression to the data. The original relationship was discovered for the C IV λ1549 line, and later studies showed that several other broad UV emission lines share this property, albeit with a different slope a. However, several optical emission lines, like Hα and Hβ, show very weak, if any, correlation. An equally strong or even stronger correlation is found between EW(line) and L/L_{Edd}. In fact, there is evidence that L/L_{Edd} may be more important than λL_λ in driving the correlation. This raises the interesting possibility that the line EW is related to the accretion rate and/or the BH spin (L/L_{Edd} depends on η and thus on the BH spin). There are also suggestions of another, albeit weaker correlation between EW and M_{BH}. Examples of luminosity and L/L_{Edd} correlations in Hβ and Mg II λ2798 are shown in Figure 7.16.

There have been several attempts to explain the origin of the Baldwin relationship. One idea is related to the shape of the ionizing SED of an α-disk. Accretion onto more massive BHs results in a "softer" ionizing SED (i.e., lower mean energy of the ionizing photons). This reduces the total number of photons capable of ionizing C^{+2} (or other ions) yet affects the lower energy continuum in a milder way. In this case, it is not surprising that L/L_{Edd} is an important parameter since the shape of the ionizing SED is strongly dependent on L/L_{Edd}.

Another suggestion correlates source luminosity with BLR covering factor, that is, a smaller covering factor leads to a weaker line for the same continuum luminosity. This idea is related to the basic RM result that shows larger BLR dimensions in higher-luminosity sources. Consider a source with a certain mass BH that goes through phases of high and low luminosity. The level of ionization in their BLR would be higher during high phases and lower during low phases. This would give a larger BLR in a sense that most emitted $H\beta$ photons originate farther away from the source during the high phase. For a limited mass BLR, larger dimensions can be associated with a smaller effective covering factor. This process cannot be applied to the entire population since the overall luminosity range in AGNs far exceeds the luminosity range of a single source. It could be viewed as a modulation of the RM relationship, which is more significant in a sample with a restricted AGN luminosity. This idea is consistent with the suggestion of a strong correlation between EW(line) and L/L_{Edd}.

A third possibility is that the disk inclination plays an important role in determining the EW distribution. A thin-disk radiation field is anisotropic and depends on the factor $\cos\theta(1 + b(\nu)\cos\theta)$ (Equation 4.54). Most emission lines are thought to be emitted more isotropically (this must be checked for individual lines; see § 7.1.1), and to a first approximation, the equivalent width (EW) of such a line relative to the disk-emitted continuum will depend on the viewing angle of the disk through the factor $1/\cos\theta$. The probability distribution of the EWs in a randomly selected sample of objects is

$$P(EW)dEW \propto EW^{-2}dEW. \tag{7.30}$$

This will result in a Baldwin-type relationship.

Unfortunately, the selection of the sample can dramatically affect this prediction. In particular, cosmological studies like the ones discussed in Chapter 9 show that the number of faint AGNs is considerably larger than the number of bright AGNs at all redshifts. This means that many more AGNs are concentrated near the flux limit of a flux-limited sample. Any study of the population properties involving source luminosities, including line EWs, will bias the distribution toward the properties of the lowest-luminosity objects. This must affect the Baldwin relationship. Indeed, there are conflicting results about different lines in different samples, regarding the exact form of this correlation.

Line width and EW

We can also test various properties related to the emission gas location and velocity. In particular, we can check the correlation of line width and EW. Using the SDSS type-I sample, there is no correlation between FWHM(Hβ) and EW(Hβ) but a significant correlation between FWHM(Mg II λ2798) and EW(Mg II λ2798) in a sense that broader lines also have larger EW. This is entirely consistent with the Baldwin relationship and the idea that, within a narrow range of M_{BH}, higher luminosity will result in larger BLR and, hence, smaller mean velocity of the gas. Here again there is a difficulty to explain why the Mg II λ2798 and Hβ lines, which are thought to have similar emissivity distributions across the BLR, show such a different relationship.

L/L_{Edd} distributions

The ability to estimate L_{bol} and to measure masses of thousands of BHs provides a valuable tool to investigate the population properties over a large redshift range. In particular, it allows us to probe the growth rate of BHs of different masses at different times, which is perhaps the most important observational data used in constructing the redshift-dependent BH mass function (Chapter 9). One example pertaining to low-redshift AGNs is shown in Figure 7.17. The diagram shows the L/L_{Edd} distribution over a very large range of values from about 0.001 to 1. It shows a clear peak whose exact location depends on the way the stellar background contribution to the local continuum was calculated. Studies at redshifts up to about 2 show similar distributions. At very large redshifts, the samples are smaller and selection effects, combined with flux limits (and some real changes too), force the peak to move toward higher L/L_{Edd}.

7.1.12. BLR models

Perhaps the most serious challenge in modeling the broad-line emission is to construct full, time-dependent line intensity maps that can explain the results of multiline RM experiments. This would mean modeling *several* emission-line intensities as they vary in response to the varying ionizing luminosity. One approach is to follow the method described in § 5.7. A full cloud model of this type must include specific values of s, p, and the initial column density, N_{col}. This should be combined with photoionization calculations of all clouds at the various locations in the BLR. The input ionizing radiation is assumed to change with time in a way

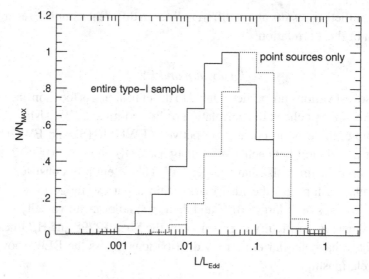

Figure 7.17. L/L_{Edd} distributions for $z < 0.2$ type-I SDSS AGNs based on $L(H\alpha)$. The solid line histogram shows the distribution for all AGNs. The dotted line is the corresponding histogram for all SDSS objects defined as point sources, that is, those likely to have a larger $L(AGN)/L(galaxy)$. As expected, this biases the distribution toward larger L/L_{Edd} (courtesy of J. Stern and A. Laor).

similar to the observed continuum variations. This affect clouds at different locations in different times. The model results are a series of predicted single-cloud line intensities that are then combined according to the model assumptions and given the velocities according to the cloud locations and the BH mass. Another approach is to try to reconstruct the time-dependent line intensity maps with the LOC model.

One modeling attempt based on the cloud model and applied to the best studied case so far, NGC 5548, is shown in Figure 7.18. It demonstrates that some observed emission-line variations are well explained by a certain cloud model with reasonable specific values of s, p, and the initial column density. However, some lines, most notably the low-ionization line of Mg II $\lambda 2798$, are not in agreement with the observations. This is likely be the result of either wrong assumptions about the gas distribution in the outer part of the nucleus or, more likely, some missing physics in the modeling of low-ionization lines.

7.2. The narrow-line region

Next we consider smaller column density ($\sim 10^{20-21}$ cm^{-2}), low-density ($\sim 10^4$ cm^{-3}) gas at a location where $L/4\pi r^2 \simeq 10^2$ erg s^{-1} cm^{-2}, that is, about 3 kpc for a very luminous AGN. We also assume a small covering factor of order of a few percent. The ionization parameter for this gas is similar to the one in the BLR,

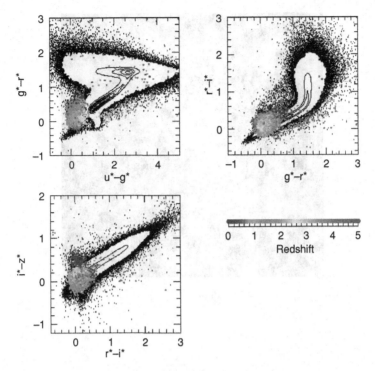

Plate 1.9.

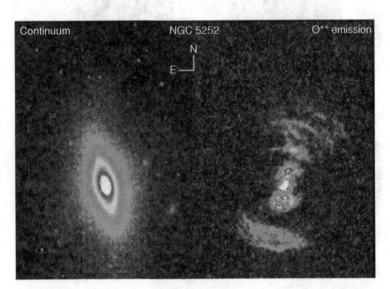

Plate 6.4.

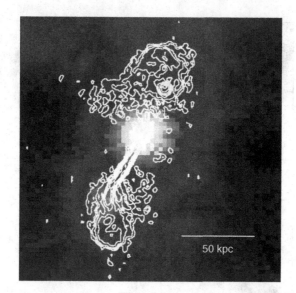

Plate 6.5.

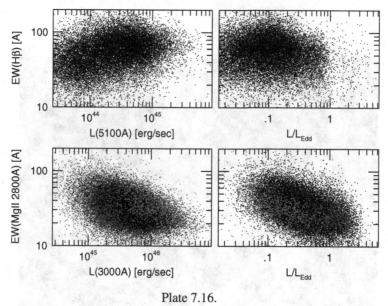

Plate 7.16.

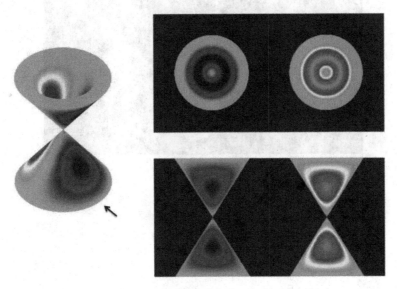

Plate 7.24.

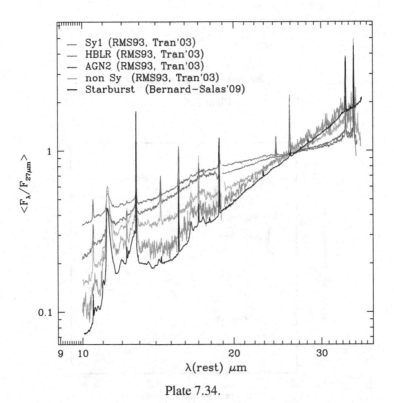

Plate 7.34.

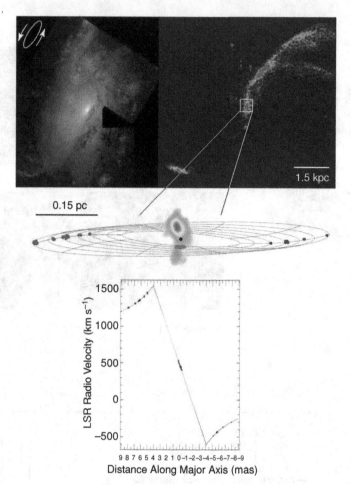

1.5 kpc

0.15 pc

Plate 7.37.

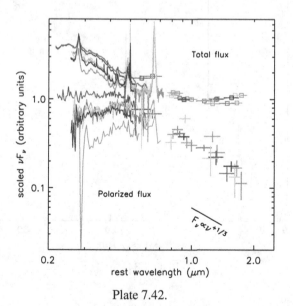

Plate 7.42.

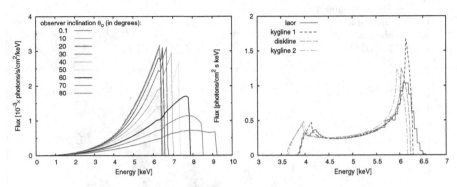

Plate 7.44.

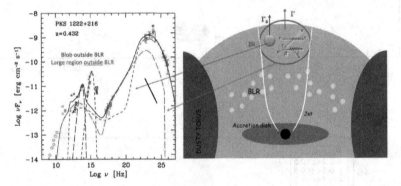

Plate 7.48.

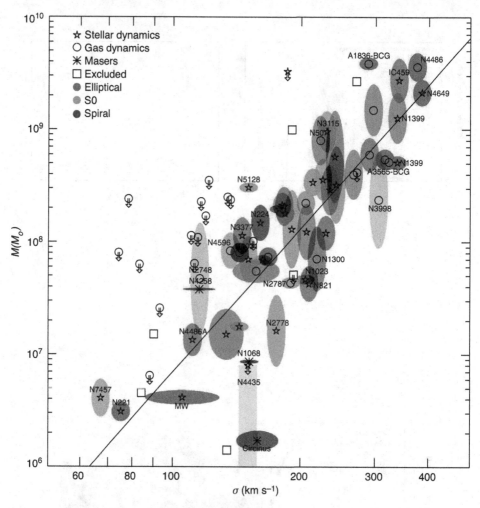

Plate 8.3.

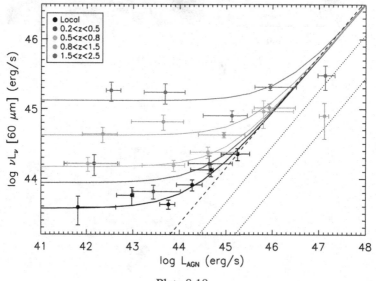

Plate 8.19.

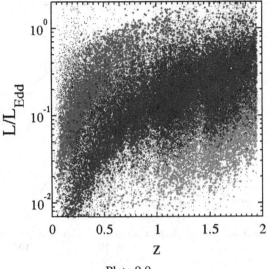

Plate 9.9.

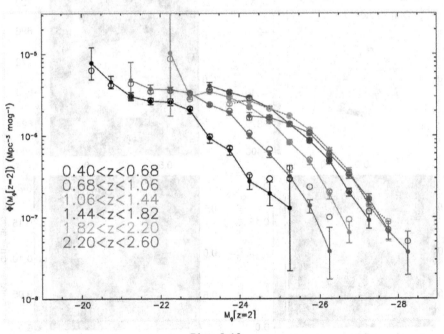

Plate 9.13.

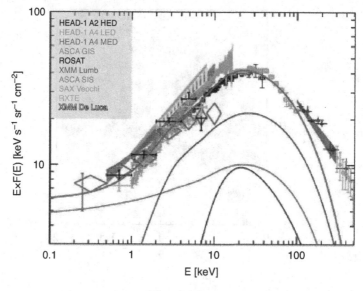

Plate 9.16.

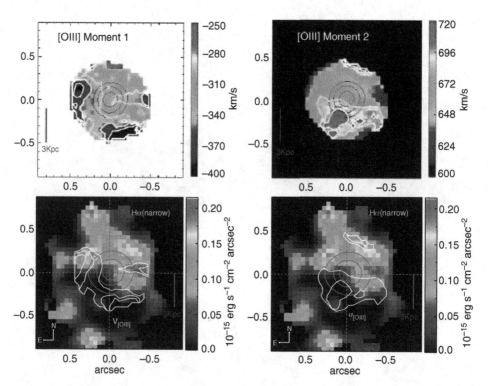

Plate 9.21.

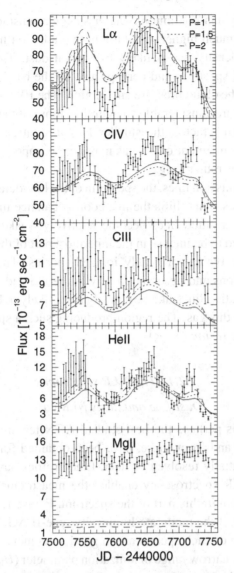

Figure 7.18. Model fit to the variable emission-line intensities in NGC 5548 (adapted from Kaspi and Netzer, 1999).

and the typical velocity, assuming the gravitational potential is controlled by the mass of the galaxy, is of order 300 km s^{-1}. There are various ways such gas can be arranged. We consider first "gas clouds" without defining exactly what this means. Later we discuss the LOC configuration and other possibilities.

Despite the similar level of ionization, the physical conditions in this region are considerably different from those in the BLR. The column density in some of the clouds is small enough to make them optically thin in the helium and possibly also the hydrogen Lyman continuum. Under such circumstances, the mean level of ionization can be much higher. In addition, the radiation field is much weaker, and hence the low-density gas is likely to contain dust. This will have a large effect on the emitted spectrum and perhaps also on the role of radiation pressure force in controlling the cloud pressure and motion. The EWs of the emission lines originating from this are considerably smaller than those of the broad lines because of the smaller covering factor, the smaller Lyman continuum opacity, and the presence of dust. Confinement of the gas may not be important since the lifetime of such large clouds is rather long.

Unlike the broad emission lines, the spectrum of the low-density gas must include intense forbidden lines. This shifts the line cooling balance in such a way that the semiforbidden and permitted lines become relatively weaker. Another group of lines that are predicted to be intense in the innermost part of this region are coronal lines. Such lines are typical of highly ionized species. They are produced by fine-structure transitions and observed mostly in the red and infrared parts of the spectrum. Low-ionization lines like Fe II or Mg II are likely to be very weak even if dust is not present in this gas. The region producing such a spectrum qualifies for the name *narrow-line region* (NLR).

7.2.1. The NLR spectrum

Dust-free and dusty NLR gas

The NLR spectrum is perhaps the best studied AGN feature. Optical–UV spectra, from the ground and from space, have been obtained for numerous sources, occasionally in a spatially resolved manner. Line profiles have been investigated in great detail, and IR spectroscopy enabled the measurement of many coronal lines that are prominent in this part of the spectrum. Figure 1.3 shows an example of the NLR spectrum of the moderate-luminosity type-II AGN NGC 5252. Observations and modeling of such spectra are in good agreement. For high-ionization AGNs, they indicate a narrow range of ionization parameter ($U_{\text{hydrogen}} \sim 10^{-2}$, hereinafter U) and a smaller ionization parameter, by almost an order of magnitude, in LINERs.

We can infer the physical conditions in the NLR using standard line ratio techniques and various diagnostic diagrams. This is helped by the fact that for nearby AGNs, there are more than 100 emission lines that can be measured in the NIR–optical–UV part of the spectrum. Moreover, the Balmer lines are easy to detect, and their ratio (the Balmer decrement) can provide a reliable way to correct for foreground extinction. Understanding dusty NLR gas, and the general role of *internal dust*, is more challenging. Here the most important processes are grain formation and destruction, depletion, and the destruction of line photons by the grains (see Chapter 5).

As explained in § 5.10, radiation pressure force on dust grains can affect the internal structure of dusty gas clouds that are exposed to the central AGN continuum. To better understand this issue, we consider it here in more detail. Dust dominates the absorption of ionizing photons in highly ionized gas (Equation 5.86) with $U \geq 10^{-1.5}$. At such high levels of ionization, the dust opacity in the hydrogen Lyman continuum exceeds the gas opacity, and hence a large fraction of the ionizing radiation is absorbed by the dust and reemitted at IR wavelengths. The HII part of the cloud shrinks and the line emission is reduced. At much lower values of U, most of the ionizing radiation is absorbed by the gas, and the effect on the emergent spectrum is not as large.

We can imagine a situation in which the value of U at the illuminated face of the cloud is above the critical value and the dust opacity dominates the attenuation of the ionizing radiation field. For this value of U, the radiation pressure on dust grains, and through this on the gas particles, is larger than the internal gas pressure. This results in the compression of the gas close to the illuminated surface and a reduction in U. Deeper into the gas, U decreases further because of the increased dust and gas opacity. Radiation and gas pressure will balance at a specific depth corresponding to a specific U. This value of U is preferred over other values regardless of the value of U at the illuminated surface. A stronger external radiation field (larger U) will increase the radiation pressure, provide more compression, and bring the internal U to the same canonical, pressure balance value. This process controls the conditions in the part of the cloud where the gas is highly ionized, for example, where most oxygen is O^{+2}, and results in similar line ratios regardless of the location of the cloud. It is less important for lines that dominate the cooling beyond the hydrogen ionization front, for example, [O I] $\lambda6300$. This mechanism was proposed to explain the very uniform value of U inferred from spectroscopy of many type-II AGNs.

The merit of dusty versus dust-free NLR models can be evaluated by considering various line diagnostic diagrams (§ 6.1.2). One such diagram is shown in Figure 7.19. It contains LINERs and high-ionization AGNs and shows the results of various photoionization calculations applied to a single-component (single cloud) dusty gas with various SEDs: a 40,000 K blackbody and two typical AGN SEDs that differ in their α_{ox}. The blackbody spectrum is calculated for a solar metallicity gas and the AGN cloud for solar and 3 times solar metallicity gas. Some of the curves clearly explain the range of line ratios presented in the diagram in both low- and high-ionization AGNs. Such theoretical diagrams have been constructed for dusty NLR gas and different line ratios. They seem to explain many observed line diagnostic diagrams and give support to the dusty, radiation-pressure-supported model.

NIR and MIR emission lines provide useful diagnostics for the conditions in the NLR, especially in those cases where large-aperture observations include both the NLR and several bright SF regions. This part of the spectrum contains lines

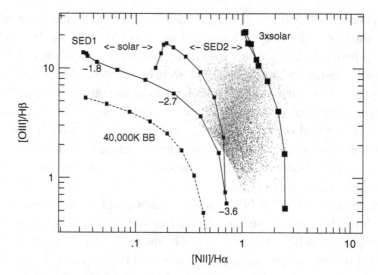

Figure 7.19. A diagnostic diagram showing the comparison of various photoion-ization calculations with observations of high-ionization type-II AGNs (red points) and LINERs (black points) from the SDSS sample. There are three different SEDs: a stellar SED (a 40,000 K blackbody), high-luminosity AGN SED (SED1 – a com-bination of an accretion disk and an X-ray power law with $\alpha_X = 1$ and $\alpha_{ox} = 1.4$), and low-luminosity AGN SED (SED2 – same components but $\alpha_{ox} = 1.1$). All models assume solar metallicity dusty gas with constant hydrogen number density of 10^3 cm^{-3}, except the model on the right, which assumes 3 times solar metallicity dusty gas.

that are clearly due to HII regions (e.g., [Ne II] 12.8 μm), lines that can originate both in HII regions and in the NLR (e.g., some Ne III lines), and high-ionization lines that can only be excited by an AGN continuum (e.g., [Ne v] 14.3 μm). A main advantage of such observations is the much-reduced extinction that allows more reliable measurements of several line ratios that provide good density and temperature diagnostics. Several IR diagnostic diagrams have been used in such studies; one of them is shown in Figure 7.20.

There are several shortcomings of the dusty, constant-pressure NLR model. They are apparent at both the high- and low-ionization parameter ends. The first applies mostly to the coronal-line region and the second mostly to the spectrum of LINERs.

Coronal lines

There are many examples of high-ionization, narrow emission lines that require a very large value of U. Examples are several mid-IR emission lines, for example, [Ne VI] 7.66 μm and [Si VI] 1.9 μm, and several optical lines due to Fe x, Fe XI, and Fe XIV. Such lines are clearly seen in many but not all AGNs. They are referred to as *coronal lines*, and the part of the NLR where they are emitted is the *coronal-line region*.

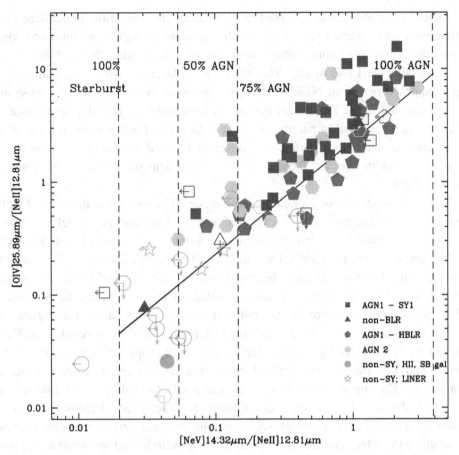

Figure 7.20. IR line diagnostics diagram (from Tommasin et al., 2010; reproduced by permission of the AAS) showing the locations of various types of objects in the [Ne v] 14.3 μm/[Ne II] 12.8 μm versus [O IV] 25.9 μm/[Ne II] 12.8 μm plane. The [O IV] 25.9 μm and [Ne v] 14.3 μm lines are the result of ionization by a nonstellar AGN continuum, and the [Ne II] 12.8 μm line is dominated by emission from SF regions. Various amounts of mixture between the two can explain the trends in the diagram. In the case shown here, AGN2 stands for "pure type-II AGN," that is, objects showing no indication of broad emission lines in reflection.

Most of the coronal lines are emitted by gas located between the NLR and the BLR. Some of the lines are very intense over very large distances, 10–200 pc in some cases. Most of the [O III] λ5007 emission in such objects is from distances of 100–500 pc. Other lines, usually those of the highest ionization species, originate mostly in regions that are several pc from the central source. Despite the very high ionization parameter, it is clear that dust has little or no effect on the spectrum of the coronal-line region. For example, there is no indication of metal depletion, which would make all iron lines much weaker than observed (iron depletion in the ISM is greater than 90%).

Given the large range of observed properties, and the similarities and differences between the lower-ionization NLR and the coronal-line region, we must consider again the various photoionization models constructed to explain the NLR spectrum, starting with the LOC model. This model is quite successful in explaining the general properties of the NLR spectrum in high-ionization AGNs that do not show strong coronal lines. The model assumes a large range of densities and hence a large range of ionization parameter *at every location*. Each ionization parameter (i.e., each density) corresponds to a different level of ionization, and the most intense forbidden lines are those that have critical densities similar to the density this location.

The LOC model is not a single location model. The entire volume is filled with gas with a large range of densities. Internal dust would change the model properties and drive the ionization parameter of all components with density below a certain value, and U above a certain value, into a small range of effective U. This cannot explain why the lowest-density, highest-U gas is dust free and produces intense iron coronal lines. The only explanation is that for reasons that are not yet clear, the conditions are not favorable to grain formation in some parts of this region.

The cloud model provides a somewhat different view of the coronal lines. This model is characterized by stratification. The range of properties at each location is small, and line emission is identified by a certain density and a certain ionization parameter. The suggestion is that the typical gas density drops with distance as r^{-s} such that $s < 2$. This gives an ionization parameter that decreases with distance from the center. Direct high-spatial-resolution observations of nearby sources strengthen this idea. The intensity contrast between high- and low-ionization lines in every location in the cloud model is larger than in the LOC model. Like the LOC model, the cloud model cannot provide an explanation to the absence of dust in the coronal-line region. Perhaps such clouds are the remnants of shocked dust-free gas in this region.

There are other suggestions that explain unusually high levels of ionization, yet they do not solve the missing dust problem. One of these involve different ionizing SEDs in various parts of the NLR due to line-of-sight obscuration. Such a situation can arise if the line of sight from some clouds to the central source passes through a low-column-density cloud that alters the ionizing SED. In particular, it preferentially removes from the spectrum photons around the He^+ ionization edge at 4 Rydberg, where the opacity is the largest. This produces a dip in the SED and lowers the effective U. Other NLR clouds at the same distance do not have such a line-of-sight obscuration and will have higher ionization ions. In this scenario, most NLR clouds see a modified SED, while the coronal-line-region clouds are exposed to the full SED. Another variant of the model assumes that the coronal-line region is the obscuring clouds themselves. This is more in accord with the smaller observed size of the coronal-line region.

Finally, we note a couple of other ideas related to the gas temperature and kinematics. There were several suggestions that the temperatures and densities in the coronal-line region are high enough, due to fast shocks, to enable collisional ionization of this gas. Shocks are appealing since they are known to destroy dust grains, thus avoiding the consequences of the constant-pressure dusty model in the coronal-line part of the NLR. Detailed models of this type are inferior to simple photoionization models in explaining the overall spectrum. In particular, they predict the presence of several high-excitation lines that are not seen in the spectrum.

An additional model combines the high ionization and high velocity (see later) of the coronal-line gas. The model assumes that the origin of the coronal lines is in a radial gas outflow that originates in the putative dusty torus around the central BH. The gas is accelerated to its observed velocity by radiation pressure force, and the iron-containing grains are destroyed in the initial acceleration phase that produces no intense lines. The high-ionization lines of iron are most intense some distance away where the flow reaches its maximum velocity and the gas is dust free.

NLR stratification

The global ionization structure of the the NLR, in particular, its stratification, can be investigated by assuming that the emission-line widths are decreasing with distance from the BH. This predicts broader lines in higher-ionization species. Such a correlation was indeed found long ago for optical emission lines and is now confirmed by careful studies of MIR emission lines. Using the MIR is advantageous because there are many fine-structure lines of highly ionized species in this part of the spectrum. This reduces the dependence of line intensity on the gas temperature. In addition, as already emphasized, the reduced reddening and obscuration is less significant in this part of the spectrum, allowing a less biased view of the velocity field. Examples of these correlations are shown in Figure 7.21.

To summarize this part, we show in Figure 7.22 calculated spectra of dusty NLR clouds exposed to a typical AGN SED. All clouds contain internal dust and have the same density and column density, which is large enough to make them optically thick in the Lyman continuum. The emitted spectrum contains a large number of emission lines in the MIR–optical part of the spectrum as well as NIR–MIR dust continuum emission. As seen in § 7.5, this continuum is similar to the one observed in many type-I and type-II AGNs.

Low-ionization LINER-type emission lines

The second extreme case involves those regions of the NLR with $U < 10^{-1.5}$. Here the ionization parameter is below the critical value, and the gas is predicted to show strong low-ionization lines. Indeed, strong low-ionization lines of O II, N II, O I, and

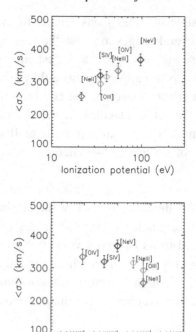

Figure 7.21. Emission-line velocity dispersions versus ionization potential (top) and critical density (bottom) for strong MIR emission lines. The line width is strongly correlated with the level of ionization of the emitting ion, suggesting that the NLR is stratified, with more ionized gas closer to the center. There is no significant correlation with the critical densities of the lines (from Dasyra et al., 2011; reproduced by permission of the AAS).

other low-ionization ions are clearly seen in the spectrum of many high-ionization AGNs. Some of the lines, for example, [O I] λ6300, are produced deep in the cloud where the gas is mostly neutral. The strength of these lines is affected by both the ionization parameter and the shape of the SED, in particular the value of α_{ox} that affects the penetration of X-ray photons to regions beyond the hydrogen ionization front. Other lines, like [N II] λ6584, are less sensitive to the exact value of α_{ox}.

LINERs are very different in this respect. Their spectrum is dominated by low-ionization lines, like [N II] λ6584 and [O I] λ6300, and higher-ionization lines, like [O III] λ5007, are considerably weaker and even absent ([Ne v] λ3426). These objects are more numerous and less luminous than the high-ionization AGNs, and the conditions in their NLRs must be considerably different.

The LINERs discussed here are all AGN ionized (to be distinguished from regions around pAGB stars and from shock-excited LINERs; see Chapter 6). Successful photoionization models with "typical" AGN SEDs indicate ionization parameters that are about a factor 3–10 lower than the typical ionization parameter in high-ionization AGNs. This must be a combination of the NLR density, the NLR dimension, the lower luminosity, and perhaps a different shape continuum.

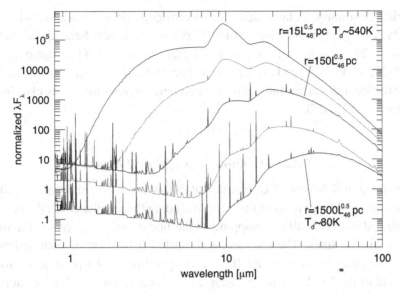

Figure 7.22. The spectrum produced by dusty AGN clouds situated at various distances, as marked, from a luminous central source with typical AGN SED and $L = 10^{46}$ erg s^{-1}. The calculations assume constant-density gas with $N_H = 10^4$ cm^{-3} and a column density of $10^{21.5}$ cm^{-2}. Galactic dust-to-gas ratio is assumed with dust properties and depletion factors similar to what is observed in galactic HII regions. The assumed gas density determines the ionization parameter at the given distance; thus clouds at the same location but with smaller (higher) gas density will show spectra of more ionized (less ionized) emission-line gas (adapted from calculations by B. Groves).

As shown subsequently, the normalized accretion rate in LINERs, as measured by $L/L_{\rm Edd}$, is some 2–5 orders of magnitude smaller than in high-ionization AGNs. The reduction is several orders of magnitude larger than the reduction in U. Thus, the NLR properties in LINERs must be different than in other AGNs.

The very low accretion rate in LINERs is likely to result in RIAFs (Chapter 4) and hence a two-temperature disk and a very "soft" SED. While the direct observational evidence for such an SED is still missing, this raises the possibility that the ionizing spectrum contains very few high-energy photons, which may explain the lack of strong lines from high-ionization ions. Photoionization models of LINERs assuming an unusually soft SED are in reasonable agreement with many observations.

LINERs without a BLR

A small fraction of LINERs show broad emission lines and are classified as type-I LINERs. However, the great majority of the locally discovered LINERs are type II with no indication of broad Hα or Hβ in their spectra. There are two observational reasons for this lack. First, the low luminosity of such sources makes their optical continua, and optical broad emission lines, difficult to detect against the stellar continuum in large-aperture spectroscopy (like the one used in several big surveys, e.g., the fiber size in the SDSS is 3 arcsec). Second, as shown in Equation 7.22, the emission lines of low L/L_{Edd} sources can be extremely broad, 10,000–20,000 km s^{-1}, and even broader. The detection of such low-contrast lines against the local stellar continuum is very challenging, resulting in misclassification of many type-I LINERs.

7.2.2. *The kinematics of the NLR gas*

Modern integral field unit (IFU) instruments allow a two-dimensional, velocity and space mapping of large parts of the NLR, including the smaller coronal-line region, in nearby AGNs. This allows mapping of the line-emitting regions in the light of certain emission lines. The (AO-assisted) spatial resolution of such instruments at NIR wavelengths is of order 0.2," corresponding to 40 pc, that is, roughly 10 percent of the NLR extent for a source at $z \sim 0.01$. Such high-resolution maps reveal complex structures with clear nonuniform gas distributions. This was already illustrated in Chapter 6 and is attributed to the uneven illumination of the gas due to a central obscuring torus. One such example is shown in Figure 7.23.

The map shown in Figure 7.23 and similar maps of other type-I sources at about the same redshift indicate a complex NLR velocity field that is a combination of rotation and outflow. The geometrical structure of the NLR and the much smaller coronal-line region are similar, suggesting the same source of ionization. The best kinematic model is made of an outflow bicone and a rotating disk coincident with molecular gas in these galaxies. In some sources, there is also a significant contribution from gas that is not rotating and is also not part of the outflow in the bicone. The presence of ionized gas in the disk, away from the central ionization cone, may be due to a clumpy central torus where some lines of sight allow the escape of ionizing photons that heat and ionized the gas in the disk. Detailed models constructed for several intermediate-luminosity type-I sources show lines of sight that are close and sometimes even outside the bicone boundaries.

An effective way to illustrate the outflow in the bicone is to construct a 3D diagram of the velocity field. A schematic of this type is shown in Figure 7.24. The picture supports the notion that in some, perhaps most, type-I AGNs, we see only one side of the bicone and thus only one-half of the total line flux.

7.2.3. *Luminosity covering factor and accretion rate*

Narrow emission lines as bolometers

Like the broad emission lines, and the various continuum bands, the intensities of the narrow emission lines provide ways to estimate the bolometric luminosity of

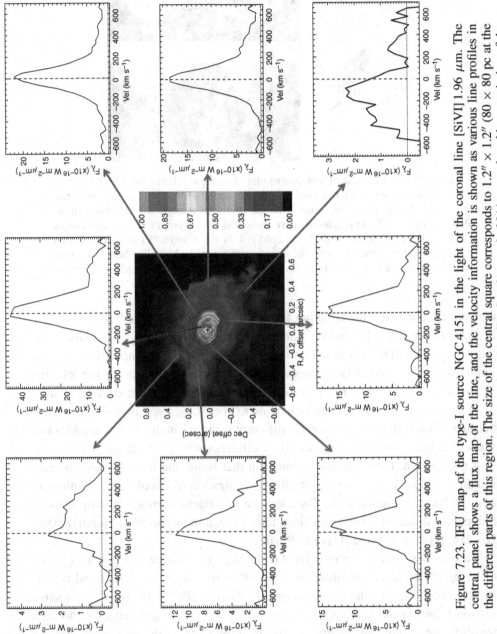

Figure 7.23. IFU map of the type-I source NGC 4151 in the light of the coronal line [SiVI] 1.96 μm. The central panel shows a flux map of the line, and the velocity information is shown as various line profiles in the different parts of this region. The size of the central square corresponds to 1.2″ × 1.2″ (80 × 80 pc at the source) (courtesy of F. Meueller-Sanchez, from Meueller-Sanchez et al., 2011; reproduced by permission of the AAS).

189

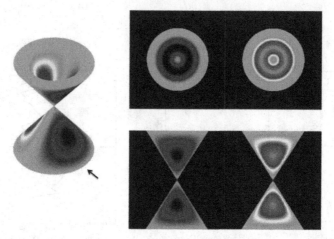

Figure 7.24. An example of a bicone model of outflow velocity field. (left) Three-dimensional structure of the model with the arrow indicating the line of sight. (top right) Front and back projections for inclination angle 90°. (bottom right) Bicone angle of 0°. The velocities are indicated by the colors, with warm colors showing the amount of redshift and cold colors the amounts of blueshift. Green represents approximately zero velocity (courtesy of F. Meueller-Sanchez, from Meuller-Sanchez et al., 2011; reproduced by permission of the AAS). (See color plate)

the central source and the covering factor (C_f) of the NLR. The basic ideas have been explained in § 7.1. Here we extend them to the NLR and type-II AGNs.

For a given SED, ionization parameter, gas density, and composition, the line luminosity in a given cloud represents a certain known fraction of the ionizing luminosity of the source. For a known covering factor, this can be used to infer L_{bol}. In reality, there is a large scatter in $L(line)/L_{bol}$, even in a single source, because of the different conditions in different locations in the NLR. An additional scatter between sources is introduced by differences in the SED and the gas and dust distribution. However, the assumption that some line luminosities represent roughly the same fraction of L_{bol} in all high-ionization AGNs of similar luminosity is not a bad assumption and allows the use of various narrow emission lines as bolometric indicators. Emission lines that are commonly used are [O III] $\lambda5007$, Hβ, Hα, [O I] $\lambda6300$, and [O IV] 25.9 μm.

There are two ways to normalize $L(line)/L_{bol}$ in a source of a given L_{bol}, one by using theoretical calculations and the other by comparing type-I and type-II sources. For example, the mean intrinsic L([O III] $\lambda5007$) is likely to be the same in type-I and type-II sources of the same L_{bol}. There are differences to do with the different geometries relative to the lines of sight to the two types that are manifested as different amounts of reddening and obscuration. Reddening can be partly accounted for by using known line ratios such as Hα/Hβ. The differences in $C_f(\text{NLR})$ from source to source are averaged out in a large sample.

A practical way to obtain the necessary calibration is to compare L([O III] $\lambda 5007$) with L_{bol} in type-I sources using a known continuum luminosity, for example, L_{5100}. This provides a line bolometric correction factor that is then used in type-II sources. Such a comparison shows that L([O III] $\lambda 5007$)/L_{5100} is larger in lower-luminosity AGNs. Typical numbers in type-II sources with $44 < \log L_{bol} < 45$ erg s^{-1} are $L_{bol}/L([O\ III]\ \lambda 5007) \simeq 3000$ for directly measured $L([O\ III]\ \lambda 5007)$ and $L_{bol}/L([O\ III]\ \lambda 5007) \simeq 800$ for reddening-corrected $L([O\ III]\ \lambda 5007)$. The corresponding number for type-I AGNs that are not corrected for reddening is about 1500, that is, reddening attenuates the line luminosity in type-II sources by an additional mean factor of ~ 2. Reddening correction in high-ionization, low-redshift type-II AGNs is easy to apply since, in many cases, $H\alpha/H\beta$ is measured to a reasonable accuracy. A similar correction in type-I sources is more challenging since the narrow $H\alpha$ and $H\beta$ lines are severely blended with the broad components of the lines and deblending introduces a large uncertainty. LINERs are very problematic because, in many cases, the $H\beta$ emission line is severely attenuated by stellar absorption lines.

Using the notation of § 7.1.5, we can write an expression for the bolometric correction factor of dereddened lines,

$$BC_{5007} \simeq 800 f([O\ III]\ \lambda 5007), \tag{7.31}$$

where $f([O\ III]\ \lambda 5007)$ takes into account the change in $L([O\ III]\ \lambda 5007)/L_{5100}$ with luminosity, most probably due to a change in the covering factor of the NLR.

The application of the $L([O\ III]\ \lambda 5007)$ method to low-ionization AGNs like LINERs is more problematic. Here the conditions in a "typical" NLR cloud are different because of the smaller U as inferred from the lower $L([O\ III]\ \lambda 5007)/L(H\beta)$. Therefore, the required bolometric correction factor is larger. To illustrate this, we show in Figure 7.25 theoretical calculations of $L(\text{line})/L_{bol}$ for three prominent narrow emission lines, [O III] $\lambda 5007$, [O I] $\lambda 6300$, and $H\beta$. The calculations represent a large range of ionization parameter. They show the very sensitive dependence of $L(O\ III)/L_{bol}$ on U and the much weaker dependence of $L(H\beta)/L_{bol}$ on U. The reason is that the [O III] $\lambda 5007$ line dominates the cooling of high-ionization gas throughout much of the HII part of the cloud. This channels an almost constant fraction of the ionizing luminosity to this line, unless the composition is very different from solar. However, in low-U gas, most of the oxygen is singly ionized, and a much larger fraction of the cooling is carried by other lines such as [O II] $\lambda 3727$ and C III] $\lambda 1909$. This is not the case for $H\beta$, which is a recombination line and as such contributes very little to the total cooling. As explained in Chapter 5, a certain fraction of all the ionizing luminosity is reradiated by the Balmer lines in a way that depends very little on the exact level of ionization. In the particular case shown here, the emission of the [O I] $\lambda 6300$ line represents an almost constant fraction of the high-energy, soft-X-ray radiation that penetrates beyond

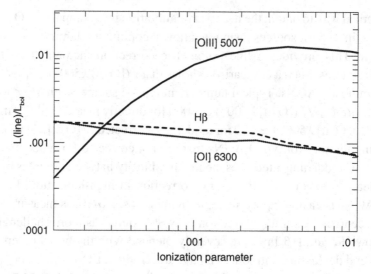

Figure 7.25. Relative line-to-continuum luminosities for strong narrow emission lines. The NLR is assumed to be made of constant-pressure dusty clouds with solar composition, $N_e = 10^3$ cm^{-3}, and full covering of the central source. The SED of the central source is characterized by $\alpha_{ox} = 1.05$ typical of low-luminosity AGNs. The [O III] $\lambda5007$ luminosity is a strong function of the ionization parameter, but the other two lines have much weaker dependences (adapted from Netzer, 2009).

the hydrogen ionization front. This constant fraction results in almost constant $L([\mathrm{O\,I}]\,\lambda6300)/L_{\mathrm{bol}}$ over a large range in ionization parameter.

The preceding considerations suggest that $L(\mathrm{H}\beta)$ is a good bolometer in both high- and low-ionization AGNs. However, $L([\mathrm{O\,III}]\,\lambda5007)$, which is a reliable bolometer in high-ionization AGNs, gives erroneous results for LINERs. Fortunately, part of this discrepancy can be corrected by noticing that the decrease in $L([\mathrm{O\,III}]\,\lambda5007)$ is compensated for by an increase in fractional cooling by [O I] $\lambda6300$. It is therefore possible to devise a bolometric correction factor that is based on the combined luminosities of the two lines. Note that the covering factor does not play an important role in all these considerations since the calibration of $L(\mathrm{line})$ is based on measurements in type-I sources with the additional assumption of a similar $C_f(\mathrm{NLR})$ in type-I and type-II sources of a similar L_{bol}.

The following expression can be used to estimate L_{bol} in *all* type-II AGNs with L_{bol} specified as earlier. They take into account a mean difference in emission-line reddening between type-I and type-II sources, probably due to the different inclinations to the line of site, and require the use of galactic extinction law prior to the calculation of L_{bol}.

For reddening-corrected $L(\mathrm{H}\beta)$,

$$\mathrm{BC}_{\mathrm{H}\beta} \simeq 6\mathrm{BC}_{5007}. \tag{7.32}$$

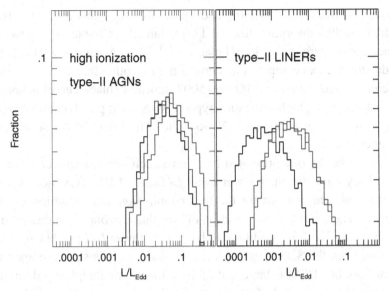

Figure 7.26. L/L_{Edd} distributions for $z < 0.1$ type-II AGNs as marked. The estimates are based on $L(H\beta)$ (red curves), $L(O\,\textsc{iii})$ (black curves), and the combined luminosity of O\,\textsc{iii} and O\,\textsc{i} (blue curves). All methods agree very well for high-ionization sources, but the O\,\textsc{iii} method underestimates L_{bol} and hence L/L_{Edd} in the low-ionization LINERs (adapted from Netzer, 2009).

For the combined O\,\textsc{iii}–O\,\textsc{i} method,

$$\log L_{bol} \simeq 3.8 + 0.25 \log L([O\,\textsc{iii}]\,\lambda 5007) + 0.75 \log L([O\,\textsc{i}]\,\lambda 6300). \quad (7.33)$$

All estimates listed here apply to the limited L_{bol} range specified earlier. This range contains the majority of local, $z < 0.1$ high-ionization AGNs in the SDSS sample. They depend on the amount of line photon destruction by internal dust, which is already included in the photoionization calculations of Figure 7.25 and accounts for a factor of 1.5–2 out of the total extinction, which amounts to a factor of ~ 4 for Hβ in high-ionization type-II AGNs. This gives a mean NLR covering factor of 4–8 percent for these sources. The estimates are based on the assumption of little or no extinction of the optical continuum in type-I AGNs, which, in itself, is subjected to some uncertainty.

L/L_{Edd} in type-II AGNs

Having estimated L_{bol} in type-II AGNs of high and low ionization, we can now use this method to investigate various line and continuum properties, including L/L_{Edd}.

Figure 7.26 shows a comparison of L/L_{Edd} distributions for a large number of high- and low-ionization AGNs in the SDSS sample. The objects were classified into the two groups, LINERs and high-ionization AGNs, by using line diagnostic diagrams. The BH mass was estimated by the M–σ_* method, which is based on the width of stellar absorption lines and is explained in Chapter 8. The two groups combined cover a large range in BH mass, $10^{6.3-9} M_\odot$. Assuming $L(H\beta)$ is the best L_{bol} indicator, we can compare the merit of the two other indicators that are based on different oxygen lines. The [O III] λ5007 indicator (black curve) is appropriate for the fast-accreting, high-ionization type-II AGNs (left panel) but underestimates L/L_{Edd} in the slower-accreting LINERs. The O III+O I indicator seems to agree much better with $H\beta$.

Diagrams like the one presented here have been constructed for several AGN samples. They show the much lower mean L/L_{Edd} in LINERs, which may explain part, but not all, the reasons for the lower ionization and excitation of the line-emitted gas. There are several selection effects that can bias such distributions. In particular, many LINERs are likely to have even lower L/L_{Edd}, but such sources are missing from the SDSS (used here) and other samples since the aperture size ($3''$ in the case of SDSS) is large and fainter objects are completely dominated by stellar lines and continua. In fact, a very detailed study of fainter nearby LINERs clearly shows that many of these sources have $L/L_{Edd} \sim 10^{-5}$ and even smaller. In addition, many AGNs are likely to have significant contributions to their narrow emission lines from SF regions in the host galaxy. This contribution, which is discussed in Chapter 8, can affect the location of LINERs and other AGNs in the various diagnostic diagrams. In fact, a very large number of narrow-emission-line objects are "composite sources" with significant contributions to the lines from both AGN-ionized and stellar-ionized gas.

7.2.4. The NLR size and extended narrow-line regions

What is the extent of the NLR, and how far can we observe low-density gas that is exposed to the direct radiation field of a luminous AGN? Given today's high-sensitivity instruments, this distance can exceed the size of the host galaxy. In fact, emission-line regions with spectra that are typical of AGN-ionized gas are seen, in some cases, to distances of 100 kpc and beyond. Such regions are referred to as *extended narrow-line regions* (ENLRs).

There were several attempts to measure the correlation between the NLR size and L_{bol} by mapping the size of the [O II] λ3727 and other line emission zones. One expression, which is based on observations of intermediate- and high-luminosity AGNs, is

$$\log \frac{R_{\mathrm{NLR}}}{\mathrm{pc}} = 0.52 \log L_{[\mathrm{OIII}]} - 18.5, \tag{7.34}$$

which, with $L_{[\mathrm{OIII}]}$ in erg/sec, is not corrected for extinction. There are expressions constructed separately for type-I and type-II AGNs, taking into account the different inclination angles. Such estimates are problematic since they do not define, uniquely, the boundaries of the NLR and hence R_{NLR}. In particular, some of these distances are measured from very low surface brightness [O III] $\lambda5007$ maps far outside the region where most of the line emission takes place. A more meaningful measurement must be specified by a certain fraction of the line luminosity in question. For example, the NLR size can be defined by the distance from the center that includes 90 percent of the line flux.

The preceding expression cannot hold for all AGNs for other reasons. In the highest-luminosity AGNs, at high redshift, $L_{bol} \sim 10^{48}$ erg s^{-1}. Using Equation 7.34, we obtain an NLR size of ~ 50 kpc, far beyond the host-galaxy size. While some residual line emission is indeed seen at large distances, the NLR as defined by observed properties of low-redshift sources is bound to the host galaxy. The conclusion is that very high luminosity AGNs may not contain an NLR. Indeed, the observed EW([O III] $\lambda5007$) in many high-luminosity objects is very small, and in some cases, the line is completely absent. There are, however, cases with relatively strong [O III] $\lambda5007$ lines and very large L_{bol}. In this case, the conditions in the NLR must be very different from those discussed here. In particular, such regions may have larger density and smaller dimensions. The estimated R_{NLR} in such cases must be different from the one given by Equation 7.34.

Returning to the ENLR, such regions are clearly seen in a number of high-luminosity AGNs. Several of those are in type-II radio-loud sources seen roughly edge on. The line-emitting gas in such cases is in the same general direction as the radio jet and the ionization cone. Spectroscopy of such regions reveals, in some cases, the nature of the obscured central source, which is seen by reflection and polarization of the central continuum. In some cases, the entire SED is seen in reflected light. Such ENLR regions are bright enough to allow spectroscopy of the low-density gas. On average, they show a very high level of ionization with strong lines of He II $\lambda1640$, C IV $\lambda1549$, N V $\lambda1240$, and other high-ionization lines. The spectra, when available, indicate an ionization parameter that is considerably higher than the mean value observed in low-redshift, high-ionization AGNs.

7.3. The highly ionized gas

7.3.1. HIG spectrum and outflow

Next consider the region between the NLR and the BLR, 0.1–10 pc from the center. Assume intermediate to large column densities, 10^{21-23} cm^{-2}, and a density that results in an ionization parameter 10–100 times larger than the one in the BLR. For such gas, $U_{oxygen} \sim 0.02$, and we expect strong absorption and emission features in the X-ray part of the spectrum. Similar temperatures and levels of ionization can be found in gas with a much lower density that is distributed inside the NLR.

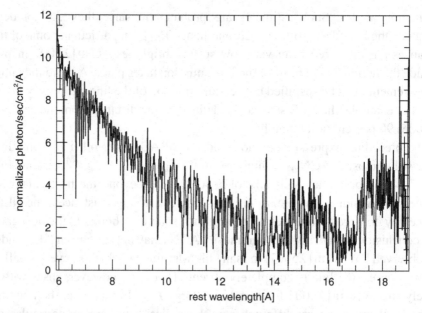

Figure 7.27. Chandra high-resolution spectrum of NGC 3783 showing hundreds of absorption lines in the 6–20 Å range (adapted from Netzer et al. 2003).

A proper name for such a component is *highly ionized gas* (HIG). In the X-ray literature, it is often referred to as "warm absorber."

For scaling purposes, we note that the density of a typical HIG cloud is

$$N_{\text{HIG}} \simeq 2 \times 10^4 U_{\text{oxygen}}{}^{-1} L_{44\,\text{oxygen}} \ R_{\text{pc}}^{-2} \ \text{cm}^{-3}, \tag{7.35}$$

and its mass is

$$M_{\text{HIG}} \simeq 10^3 \ N_{22} C_f R_{\text{pc}}^2 \ M_\odot, \tag{7.36}$$

where C_f is the covering factor of the HIG, N_{22} is the hydrogen column density in units of 10^{22} cm^{-2}, and $L_{44\,\text{oxygen}}$ is the integrated 0.54–10 X-ray luminosity in units of 10^{44} erg s^{-1} (0.54 keV is the K-shell threshold ionization of neutral oxygen). Note that for low-luminosity AGNs, like NGC 5548 and NGC 3783, $L_{44\,\text{oxygen}} \simeq 0.05$, and for high-luminosity quasars, $L_{44\,\text{oxygen}} \simeq 1$.

X-ray observations of type-I AGNs show the clear signature of the HIG. The strongest spectral features are numerous absorption lines of the most abundant elements, strong bound–free absorption edges due mostly to Ovii and Oviii, and several emission lines. A good example of this type is the 900 ks spectrum of NGC 3783 obtained by Chandra HETG. This spectrum was observed in two states of the source, high and low, that differ by about a factor of 1.5 in the flux of the hard X-ray continuum. Part of the low-state spectrum is shown in Figure 7.27.

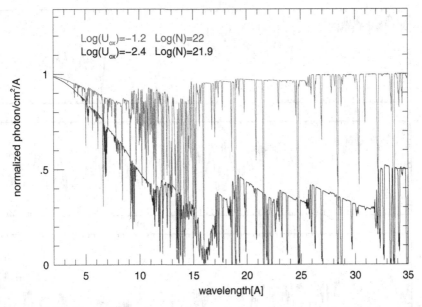

Figure 7.28. Two of the three components required to model the low-state spectrum of NGC 3783 shown in Figure 7.27 (adapted from Netzer et al., 2003).

Observations like the ones shown here are accurate enough to allow a detailed study of the physical conditions, including the determination of the absorbing columns of many ions and the exact (somewhat SED-dependent) value of the ionization parameter. Unfortunately, in most cases, the densities are not high enough to show a clear response of the gas to observed continuum variations. Thus, the density cannot be directly determined, and there are only upper limits that translate to lower limits on the distance. Typical numbers suggest $r_{HIG} \gtrsim 1$ pc, although there are claims for much smaller dimensions.

Figures 7.28 and 7.29 show the results of an ambitious attempt to fit a 900 ks Chandra spectrum of NGC 3783. An interesting result of the modeling of this spectrum is the need to introduce several HIG components. The components differ substantially in their ionization parameters (more than a factor 10) and absorbing columns. Models of two of the three components are shown in Figure 7.28, and the full, three-component fit to the spectrum is shown in Figure 7.29.

A very interesting feature of the three-component HIG in NGC 3783 is the fact that the product $N_H T$ is about the same in all three components. This means that the three can coexist in the same volume of space in pressure balance. In other words, the highest-temperature, lowest-density component can provide the confining pressure for the two other components. An inspection of the stability curve relevant to this case shows that, indeed, there are three stable solutions corresponding to the three observed components. This is shown in Figure 7.30.

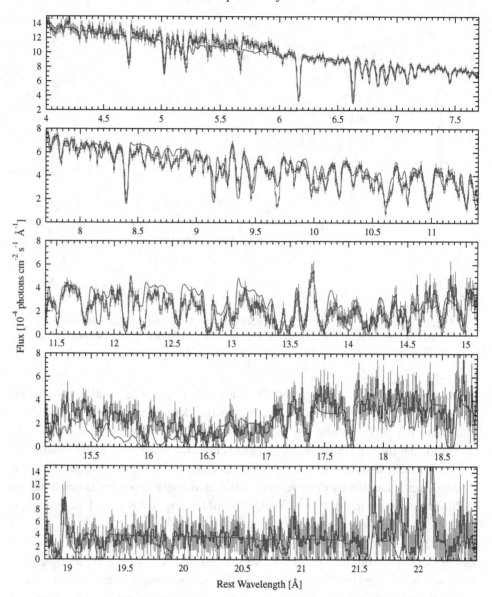

Figure 7.29. A three-component fit to the spectrum shown in Figure 7.27 (adapted from Netzer et al., 2003). The smooth solid line is a model fit to the data.

Studies of HIG systems in other type-I AGNs indicate that two or three components in pressure equilibrium are common in other cases too.

The X-ray spectrum of type-II AGNs is completely different because of the large obscuration in this type of source. It shows prominent X-ray emission lines that indicate similar properties to the gas producing the strong absorption in type-I AGNs. In some of the sources, the X-ray emitter is resolved, and its dimension can

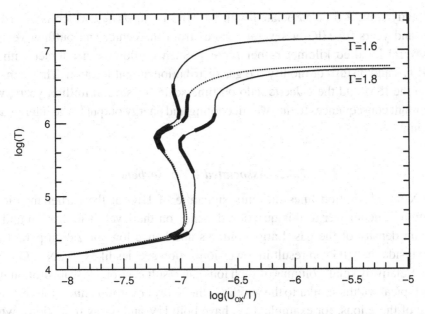

Figure 7.30. Various stability curves for the HIG in NGC 3783 with different hard X-ray slopes Γ. Each curve contains three regions, marked with thick lines, allowing three different stable solutions for gas in thermal equilibrium with different levels of ionization. Two of these components are shown in Figure 7.28 (adapted from Netzer et al., 2003).

be directly measured. This indicates an X-ray emission region of several hundred parsecs in diameter. It is not yet clear whether the type-I absorbers are of similar dimension. In fact, some well-studied cases indicate much smaller absorption regions, which raises the possibility that some HIG regions extend over much larger distances than indicated by the analysis of the absorption spectrum.

The HIG gas is transparent to the UV and soft X-ray radiation and is thus subjected to a strong radiation pressure force, compared with the opaque BLR gas. This results in mass outflow from the nucleus. Such outflows have been observed in several well-studied cases, and it is interesting to consider their effects on the general AGN environment.

The mass outflow rate of the HIG can be inferred from its observed velocity and from the known (or guessed, in some cases) location. It is given by

$$\dot{M}_{\mathrm{HIG}} = 4\pi r^2 C_f \rho v_{\mathrm{HIG}} \epsilon_{\mathrm{HIG}} \tag{7.37}$$

where C_f is the covering factor and ϵ_{HIG} is the filling factor of the HIG. In most such cases, the column density and the ionization parameter are known. However, the density and filling factor are still missing. The resulting uncertainty on the mass outflow rate is very large. For example, estimates for the well-studied case of NGC 3783 mentioned earlier range from 0.1 to more than 10 $M_{\odot}$ yr^{-1}.

The dynamical time associated with the HIG is r_{HIG}/v_{HIG}, which is of order a thousand years for HIG at several parsecs from the center and outflow velocity of several hundred kilometers per second. Such outflows can interact with the NLR gas and disturb or modify the gas distribution at that location. They can also reach the ISM and the galactic halo on time scales of several million years, with important consequences to the overall cooling and energy output to the intergalactic medium.

7.3.2. *Associated UV absorbers*

Are X-ray absorption lines the only signature of HIG at the earlier mentioned location? The answer to this question depends on the level of ionization and the column density of the gas. Large columns and hence low-ionization parameters deep inside the HIG can result in lower-ionization species like C IV, N V, O V, and O VI. Relatively small columns of such ions can result in strong UV absorption lines with typical widths similar to the widths of the X-ray absorption lines. Interestingly, some of these ions, for example, O V, have both UV and X-ray transitions, which can be compared to test the idea that the UV and X-ray absorption lines originate in the same component. Some studies of this type show clear-cut cases in which the UV and X-ray lines share the same velocity and are likely to originate from the same location. Other cases show X-ray absorption but no similar velocity UV absorption. Figure 7.31 shows a typical example of UV absorbers in an AGN whose X-ray spectrum shows the clear absorption signature of HIG.

The mutual existence of UV and X-ray lines can be the result of a large-column-density absorber where the high-ionization species at the illuminated face produces the strong HIG X-ray lines, and residual low-ionization species deeper in are responsible for the UV lines. An alternative, more complex scenario suggests a case where the kinematics and the microphysics of the gas are only loosely connected. For example, in a clumpy coherent flow, every location is associated with a certain velocity. The higher-density clumps are of lower ionization and can give rise to UV absorption lines. The lower-density gas at the same location is more highly ionized, which can explain the X-ray HIG lines.

7.4. Broad- and narrow-absorption-line regions

7.4.1. *Broad absorption lines (BALs)*

The class of broad-absorption-line (BAL) AGNs was described in Chapter 6. The broad (up to $\sim 0.2c$) blue-shifted absorption troughs that characterize this group of sources are observed in 10–20 percent of all high-luminosity, type-I AGNs. They are hardly ever observed in low-luminosity, local sources. Other characteristics of such objects include significant reddening of the continuum, a high degree of

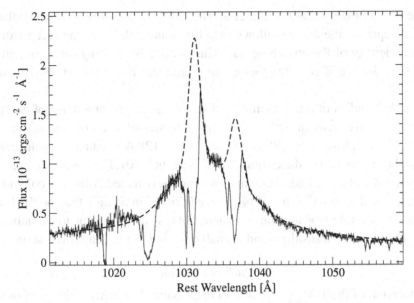

Figure 7.31. The FUSE spectrum of the quasar MR 2251-178 showing narrow
O VI absorption lines superimposed on emission lines of the same ion. The derived
outflow velocities are in the range 300–600 km s^{-1}. The source also shows time-
dependent X-ray absorption lines.

polarization, weak X-ray emission compared with AGNs of similar L_{bol}, and weak
radio emission. Here we assume that the features are part of a general BAL region
that may be common to some, but not all, AGNs.

High-quality spectroscopic observations allow a systematic study of many BAL
features. Some examples were shown in Figure 6.8. The velocity field has been
mapped to great accuracy, and there are observations of resonance lines as well
as lines from excited levels. The very high velocity field is complex and must
include nonradial components. It is also clear that part of the outflowing gas is not
covering the entire continuum source, which complicates the analysis. A realistic
dynamical model must combine column density of all outflowing species with
their geometry. Since different levels of ionization are associated with different
locations, the covering factor is not identical for all the ions.

Column density and covering factor

The kinematics of the high-velocity outflowing gas is complex. In some cases,
the seemingly broad troughs are clearly split into several narrower components. In
others, the troughs remain broad, suggesting a smooth, windlike flow. Perhaps the
most useful lines that are used for modeling such flows are the various doublets of
Li-like ions, C^{+3}, N^{+4}, and O^{+5}. In such cases, the relative absorbing column of
the two components of the doublet is set up by the atomic physics to be 1:2 because

of the ratio of the oscillation strengths. The ability to measure, separately, the two optical depths in the doublet allows us to disentangle the covering factor from the column density of the absorbing gas. This is done by looking for a combination of optical depth and covering factor that reproduces the observed absorption line ratio.

Detailed studies of a large number of BAL spectra reveal a range of obscuring columns, mostly between 10^{21} and 10^{22} cm^{-2}. In several cases, there is a clear detection of a broad phosphorus (PV) line at $\lambda\lambda 1118$, 1128 Å. Assuming a solar composition, the expected abundance ratio of P/C is about 0.001. Even if P/C is larger than solar an order of magnitude, such lines will not be detected unless the column density is larger than $\sim 10^{22}$ cm^{-2}. This is an additional indication that the flat-bottom profiles of several resonance lines in many BAL systems are due to a combination of highly saturated transitions and a small covering of the central UV source.

Outflow location

The location of the BAL gas is still an open issue. The BAL width is of order of the typical FWHM of the broad emission lines, which suggests similar dimensions. Indeed, a location just outside the BLR is one plausible explanation, especially in those cases where the broad emission lines are observed to be absorbed by the BAL gas.

The distance of the bulk of the outflow can be directly derived in a handful of cases where lines from both the ground level and an excited fine-structure level of a certain ion are observed in the same spectrum. Photoionization modeling allows us to solve for the ionization parameter of the gas using the measured column densities of the more abundant ions. This provides a good estimate of the electron temperature in the gas. Given the value of T_e, we can find the electron density N_e that gives the relative populations of the ground and fine-structure levels. This density can be combined with a dynamical model to estimate the location of the gas. An absence of absorption from a fine-structure level can be used to derive, in a similar way, an upper limit on the density and thus a lower limit on the distance. Several such studies suggest a typical distance of order 1–10 pc but there are some that suggest much larger distances, up to a kiloparsec.

A different estimate of the location of the outflow comes from variability studies. Variations in line depths are seen on time scales of days to years. The BAL samples that were monitored systematically are small and do not allow very good statistics. However, the number is large enough to show that spectral variations are very common and occur in about 50 percent of the sources. Several examples were presented in Figure 6.8. The more common observed variation is one in which the entire line moves up or down relative to the continuum, which results in a change of the line EW. Such variations are not associated with acceleration or deceleration of the flow.

There are two possible reasons for the observed variations: changes in the level of ionization due to variations of the ionizing flux and changes due to the motion of the absorber across the sight line. The latter are easier to spot in lines with very large optical depth. Such lines, if fully covering the UV continuum, would show completely saturated, black bottom (zero-intensity) profiles. If the covering factor is smaller than unity, they will show a saturated profile but larger than zero flux at the bottom of the trough. Motion across the line of sight will result, in many cases, in a change of the covering factor, which will manifest itself as up or down motion of the flat part of the profile relative to the unabsorbed continuum. The combined estimate of the dimension of the UV source (the part of the disk emitting the UV continuum) and the velocity perpendicular to the line of sight lead to an estimate of the distance. Typical numbers are about 1 pc, which, for such luminosity sources, is comparable to the BLR size. There are a handful of sources with indications of even smaller dimensions. Variations in absorbing column are more difficult to spot and are less common. In fact, there are documented cases of large changes in the continuum flux that *did not* result in line variations.

Regarding the orientation to the central disk and/or torus, there are indications that the flow may be at inclination angles that are close to the inclination angle of the central accretion disk. The continuum reddening, the high level of polarization, and the attenuation of the central X-ray source all point in this direction. There are several clear counterexamples with flows that are perpendicular to the plane of the disk. Several such cases are seen in radio-loud BAL AGNs that are much less common among BAL AGNs.

Outflow velocity

Acceleration by radiation pressure force (§ 5.9.2) is the most efficient mechanism to drive AGN gas to high velocities. This mechanism is most efficient (the largest force multiplier; see Figure 5.11) when the force is due to the absorption of continuum photons by absorption lines in the flow (bound–bound transitions). The force multiplier due to bound–free transitions is less efficient and that due to electron scattering is even smaller. In highly ionized gas, most of the species with strong resonance lines are highly ionized, and the force multiplier is low. Because of this, it is difficult to accelerate such gas to high velocities.

The large X-ray-obscuring columns observed in BAL AGNs provide a motivation for a model that can naturally explain the extreme velocities. The model is based on the idea of "shielding gas," which is a large-column-density material located between the strong UV and X-ray radiation sources in the inner accretion disk and the high-velocity outflowing gas further out (see Figure 7.32). The gas in the shield is highly ionized at the illuminated face, resulting in a small force multiplier. Such gas prevents the ionization of the gas layer just outside of it that is not exposed to the strong ionizing radiation. The level of ionization in this component is low and

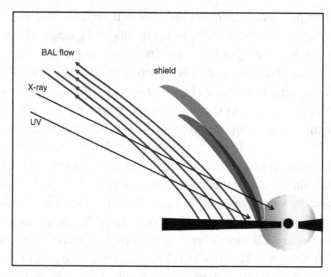

Figure 7.32. A thick shield between the inner disk and X-ray-emitting corona (spherical region around the BH) prevents most UV-ionizing photons from reaching the gas that is lifted from the disk farther away. As a result, the force multiplier in this gas is very large, allowing it to reach a terminal velocity of up to $0.1c$.

very different from the conditions in the shield. The nonionizing photons with $E <$ 13.6 eV can reach this component, absorbed by the numerous resonance transitions typical of low-ionization gas, and increase the force multiplier considerably. Such conditions can result in efficient acceleration and terminal velocities approaching $0.1c$.

The geometry presented in the diagram depicts only one possibility where both components are closely related to a disk-produced wind. There are other possibilities such as a set of concentric shells with small global covering factor but large covering in some preferred directions.

The terminal velocity of radiation-pressure-driven gas is of order of the escape velocity at the location where the flow is launched. For a $v = 0.1c$ outflow, the location is roughly at $100r_g$. For a standard α-disk around a massive BH, this is just outside the location where most of the ionizing radiation is emitted (Chapter 4).

Emission lines from BAL outflows

The observed BALs are the result of absorption of continuum photons by outflowing gas along the line of sight to a remote observer. The physics of this process was discussed in Chapter 5 under the subject "line fluorescence." The subsequent emission of the absorbed radiation contributes to the emission line as observed from other directions. In a complete spherical shell of outflowing material, the number of absorbed and emitted photons is the same (unless there is another escape route, i.e., another transition from the same upper level), thus the EWs of the absorption and emitted lines are the same. Such a configuration results in the canonical P-Cygni profile. In fact, detailed HIG observations suggest that the process contributes significantly to several strong X-ray emission lines.

Observations show that in many BAL AGNs, EW(absorption)>EW(emission), which is yet another indication (on top of the small covering factor deduced from flat-bottom lines that are not black) for an incomplete shell. However, the comparison of the two EWs cannot be used to estimate the covering factor since there must be other sources of line emission such as emission by gravitationally bound BLR clouds. The process must be considered when calculating photoionization models for the line-emitting gas since, in such cases, the contribution from line fluorescence can be significant.

7.4.2. Narrow absorption lines

Narrow absorption lines (NALs) are seen in about 50 percent of all nearby intermediate-luminosity AGNs. Much like the BALs, the strongest lines are due to the resonance transitions of C IV λ1549, Si IV λ1397, N V λ1240, and O VI λ1035. The typical velocities are several hundred to several thousand kilometers per second, and the typical widths are a few hundred kilometers per second. There is confusion between NALs and what was referred to earlier as "associated absorbers." If fact, there is no physical reason to differentiate between the two groups.

Many NAL systems show clear variations over time scales of weeks to months. Their density and level of ionization are known to a good accuracy from careful, time-dependent spectroscopic observations and from detailed photoionization modeling. Here, again, the column density and the covering factor play important roles in influencing the observed properties. The location of such systems is most likely in the inner NLR. In fact, in a handful of well-studied cases, the absorption features are identified with specific emission features of the lines in question.

Astronomers are fond of acronyms. Since BAL and NAL leave a small gap in velocity (and terminology), there are suggestions to add the name "mini-BAL" to characterize systems that are narrower than the typical BALs but broader than most NALs, of order 1000 km s^{-1}. Despite the continuous change of properties, it is not at all clear that BALs, mini-BALs, and NALs represent the same phenomenon.

7.5. The central torus

7.5.1. Obscuring structure in the inner 100 pc

Next we consider the region between 0.1 and about 10 pc from the central BH. Simple interpolation between the BLR and NLR properties suggests gas density of about 10^4–10^7 cm^{-3}; Keplerian velocities of order 1000 km s^{-1}, depending on the BH mass and sphere of influence; and a large range of possible column densities and levels of ionization. The entire region is beyond the dust sublimation radius and is likely to contain dust and molecular gas. The related opacity, at all wavelengths, except for the infrared even for small column densities, must be large, and hence the central source is likely to be obscured from some directions. Optical–UV observations show that in most local AGNs, the material in this region completely obscures the emission of the central accretion disk and the BLR. X-ray observations allow a good estimate of the obscuring columns in many of these sources. The range is very large, from about 10^{22} cm^{-2} to more than 10^{24} cm^{-2}.

The study of the 0.1–10 pc obscuring structure has become a major area of research, leading, among other things, to better understanding of the dichotomy between the two generic types of AGNs. The main assumption, which is supported by many observations (and which was discussed in Chapter 6 in connection with AGN unification) is that the main component in this region is a flat, thick structure made of a combination of dusty atomic and molecular gas. This structure assumes the shape of a torus, or a donut, with a small central opening and a much larger outer dimension.

There are two direct ways to measure the size of the torus. The first is the dust RM, which was explained in § 7.1.9. This is a measure of the lag between the V- and K-band light curves, which was measured in about a dozen sources. It can be interpreted as reverberation of the dusty, innermost part of the torus in response to the optical–UV continuum variations. The observations show that the lag corresponds to a size that is about $3R_{\mathrm{BLR}}(\mathrm{H}\beta)$. The measurements are shown in Figure 7.33.

The second is based on N-band (~ 12 μm) interferometry that is capable of resolving structures of less than 1 pc in nearby AGNs. Here, again, there are direct measurements in about 20 type-I and type-II AGNs, this time in the light of cooler dust. The measurements confirm a size that is about an order of magnitude larger than the size inferred from the V versus K reverberation, in rough agreement with model predictions. These results are also shown in Figure 7.33. All these measurements show sizes that are roughly proportional to $L_{\mathrm{bol}}^{1/2}$.

7.5.2. Dusty torus models

Several types of torus models have been constructed and studied in great detail. The first is a continuous density distribution structure (*continuous torus*) and the second a clumpy structure (*clumpy torus*). The two agree with some of the properties but disagree substantially with others. The result is a significant difference in the predicted emergent spectrum that can be put to an observational test.

The continuous, smooth distribution torus model assumes gas that extends all the way from the inner torus wall to very large distances. Since dust is present at all locations, the grain temperature is a monotonic decreasing function of the distance. The hottest dust is at the inner walls. Its emission can be directly observed

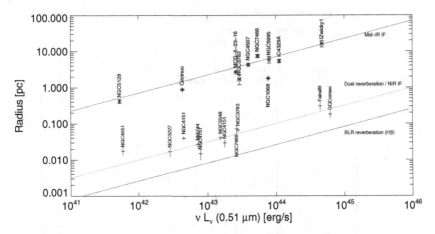

Figure 7.33. Two methods to measure the torus size in AGNs: (1) V versus K RM that measures the innermost part of the torus (middle correlation); (2) N-band interferometry (top line) that measures the region where most of the 12 μm flux is emitted. The RM results are represented by the lowest line. All sizes are roughly proportional to $L_{bol}^{1/2}$ (courtesy of L. Burtscher and the MIDI team).

in type-I sources and may explain their strong NIR emission. The temperature of the farther gas is very low because of the very low flux reaching these regions. Since type-II AGNs are seen only at large inclination angles, they are associated, according to this model, with colder dust emission. The model also suggests strong silicate absorption features in type-II AGNs. Type-I AGNs are predicted to show much hotter dust emission and strong silicate features in emission.

The NIR–MIR observations of type-I and type-II AGNs are not in accord with the prediction of the continuous torus model. They show a general similarity in the MIR SED of the two types and do not show strong silicate absorption features in type-II spectra. In fact, the observations show a gradual change of properties with a gradual change of MIR slope, from type-I AGNs to SF galaxies (Figure 7.34), with some indications that "pure type-II AGNs," those without broad emission lines, have a slightly different shaped MIR continuum.

The alternative clumpy torus model seems to be in better agreement with many observations. The main difference in this model is the removal of the 1:1 correlation between dust temperature and distance. Because of the clumpy structure, some clumps that are much farther away from the center compared with the torus inner wall are still exposed to the main source of radiation. The illuminated face of such clumps can attain high dust temperature and be a source of shorter wavelength MIR emission. The back sides of such clumps are also sources of dust emission, which, depending on the column density of the clump, can be similar or different from the illuminated side temperature. Both sides of the such clumps can be seen from various inclination angles, which explains the similarity between the MIR SED of

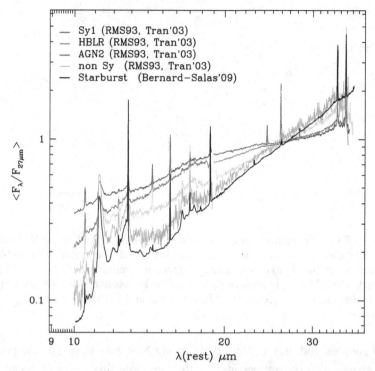

Figure 7.34. Composite MIR continuum for various types of low-redshift AGNs and SF galaxies (from Tommasin et al., 2010. Reproduced by permission of the AAS). The MIR slope changes gradually between type-I AGNs and SF galaxies with indications that the pure type-II sources, those without broad lines in reflection, have a somewhat different shape than other AGNs. (See color plate)

type-I and type-II sources. The important parameters of the model are the cloud column density and distribution in both directions, along the plane of the central disk and perpendicular to it. An additional interesting possibility is that under some (albeit extreme) conditions, the central source can be directly observed through the clumps even for very large inclinations where the torus is seen edge on.

Medium-resolution MIR spectroscopy by ISO and Spitzer enable a direct comparison between torus model predictions and AGN observations. One such example is shown in Figure 7.35. In this case, the predictions of the clumpy torus model are compared to the observations of more that a hundred type-I AGNs over a large luminosity range. The general conclusion of this work is that present-day (2011) clumpy torus models cannot explain the observed NIR–MIR SED. They give a reasonable fit to the 5–15 μm spectrum but fail to explain the NIR emission, at 1–5 μm, and predict too weak silicate emission features at rest wavelengths centered on 10 and 18 μm.

The strong NIR emission of type-I AGNs is a distinct spectral feature. The integrated 1–3 μm luminosity is roughly 20 percent of the bolometric luminosity

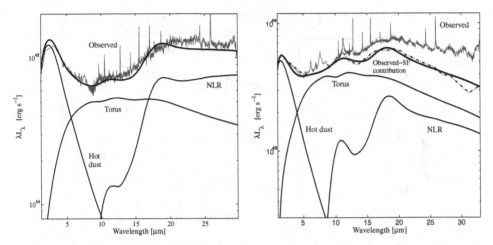

Figure 7.35. Three-component model fits to the spectra of Mrk 705 and PG 2349, both luminous type-I AGNs. The models includes hot (pure graphite) dust, clumpy torus, and NLR dust, as marked. A SF contribution was subtracted from the observed spectra prior to the fits. This contribution is much larger in the case of Mrk 705 (adapted from Mor and Netzer, 2012).

of many type-I sources. The inferred dust temperature is higher than 1500 K, the mean sublimation temperature of silicate grains, and pure graphite dust has been proposed to explain this emission (see Figure 7.36). The clumpy dust torus used in the modeling does not include such dust and hence cannot explain this emission. The too weak silicate emission features predicted by the model can be explained by noting that the observed spectrum comes from regions that cover a large part of the NLR in most of the sources. This suggests that NLR dust emission, like the one shown in Figure 7.22, must be included in the fitting. Indeed, the addition of such dust vastly improves the agreement with the model prediction.

Models like the one shown here can be used to estimate the relative contributions of the various dusty components to the overall AGN dust emission. A study of more than 100 sources suggests relative contributions of about 40:40:20 for hot graphite dust, clumpy torus with silicate dust, and NLR dust, respectively. The combined covering factor of all three components in low- to intermediate-luminosity AGNs is about 50 percent, that is, the integrated dust emission over the 1–40 μm is roughly 50 percent of L_{bol} for low-redshift intermediate-luminosity AGNs. In high-luminosity sources, that fraction can be much smaller with some indications that C_f(torus) is inversely correlated with L_{bol}.

7.5.3. The inner torus and the HIG connection

The inner part of the torus is directly exposed to the central radiation field. The temperature and ionization of the gas in this location depend on the ionization

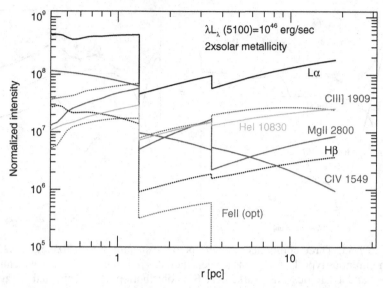

Figure 7.36. Emission-line emissivities in the outskirts of the BLR. The graphite sublimation radius in this case is at 1.3 pc and the silicate sublimation radius at 3.5 pc. The line emissivities drop dramatically beyond these radii due to the absorption of most ionizing photons by the dust (adapted from Mor and Netzer 2012).

parameter. There are two extreme possibilities with different observational implications. The first is the natural extension of the BLR. This involves a system of high-density ($\sim 10^8$–10^9 cm^{-3}), large-column-density clouds that are beyond the dust sublimation radius and therefore contain dust grains. The second is a much lower-density dusty gas with $N_H \sim 10^5$–10^7 cm^{-3}, much higher ionization parameter, and very high temperature. Obviously, the real situation may involve both. Nevertheless, the following discussion considers them as two different scenarios.

The physical conditions inside a dusty gas cloud depend on the gas density, the grain composition, the dust-to-gas ratio (related to the gas metallicity), and the ionization parameter. As explained in Chapter 5 and § 7.2, the latter determines the conditions under which the dust opacity dominates the ionization structure by absorbing most of the ionizing flux. Standard dust-free models of the BLR, like the ones discussed in § 7.1, can be extended to include the dusty gas. One such example is shown in Figure 7.36. The diagram shows emissivity as a function of distance for the strongest lines in the BLR spectrum. This is part of a BLR cloud model that is designed to mimic the observed spectrum, the line profiles, and line intensities for a typical high-luminosity type-I AGN.

In the case shown here, there is no outer limit to the extent of the structure, and the cloud distribution reaches well beyond the dust sublimation radius. The dusty gas has two distinct regions. The one closer to the center contains only pure graphite grains that can survive at higher temperatures, up to about 2000 K, depending on the grain size. Further out, the dust temperature drops below 1500 K, which allows silicate grains too. The model shows a very large drop in all line intensities at the first, graphite sublimation radius reflecting the fact that further from this location, most of the ionizing radiation is absorbed by the grains. Further out, there is an additional drop due to the appearance of silicate grains. Considering the total line intensities, there is very little emission beyond the first limit. The graphite limit is a factor of about three outside the BLR size inferred from Hβ RM (Equation 7.18). This behavior can naturally explain the RM size of the BLR and its dependence on the source luminosity.

Interestingly, the model predicts that the intensity of several low-ionization lines, in particular, Mg II $\lambda 2798$ and He I $\lambda 10830$, can increase significantly even beyond the graphite edge, which, in principle, can be tested by RM of these lines. The dependence of all the line intensities in this and similar models depends on the change of density with distance.

The second possibility is related to the presence of much lower density gas in the inner parts of the torus. Such gas may be part of the general intercloud (interclump) medium inside the torus or material that boils off the surface of the higher-density clouds. Its level of ionization depends on its density and distance. Under plausible conditions, it can be similar to the general level of ionization of the HIG, which suggests a general association of the torus and the HIG. We can estimate the properties of the gas using the previously defined $L_{44\,\text{oxygen}}$ and assuming a typical X-ray SED. This gives

$$U_{\text{oxygen}} \simeq 0.2 L_{44\,\text{oxygen}} r_{\text{pc}}^{-2} N_5^{-1}, \tag{7.38}$$

where $N_5 = N_H/10^5 \,\text{cm}^{-3}$ and r_{pc} is the distance in parsecs. Assume, for example, $L_{44\,\text{oxygen}} = 0.1$ (intermediate-luminosity AGN), $r_{\text{pc}} = 1$, and $N_5 = 0.1$; we get $U_{\text{oxygen}} = 0.2$. Such gas could fill a large volume and can produce strong high-ionization X-ray lines from the torus cavity and outside of it (the torus inner walls for such a source are at about 0.1 pc).

Because of the large column density, the torus walls (or the torus clumps in the clumpy torus model) can develop a highly ionized skin that is a very efficient X-ray reflectors. This will scatter the incident optical light and can be related to the observed broad emission lines seen in polarized light in many type-II sources. It can also reflect the X-ray continuum radiation, thus explaining the strong iron Kα in type-II AGNs. Given such conditions, the inner parts of the torus can become the reservoir for X-ray-driven winds seen in HIG outflows. Interestingly, part of this gas may have little angular momentum and thus a potential source of spherical accretion onto the BH.

7.5.4. The origin of the torus

Detailed studies of nearby AGNs in disky galaxies indicate that the orientation of the torus, which can be inferred from the NLR geometry, is similar to the larger-scale molecular disks observed in these sources. It is reasonable to assume that the origin of the torus material is gas inflow from the galaxy along with larger-scale flows in this general plane. While the details of the mechanisms causing the flow are not fully understood (see Chapter 9), the torus may well be part of the general flow that continues all the way to the central BH via the central accretion disk. In this scenario, the torus gas finds its way to the central accretion disk and, eventually, the central BH. It is interesting to note that the total mass in a clumpy torus can be derived from the model and is only a small fraction of the BH mass. Among the open questions in the model are the origin of the dust in the torus, the connection with stars and star clusters in this general region, and the frequency and magnitude of SF in the torus.

An alternative scenario relates the inner accretion disk and the central torus to large-scale winds that originate from a massive central disk, which extends well beyond the central accretion disk in the same general plane. While the origin of such winds is not clear, it is speculated that they may be related to large-scale magnetic fields and may be driven by radiation pressure force. According to some models, the wind is clumpy and is the origin of the clumpy, dust-free BLR clouds and the dusty clumps in the torus. In the wind scenario, the BH is fed by the central disk and the torus, and BLR material is outflying from the center.

Finally, all these scenarios do not exclude somewhat different geometrical shapes. For example, the central torus may be replaced by a similar size structure, for example, a warped disk that joins the main accretion flow a few parsecs from the BH. The warped structure provides the needed covering factor and the observed column density. It may be related to large-scale inflow that, at large distances, is not along the plane of the disk. All these scenarios can be tested by observations: the covering factor by the relative number of type-I and type-II sources, the column density by X-ray observations, and the overall structure by fits to the NIR–MIR spectrum.

7.5.5. AGNs without a torus?

While AGN tori became a cornerstone of our understanding of AGN unification, there is evidence that a certain fraction of high-luminosity objects, and a large fraction of low-luminosity AGNs, do not have such a structure.

The evidence regarding the high-luminosity AGNs comes from large NIR surveys like the one conducted by the IR satellite WISE. The surveys cover a large wavelength band and allow the observation of hot dust in AGNs up to a redshift of ∼2. The WISE survey overlaps with the SDSS and the sensitivity is high enough to look for such dust in a large fraction of luminous SDSS type-I sources. A surprising result that came out from the observations is that ∼15 percent of the sources do not show the signature of hot dust, while others of similar luminosity, and similar emission-line properties, show it very clearly. Several ideas were proposed to explain the lack of dust emission, including lack of central torus or a geometrically thick central disk, with a central funnel, which hides the torus from the direct view of the central radiation source. A thick accretion disk with a funnel is a natural consequence of very fast accretion onto the BH (Chapter 4). It may also be related to radio jets as some models require such structure to efficiently collimate the jet.

Many LINERs that are at the other extreme of the luminosity scale, with L_{bol} as small as 10^{42} erg s^{-1}, also do not show the NIR–MIR emission signature of a torus. Part of this may be due to the difficulty in detecting this emission because of strong stellar emission at NIR wavelengths and SF emission at MIR wavelengths that outshine the torus emission. Another possibility is that the entire torus structure disappear at very low luminosity.

7.6. The megamaser molecular disk

Water maser emission has been detected, by radio observations, at a wavelength of 1.35 cm, in a number of nearby type-II AGNs. Such sources are referred to as megamasers because of their very high power compared with galactic masers. The measurements are unique because of the extremely high precision in determining the velocity and width of the megamaser lines, which are extremely narrow due to their intrinsic monochromaticity.

AGN megamasers are localized regions in a large molecular disk at typical distances of about 0.1 pc from the central BH. Like all masers, their amplification factor varies exponentially with the path length, and the emission is seen from those regions where the path has large velocity coherence. For an edge-on thin molecular disk around a central radio source, this occurs at three points: directly in front of a radio continuum source, where the velocity spread is very small and the maser amplifies the central source emission, and at the two tangent points on the disk where the radial velocity is close to its maximum and the projected velocity gradient across the disk is a minimum allowing self-amplification. In a disk made of concentric rings, and a velocity field that is entirely dominated by a central BH, we expect several tangent points on each side of the center, where the Keplerian velocities for the rings are maximum. Megamasers that lie in the same plane provide the most accurate way to determine the differential motion in the disk and hence the BH mass.

Of the discovered megamasers in AGNs, the most accurate measurements, so far, are the ones obtained for NGC 4258. This object is a type-II LINER possessing

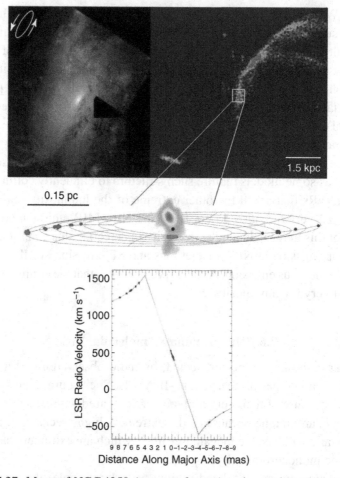

Figure 7.37. Maps of NGC 4258 (two top frames) and an enlarged view of the center source with a wrapped disk around a central 22 GHz radio source and several maser sources in the disk. The bottom part shows the velocity curve of the maser sources and a fit to a Keplerian disk (by permission of NRAO). (See color plate)

a highly obscured central X-ray source with a column density of $\sim 7 \times 10^{22}$ cm^{-2} and $L(2\text{--}10$ keV$) \simeq 4 \times 10^{40}$ erg s^{-1}. The hidden nuclear continuum and several NLR lines are seen in reflected, polarized optical light. The galaxy harbors nuclear megamasers that are situated in a nearly edge-on, extremely thin, slightly warped Keplerian disk (see Figure 7.37). The masers extend from 0.13 to 0.26 pc, and the Keplerian rotation curve indicates a central BH mass (assuming no other mass within 0.13 pc), with mass of $\sim 3.9 \times 10^7 M_\odot$. Interestingly, the velocities in this source are so accurately measured that the centripetal acceleration is also known, which allows an independent measurement of the maser–BH distance and hence an independent measurement of the Hubble constant.

The VLBA observations of NGC 4258 reveal a sub-parsec-scale jet oriented along the disk axis. The velocity centroid of the disk agrees well with the systemic velocity of the galaxy, and the rotation axis of the disk is almost perfectly aligned with the inner portion of large-scale twisted jets.

The dimensions of the molecular disk in NGC 4258 correspond to about 10^5 gravitational radii, outside the typical dimension of a standard α-disk. They correspond, much better, to the expected dimension of the dusty torus and may well be part of the same structure. However, the thickness of the maser disk must be much smaller than the expected thickness of the torus and is estimated to be few $\times 10^{14}$ cm, that is, $H/r \sim$ few $\times 10^{-4}$.

The central accretion disk in NGC 4258 must have very low accretion rate, as inferred from the X-ray flux, much like in other LINERs. As explained in Chapter 4, such disks are likely to be advection dominated and are classified as RIAFs. It is not clear whether such a two-temperature RIAF goes all the way from the center to the region of the masers or, perhaps, there are two distinct coaligned structures.

Measuring M_{BH} using megamasers is the most accurate direct measure we have so far in AGNs. The radio observations allow much better determination of the orbital velocity, and the BH–maser distance can be measured to a great precession. Also, the megamasers are well within the BH sphere of influence, thus the stellar mass is not an important contribution to the measured mass. Megamasers have been detected in more than a hundred sources. Only about a dozen of those are good enough to attempt real M_{BH} estimates. Except for NGC 4258, all other suitable velocity curves are more complex. In some sources, the agreement with Keplerian motion is not very good, and there are cases that are probably affected by molecular outflow and by the self-gravity of a central massive disks.

While all megamasers are in type-II high-ionization AGNs, or type-II LINERs, there is direct evidence, from polarization measurements, for hidden BLRs in some of the sources, for example, NGC 1068, NGC 4388, and the Circinus galaxy. In such sources, we can measure FWHM($H\beta$) from the reflected light and estimate the velocity in the BLR. Moreover, these are all strong X-ray sources that allow an estimate of the BLR size (§ 7.1). We can thus make a direct comparison between the two estimates of M_{BH}: the megamaser method and the RM method. The comparison shows good agreement with a larger uncertainty on the RM-based estimates. This provides another confirmation that the RM method is, indeed, a good way to measure M_{BH}.

Megamasers in AGNs, at least the ones showing BLR lines and nonstellar continua in reflection, are likely to contain both an inner standard thin accretion disk and an outer cold molecular disk. The two are not necessarily part of the same structure and may well be completely detached. Moreover, there is evidence that the mean obscuring column of megamaser tori is larger than the mean column density of type-II AGNs. In fact, a very large fraction of AGNs with detected megamasers are Compton thick.

7.7. Stars and starburst regions

7.7.1. Stars as emission-line clouds

Like any gas cloud, stars in the (hypothetical) nuclear cluster absorb and reprocess the incident AGN radiation and are hence potential sources of emission lines. Main sequence stars in this location are not likely to show any spectral signature because of their high-density atmospheres and their small cross section (i.e., small covering factor). Emission from such stars is dominated by their stellar properties. However, some stars may develop extended winds, or envelopes, as a result of the interaction of their atmosphere with the hard, external radiation field. This will lower the density and increase the cross section for absorbing the ionizing radiation. Such extended envelope stars have been named *bloated stars*. It has been speculated that star–star collisions may also produce bloated stars.

Bloated stars' extended envelopes are similar, in many ways, to the low-density gas discussed so far. Close to the center, they will show a typical BLR spectrum and, further away, a typical NLR spectrum. Such envelopes or winds are likely to have large density gradients that, far enough from the star, could have the properties of the HIG. Thus, the presence of giant envelope stars can bridge the gap between several of the observed components of AGNs with different densities. They can also provide another reservoir, or source, for the emission-line gas.

Photoionization models of bloated star envelopes result in emission-line spectra that are not very different from the one produced by other cloud models. This is not surprising since the general properties like density, column density, and covering factor are very similar, and so is the assumed location. There are two other properties that require more investigation. First, giant stars in a nuclear cluster must be accompanied by a much larger number of stars that are less massive but determine (depending on the exact IMF and cluster evolution, which may be different in this environment) the global mass. The conditions that the cluster mass is smaller than the BH mass results in a small number of giant stars of the order of few $\times\ 10^5$ near a BH of $10^8\ M_\odot$. This small number raises again the issue of the very smooth observed line profiles that suggest a very large number of emitting clouds. There is also the issue of the frequent giant star collision that is predicted in such cases and the lack of empirical evidence for such events.

Another interesting prediction is that bloated stars exposed to the strong central radiation field and moving through any other medium will develop extended, lower-density cometary tails. The direction of the tail depends on the direction of motion, and its large dimensions can make it the main source of line emission. Detailed models of such tails are not yet available. However, new results concerning the hard

X-ray variations in several low-luminosity AGNs suggest the occasional occultation of the central source by clouds that are consistent, in terms of general dimension, column density, velocity, and structure, with such hypothetical cometary tails.

7.7.2. Starburst regions in AGNs

Starburst regions in AGN host galaxies are the topic of § 8.5. These are different from other AGN components because of their dimension, their location, and, most important, their main source of energy. The size of such regions in large SF galaxies can approach several kpc. As discussed later, such regions are clearly observed in many AGN hosts at low and high redshift. Here we address the high-energy spectrum of the gas in SF regions because some of it is occasionally confused with AGN excited gas. In Chapter 8, we give more information about the general physics and the longer wavelength emission.

The main ionization source for the gas in SF HII regions is the radiation of young massive stars. The main sources of mechanical energy are supernovae explosions and fast stellar winds. The most active phase of a short-duration SB lasts only 10–20 Myr. Continuous SB events can last much longer – 100 Myr and more. The observed optical signatures of SF regions are strong recombination and forbidden emission lines. Diagnostic diagrams like the ones presented in Chapter 6 are very useful in identifying such regions by ways of line ratios like [N II] $\lambda6584$/Hα and [O I] $\lambda6300$/Hα that are significantly different from those in AGN ionized gas. The typical X-ray signature of the ionized SF gas is that of a multitemperature hot plasma that contains strong collisionally excited emission lines.

High-resolution X-ray observations of AGNs show spectra that are very different from the X-ray spectra of typical starburst galaxies. An example is given in Figure 7.38, which compares the X-ray spectrum of a starburst galaxy, M 82, with the X-ray spectrum of the "classical" type-II AGN, NGC 1068. The spectra are very different and indicate different excitation mechanisms. The AGN gas is much cooler, probably less than 10^5 K. It is excited by the hard X-ray continuum, and the most intense lines are due to recombination, mostly of H-like and He-like species (e.g., O^{+7}). The starburst gas is much hotter, with line ratios that indicate a temperature of about 10^7 K. Such temperatures are typical of collisionally ionized gas whose source of energy is SN and stellar wind-driven shocks.

Hot starburst gas in AGNs, when present, is not necessarily expected to show the spectrum of a "pure starburst" gas. The reason is the strong radiation field of the AGN that can affect the hot starburst gas even at large distances. In particular, the hard X-ray AGN radiation can penetrate to large columns, even to the inner part of large molecular clouds, where new stars are born. Depending on the gas density and continuum flux, the penetrating radiation can provide significant additional ionization to the starburst gas, thus changing its ionization and emergent spectra.

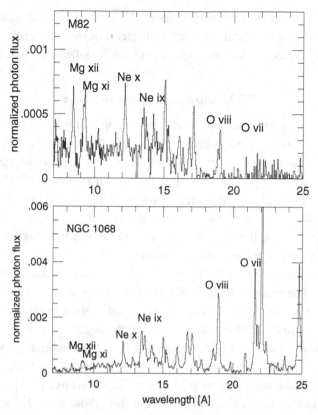

Figure 7.38. A comparison of X-ray photoionized gas (NGC 1068, a type-II AGN with an extended X-ray nebulosity) and X-ray collisional gas (M82, a "classical" starburst galaxy). Both spectra were obtained by the RGS on board XMM-Newton and were retrieved from the XMM archive.

Such starburst regions will show a mixed-type spectrum, which can differ from the spectra of starburst regions that are not exposed to active nucleus radiation.

An interesting example is NGC 6240. The galaxy belongs to the subgroup of ultraluminous infrared galaxies (ULIRGs), which are described in § 8.5. Such systems contain two or more interacting galaxies that are observed several hundred million years before the final stage of the merger event. Such interactions are associated with intense SB activity. Later on in the process, one or more of the participating massive BHs can start accreting, which will show as a young AGN. In the case of NGC 6240, the two merging BHs are active, which results in a double X-ray source. The separation of the two nuclei is about 1.4 kpc, and the entire region surrounding the merger is full of IR bright SB regions. The MIR spectrum of NGC 6240 shows very weak high-ionization, high-excitation emission lines that are typical of AGNs. However, the X-ray continuum is very similar to many type-II spectra with a large

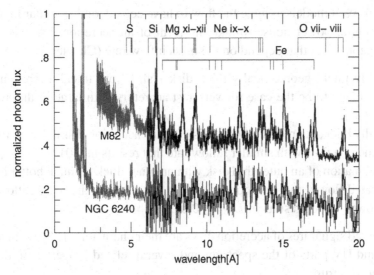

Figure 7.39. A comparison of the X-ray spectra of the SB galaxy M 82 and the double-nucleus AGN ULIRG NGC 6240 (adapted from Netzer et al., 2005). The similar soft X-ray emission-line spectra of the two sources are typical of collisionally ionized plasma. The only exception is the strong Kα line at around 1.9 Å, which is seen only in the spectrum of the AGN.

obscuring column ($\sim 10^{24}$ cm^{-2}) and strong hard X-ray continua. Judging by the hard X-ray spectrum, at energies above about 3 or 4 keV, this is a clear type-II AGN.

Figure 7.39 compares the X-ray spectra of NGC 6240 and M 82. Unlike the previous example, and despite the huge difference in luminosity (the 0.5–5 keV luminosity of NGC 6240 is some 50 times larger than in M 82), the two spectra are very similar, in their soft X-ray emission indicating that most of the X-ray lines at those energies are due to SB excitation. The star-forming regions in this galaxy are probably shielded from the central X-ray source. The X-ray spectrum also shows a strong iron Kα line. This is probably from the innermost "real AGN" region and is not associated with the SB region. NGC 6240 is thus a clear example of a combined AGN–SB system.

7.8. The central accretion disk

Most of the radiation emitted by the central power house is due to accretion flow onto the central massive BHs. The loss of gravitational energy by the inward-spiraling gas is converted to electromagnetic radiation or kinetic energy of a wind. The central power house, or a large part of it, must be a compact high-density, large-column-density structure with typical dimensions of $\sim 1000 r_g$. The more successful AGN models assume the flow to have considerable angular momentum and the central power house to take the form of an accretion disk with various possible geometries that depend on the mass inflow rate (Chapter 4):

1. Optically thick, geometrically thick disk ("slim" or "thick" accretion disks): This is likely to be the case for very fast accretors with large $\dot{m}$ that results in $0.5 \leq L_{Edd}$.
2. Optically thick, geometrically thin disk ("thin" accretion disk): This is the likely outcome of a flow with smaller inflow rate that results in $0.01 \leq L/L_{Edd} \leq 0.5$.
3. A combination of an outer thin disk with an inner thicker, much hotter advected dominated flow ("two-component disk"): In this case, the mass inflow rate is even lower, which results in $L_{Edd} \leq 0.01$.

The observed signatures of accretion disks are their characteristic SEDs in the NIR, optical, and UV parts of the spectrum and several related properties in the X-ray part of the spectrum.

7.8.1. *The optical–UV continuum of accretion disks*

The best-understood disks are thin disks with or without X-ray emitting coronae. The majority of intermediate- and high-luminosity AGNs found in large surveys are thought to be powered by such disks.

Thin-disk theory suggests that the SED of such systems contains a broad wavelength band where the spectral slope α ($L_\nu \propto \nu^\alpha$) is in the range 0–0.5. The "classical" thin-disk slope is $\alpha = 1/3$ (Chapter 4). X-ray emission from the hot corona and X-ray reflection from the surface of the disk are additional important characteristics of such systems. Much observational effort has been devoted to the measurement of α and the characterization of the part of the continuum showing this slope. This part of the SED is occasionally referred to as the *big blue bump*. For $M_{BH} = 10^9 M_\odot$ and $L/L_{Edd} = 0.1$, the peak of this emission is predicted to be around 1000 Å.

Figure 7.40 shows an attempt to look for the signature of the big blue bump by combining the spectra of several hundred type-I AGNs with $L/L_{Edd} \sim 0.1$. The experimental 1–20 eV part was supplemented by a typical X-ray power-law continuum observed in such objects. This observed spectrum is rather different from the schematic disk spectrum of Figure 4.3 and also from the more realistic disk spectrum of Figure 4.5. In particular, in no part of the spectrum is $\alpha > 0$.

Despite the preceding, there are good reasons to believe that the optical–UV continuum of AGNs is dominated by thin accretion disk emission. First, as explained in Chapter 4, even the most sophisticated calculated spectra are rather limited, and several models show spectra that differ considerably from the schematic $\nu^{1/3}$ dependence. Second, the infrared ($\lambda \geq 1~\mu$m) part of the SED is dominated by

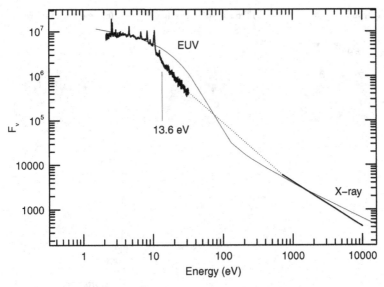

Figure 7.40. Observed SED (thick solid lines connected with a thin dotted line) compared with a theoretical model made of the spectrum of a thin accretion disk around a $3 \times 10^9 \, M_\odot$ BH with $L/L_{Edd} = 0.1$ and an additional X-ray power-law component. Note the strong blue bump (energy excess just beyond the Lyman limit at 13.6 eV) predicted by this model compared with the sharp drop in the observed spectrum.

nondisk emission, in particular, stellar emission by the host galaxy and thermal emission by warm dust, presumably in the central torus. In radio-loud AGNs, non-thermal emission can also contribute at this and even shorter wavelength bands. This obscures the part of the spectrum where the standard thin-disk theory predicts $\alpha = 1/3$. Having said all that, we must also consider the (yet hypothetical) possibility that additional processes, related perhaps to disk winds, are taking place and changing the observed SED.

7.8.2. The polarized signature of AGN disks

Thin-disk calculations suggest linear polarization of the emitted flux at a level of a few percent. Conversely, the emission from the other strong spectral components in AGNs, like ionized gas and warm dust, is mostly unpolarized. Spectropolarimetry is a well-developed technique that is already providing clues about the nature of hidden AGNs (Chapter 6). Can we use this technique to search for the polarized radiation from accretion disk atmospheres?

The various types of AGNs are associated with various viewing angles of the innermost part and, therefore, different location of the scattering region (the "mirror") and different polarization angles. In type-II AGNs, the ionization cone and the radio jet (if any) are at large angle to our line of sight. In this case, there

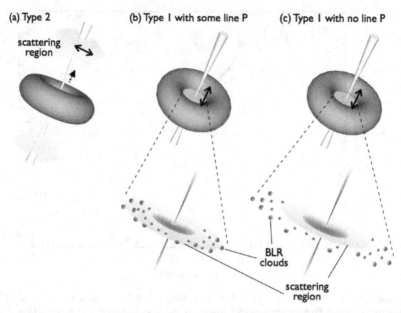

Figure 7.41. Schematics for the geometry of dominant scattering regions in AGNs. (a) Type-II AGN with a scattering region along the jet direction and outside the torus. (b) Type-I AGN with some broad-line polarization indicating a scattering region that is of the same dimension and at about the same location as the BLR. (c) Type-I AGN with no broad-line polarization. In this case, the scattering region is likely to be inside the BLR, where we expect to have an accretion disk. In each panel, the double arrow shows the position angle of continuum polarization (courtesy of M. Kishimoto).

is usually a single scattering region, which is suspected to be located along the direction of the radio jet, outside the opening of the torus. This position will result in a polarization position angle, which is perpendicular to the direction of the jet (parallel to the central disk). In type-I sources, where there is a direct view of the center, there are two likely scattering regions. One is cospatial with the BLR and results in polarization of the broad emission lines. Such a polarization, at a level of ~ 1 percent, is observed in many sources. The second is inside the BLR, where the thin accretion disk is expected to be. In this case, the broad-line flux will not be polarized, but the disk flux will be. In both cases, the polarization angle is roughly perpendicular to the plane of the disk. Figure 7.41 illustrates the three possibilities.

Figure 7.42 shows the directly observed and polarized spectra of six luminous type-I AGNs. The polarized spectra are normalized, separately, at 1 μm and show the signature of a thin accretion disk, in particular, the $\nu^{1/3}$ dependence at long wavelengths. This part of the spectrum is dominated by emission from the low temperature of the disk and, according to this interpretation, contains contributions from other sources, for example, hot dust just outside the BLR. The polarized fluxes

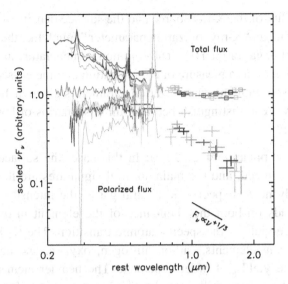

Figure 7.42. The polarized (bottom part) and unpolarized spectra of four type-I AGNs showing the "bare accretion disk" spectrum in polarized light (courtesy of M. Kishimoto). (See color plate)

show some hints for Balmer absorption edges, at 3646 Å, as predicted by some disk models.

To summarize, spectropolarimetry of several nearby type-I AGNs shows a $L_\nu \propto \nu^{1/3}$ spectrum indicative of optically thick, geometrically thin accretion disks. Adopting this interpretation, we reach the conclusion that the directly observed SED is dominated by nondisk emission at all wavelengths greater than about 1 μm. The short-wavelength part is more luminous but, in this range, the disk SED is already very different from the canonical $\nu^{1/3}$ shape.

7.8.3. The X-ray signature of AGN disks

Irradiated X-ray-emitting disks

Thin accretion disks that are irradiated by an external X-ray source provide other characteristic disk signatures. There are various possibilities depending on the disk properties and the location and luminosity of the external X-ray source (for general considerations regarding such cases, see § 4.2.2). The hot disk corona is the most likely origin of the observed X-ray power law. Observations show that the typical spread in X-ray slope is not large, with a notable exception of NLS1 galaxies, which show much steeper soft X-ray continua. The radiation extends over several orders of magnitude in energy, from about 0.2 keV to about 100 keV. Here we address in more detail the physical conditions that are thought to govern the main observed X-ray emission features.

The incident flux of the X-ray source and the surface density of the disk can be combined to define an "X-ray ionization parameter," much like the one used in the analysis of the BLR gas or the HIG. The common nomenclature in X-ray studies is to use the parameter ξ as a measure of the ionization parameter (see Equation 5.49 and the comparison with various ionization parameters given in Table 5.2). Using this parameter, we can distinguish between three scenarios of low, intermediate, and very large ξ:

Low-ionization parameter, $\xi \lesssim 100$: In this case, the surface of the disk is not highly ionized, and the main spectral signatures are fluorescence lines of relatively neutral species. In neutral gases, the intensity of fluorescence lines depends on both the abundance of the element in question and the fluorescence yield of the specific atomic transition. The K-shell lines of the most abundant elements, carbon, nitrogen, oxygen, and neon, have small fluorescence yields, of order 1 percent. The heavier elements have larger yields for the same transitions, with iron having a yield of about 30 percent. Under the conditions assumed here, the strongest expected lines are due to FeI-XVII, all at around 6.4 keV.

The incident X-ray power law results also in strong reflection from the disk surface, in particular, a strong "Compton hump" centered at around 30 keV.

High-ionization parameter, $\xi \sim 500$: In this case, the "skin" of the disk is almost entirely ionized, and the only elements that did not lose all their electrons are heavy elements like iron and nickel. The strongest predicted lines are the leading resonance transitions in the H-like and He-like sequences of magnesium, silicon, sulfur, argon, and iron. The strongest iron lines are Fe xxv and Fe xxvi, at around 6.67 and 6.97 keV, respectively.

Reflection off a fully ionized disk surface is much more efficient at lower energies, 0.1–10 keV, compared to the first case of a more neutral surface. The Compton hump is still strong but weaker relative to reflection at lower energies.

Extreme conditions, $\xi \gtrsim 3000$: Here the disk is too highly ionized to show significant emission in any line. The reflected spectrum is almost a perfect mirror image of the incident X-ray spectrum, except for a weak high-energy bump.

Figure 7.43 shows several calculated disk spectra covering the entire range, from almost completely neutral to very highly ionized disk surface. The X-ray energy slope is 1 (photon index of 2), and the ionization parameters are marked. Note the changes in the emitting species, the strong soft X-ray excess due to the conglomeration of strong emission lines at low X-ray energies, and the strong Compton hump at around 30 keV.

X-ray emission lines from the surface of an illuminated disk can form at various distances from its center. Far from the BH, GR effects are negligible, and the line

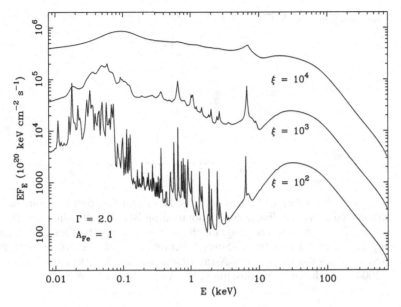

Figure 7.43. Theoretical ionized disk spectra for various ionization parameters. The disk is assumed to have solar abundance and constant gas density. The values of the ionization parameters, ξ, are marked along the curves. The strong hump at around 30 keV is due to the reflection of the incident X-ray power law (from Ross and Fabian, 2005; reproduced by permission of John Wiley & Sons Ltd.).

width will reflect the Keplerian velocity at this location. Closer in, GR effects become important, and both Doppler shift and gravitational redshift are important. The closer the emission is to the ISCO, the larger are the shift and distortion of the line profile. In particular, higher spin is associated with a closer ISCO and more extended "red" wings. Another factor that affects the line profile is the disk inclination to the line of sight. The inclination determines the Doppler shift, and thus, highly inclined disks show larger blue shift (stronger high-energy wing). Finally, the observed profile depends on the line emissivity pattern across the disk since every ring produces a somewhat different (local) profile. In many disk models, the assumed line emissivity pattern is modeled by a simple power law where the emitted line flux per unit area at an angle i and distance r from the BH is given by

$$I(i, r) = f(i)r^{-\beta}. \tag{7.39}$$

The dependence of the disk line profiles on the emissivity index β is such that larger β results in a smaller emission region closer to the ISCO. This increases the importance of the GR effects and results in broader lines that extends further into lower energies.

Blurred reflection

Present-day X-ray modeling combines the calculation of line emission from the surface of the disk and the various relativistic effects into a full spectral model.

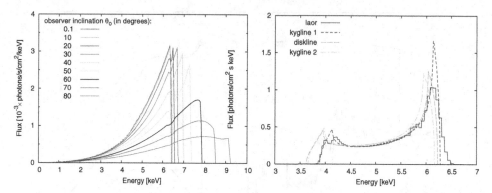

Figure 7.44. (left) Relativistic "neutral" (6.4 keV) iron line profiles from an X-ray-illuminated accretion disk around a nonrotating BH. The illumination pattern (Equation 7.39) is $\beta = 3$. The diagram shows the observed profiles at different viewing angles, as marked. (right) Several theoretical relativistic $K\alpha$ profiles that are used to model the spectrum of X-ray-illuminated AGN disks (from Dovčiak et al., 2004; reproduced by permission of the AAS). (See color plate)

All lines in such models, and all absorption edges, will be smoothed to some degree by the relativistic corrections. The level of smoothing depends on the location and is, in principle, different for every line and every edge. Since the X-ray jargon refers to the emission from the surface of the disk as "reflection," such models go under the name of *blurred reflection*.

A clear prediction of blurred reflection models is a correlation between the variations of the X-ray continuum source and the resulting emission from the surface. Such correlations have never been clearly observed, despite the large amplitude continuum variations typical of AGN disks. Another problem, or perhaps unknown, is the efficiency of reflection. For a simple flat disk and a central point X-ray source with a direct line of sight to the observer, the reflection cannot exceed 100 percent or a solid angle of 2π. However, several individual sources have very hard spectra, indicating solid angles much greater than 2π. There are several possible explanations that involve curved or twisted geometry of the disk and obscuration of the central source but clear line of sight to the reflector.

Relativistic line profiles

For low- to medium-ionization parameters, the strongest iron lines in X-ray-illuminated disk are due to Fe I–Fe XVII all very close to 6.4 keV. Figure 7.44 shows the results of modeling the 6.4 keV feature under a variety of conditions. The diagram on the left shows a case of a disk with spin parameter $a = 0$ and various inclinations to the line of sight, from pole on ($\theta_0 = 0.1°$) to almost edge on ($\theta_0 = 80°$). The emissivity pattern in this case assumes $\beta = 3$. All profiles extend to very low energies due to emission from regions very close to the ISCO. The increased inclination results in larger Doppler shift and the "bluest," pole-on spectra show emission at 9 keV and even beyond. The diagram on the right shows various Fe-Kα lines that are used to model real X-ray observations of AGNs. The profiles show a distinct double-horn shape with a steep rising high-energy wing that extends only slightly beyond 6.4 keV and a more extended red wing.

Decades of study resulted in a large number of detailed, high-quality observation of X-ray emission lines. This enables good statistics of line strength (EW) and variability. In particular, there are several observations of iron emission lines in a number of nearby type-I AGNs that are detailed enough to identify the lines and measure their profiles at great accuracy. Most of the X-ray lines in low- and intermediate-luminosity type-I AGNs were identified as low-ionization iron Kα lines with a peak intensity at around 6.4 keV. The lines are relatively weak, with typical EWs of 30–100 eV. A substantial number of sources do not show such a line, which suggests even smaller EWs. The observed EWs provide important information about the central disk. First, a small fraction of the line intensity must be the result of absorption followed by fluorescence emission in large-column-density BLR clouds. Simple calculations show that this can amount to line EW of 10–20 eV, depending on the column density and iron abundance. The remaining emission must come from the disk or its immediate vicinity. This has led to a big theoretical and experimental effort to detect the lines and to measure their profiles in attempt to discover the unique relativistic signature from the inner part of the disk.

Measuring the detailed profile of a very broad, very weak disk line is very challenging. The measurements use sensitive but low-resolution CCD detectors with a typical spectral resolution of $E/\Delta E \sim 50$ at around 6.4 keV. The energy resolution is not the limiting factor since the lines are predicted to be extremely broad. In some cases, for example, the spectrum of the NLS1 MCG-6-30-15, there are claims that an extremely broad, variable iron line has been detected with a profile that agrees with the prediction of the relativistic disk model. A fit of the observed spectrum of this source is shown in Figure 7.45. Indeed, the low-energy wing of the line seems to extend all the way down to 2 keV and even below. A major limitation in fitting theoretical disk line models to such observations is the exact placement of the underlying continuum since the total observed energy band is not much broader than the predicted width of the Kα line.

Attempts to fit many other X-ray spectra, using similar theoretical models, are controversial. According to some work, broad relativistic iron X-ray lines are very common in nearby AGNs and show in perhaps 30–50 percent of all sources. According to others, there are very few, if any, nearby AGNs where relativistic disk lines have been positively identified. The situation regarding higher-luminosity sources is even more problematic since in many of these cases, the S/N is very low and does not allow a meaningful fit to the data.

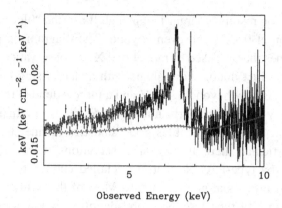

Figure 7.45. A fit to the XMM-Newton observation of MCG-6-30-15 using a relativistic disk line profile with $a = 0.99$, $\beta = 3$ and an inclination of $40°$ (from Dovčiak et al., 2004; reproduced by permission of the AAS).

An interesting, not yet fully understood point is that the handful of sources showing the most convincing relativistic iron lines, for example, MCG-6-30-15 belong to the subgroup of NLS1s discussed in Chapter 6. These are nearby, low-luminosity type-I AGNs with relatively small BH masses ($M_{BH} \simeq 10^7 M_{\odot}$) and very high accretion rates ($L/L_{Edd} \simeq 1$). Such sources are known to show very steep soft X-ray spectra with very large amplitude, fast X-ray variations. The high accretion rates make them unique and may be related to the spin of the BH. Conversely, these are also the cases where the thin-disk theory breaks down since the disks become slim or thick. The energy transport in such disks can be dominated by advection, which complicates the mass and accretion rate estimates as well as the assumed disk geometry.

Alternative models

Are such relativistic line models the only way to explain the observed X-ray features? Given the large number of model parameters, the poor spectral resolution, and the limited energy band, the answer is not very clear. At least one other type of model, which requires no iron line emission from the surface of the disk, does equally well in explaining the X-ray observations of many AGNs, including the more extreme cases discussed here. The model combines highly ionized disks (i.e., no strong disk spectral features) and several systems of large-column-density clouds that partially obscure the primary power-law sources and the reflected radiation. The model explains the overall X-ray SED and the source variability, which is mostly due to the motion of the various obscurers across the central source. The main components of one such model constructed to explain the spectrum of MCG-6-30-15 are shown in Figure 7.46.

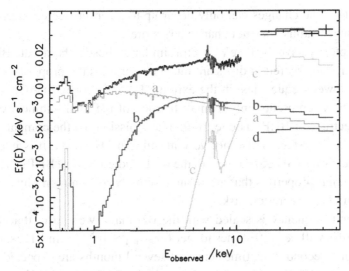

Figure 7.46. A fit to the Suzaku observations of MCG-6-30-15 that requires no relativistic Kα line (from Miller et al., 2008; reproduced by permission of A&A). The model includes several absorbers covering the (a) primary power-law continuum, (b) the partially covered power law, and (c) the reflected continuum. The fit to the data is the upper curve. The fit quality is similar to the fit by relativistic Kα lines and requires no relativistic effects.

7.8.4. Analogy with black holes in X-ray binary systems

Analogy and similarity between massive BHs and galactic black holes in X-ray binaries (black hole binary systems (BHBs)) suggests a new way to look at X-ray variability, BH mass, BH accretion rate, and the properties of the central disk.

BHBs are found in three states: soft state, hard state, and very hard state. This terminology refers to the shape of the X-ray spectrum and the 2–10 keV luminosity. In the hard state, the X-ray luminosity is low, and in the soft state, the X-ray luminosity is high. In the very high state, the luminosity is the highest, and the spectral shape is intermediate between hard and soft. The changes in luminosity indicate a rise in accretion rate as we go from hard to soft to very hard states.

The much shorter time-scale variations observed in BHBs enable a more detailed analysis of their power spectral density (PSD). Typically, the PSD in the soft state can be described as a bending power law, $P(\nu) \propto \nu^{-\alpha}$, where at high frequencies, $\alpha > 2$. Beyond a characteristic break frequency, ν_B, the shape changes to $\alpha \simeq 1$. It seems that the corresponding break time, T_B, in such systems is proportional to the BH mass and inversely proportional to the the normalized accretion rate, $\dot{M}/\dot{M}_{\mathrm{Edd}}$. One suggestion for this behavior is that larger accretion rate leads to smaller inner disk radius and hence a change in the bend frequency of the PSD, which reflects the inner radius of the disk. This proposed change is not related to the BH spin, only to the accretion rate. Indeed, spectral fittings to disk spectra in BHBs are consistent with this idea. The changes can only occur up to a certain accretion rate, beyond which the inner radius does not change any more.

AGNs typically have 2–10 keV spectra similar to those of hard state BHBs ($\Gamma \sim$ 2). However, their accretion disks are much cooler, thus the proper comparison is with much lower frequencies, in the extreme UV. Moreover, in AGNs, the 2–10 keV luminosity represents only a small fraction of the total luminosity. Much of the soft spectrum of BHBs is due to thermal emission, so the comparison of the spectra in the soft state is also not very meaningful. However, it is interesting to compare the X-ray PSD of AGNs with the well-studied PSD in BHBs in an attempt to derive similar properties that are related to the basic common physical process of accretion via an accretion disk.

If the bend frequency is scaled with the BH mass, we expect that the typical times in AGNs will be $\sim 10^7$ times longer for $M_{BH} \sim 10^8 M_\odot$. In BHBs, the time is a fraction of a second, thus time scales of several months are expected in AGNs. Fortunately, several X-ray missions, in particular RXTE, provided well-sampled days-to-years light curves for a small number of intermediate-luminosity AGNs, which enabled a proper PSD analysis. The best-studied cases show a clear bend in the PSD, from $\alpha = 2$ to $\alpha = 1$, as we go from high to low frequencies. This seems to continue the trend observed in BHBs and provides a new tool to assess M_{BH}, or perhaps $M_{BH}/\dot{m}_E$, in such systems. BH masses obtained in this way are in general agreement with the mass obtained by using the RM method. One attempt to connect BH masses and bolometric luminosities in both type BHs gives the following empirical fit:

$$\log T_B = A \log M_{BH} - B \log L_{bol} + C, \tag{7.40}$$

with $A = 2.1$, $B = 0.98$, and $C = -2.32$, where T_B is in days and M_{BH} and L_{bol} are in solar units.

There are other interesting suggestions, in particular the proposal that the $T_B \propto M_{BH}/(\dot{M}/\dot{M}_{Edd})$ relationship should lead, in a virialized system of moving BLR clouds, to the relationship $T_B \propto \text{FWHM}(H\beta)^4$. This seems to be verified by observations.

All well X-ray-sampled AGNs (about 30 as of 2011) are considered to be in the soft state. This may be a selection effect because in all sources monitored in this way, the normalized accretion rate is too high to be in the hard state (in BHBs, the transition to the hard state occurs when $\dot{M}/\dot{M}_{Edd}$ drops below ~ 0.02). Interestingly, some very high accretion rate AGNs, like NLS1s, show X-ray behavior consistent with the very high state of BHBs (e.g., in Akn 564, there are two bends in the PSD). There are some yet unconfirmed claims for quasi-periodic oscillations in the X-ray light curves of some sources.

The correlation of different mass BHs in their soft-high state is not the only connection between BHBs and AGNs. There is a different correlation that connects very low accretion rate systems, in low state, over the same range of mass. This correlation goes under the title "the fundamental plane of BHs" and is discussed in § 9.5.5.

7.8.5. *Direct detection and size measurements via microlensing*

The latest technique to detect AGN accretion disks, and to estimate their size, is based on microlensing by stars in foreground galaxies and the measured correlated (or uncorrelated) variability due to such transits. The idea is that each transit is covering a different part of the disk that has a different temperature, and their correlation can reveal the disk structure. So far (2011), there are about 20 cases in which such experiments have been conducted. The results are usually based on measurements in the (rest) UV part of the spectrum, where the continuum is bright and the broad emission lines allow an estimate of the BH mass.

The available data are based mostly on variability measurements of high-redshift luminous AGNs. The mean BH mass in this sample is of order $10^9 M_\odot$, and the mean disk size (i.e., the part of the disk emitting most of the 1500–2000 Å radiation) is 3–10 light-days. Given the BH mass, this is translated to 20–40 gravitational radii. This value is comparable, or perhaps a bit larger than the size expected to emit this radiation in thin accretion disks.

7.9. The central jet

7.9.1. *Radio optical X-ray and gamma-ray jets*

Radio-loud and γ-ray-loud AGNs were introduced in Chapter 6. They are classified by their radio morphology (FRI or FRII sources), bolometric luminosity, radio luminosity, the level of ionization of their optical emission lines, the collimation and velocity of the jet, and the inclination of the jet to the line of sight. According to the AGN unification scheme, all the preceding, and perhaps also the morphology and luminosity of the host galaxy, play important roles in the classification. This section focus on the unique physics of the power house of such sources.

Almost all powerful radio-loud AGNs show central radio core or radio jets. The launch point of the jet coincides with the location of the optical–UV and X-ray continuum source. FRII radio sources, the ones with the higher radio luminosity, typically show at most a single central jet with linear polarization perpendicular to the jet direction. Assuming synchrotron emission, it suggests magnetic fields along the jet. FRI sources are less luminous, usually contain two antiparallel jets, and are more likely to show electric vectors parallel to the jet. The radio emission of the jet in both types of sources is beamed, along the jet direction. The two lobes that are common in FRII sources are not beamed, and more pole-on radio maps show them as fainter extended contours around the central point source.

The viewing angles of compact radio sources are strongly correlated with their radio SED. One way to measure the viewing angle is to compare the radio power of the jet and the lobes. The comparison shows that the radio spectral index α_R ($L_\nu \propto \nu^{-\alpha_R}$) is smaller when the core-to-lobe ratio is larger. Thus, the distinction between steep and flat spectrum radio sources (Chapter 6), at around $\alpha_R = 0.5$, is related to the viewing angle and reflects the beaming of the radio radiation along the central jet direction. The more powerful beamed radio sources are occasionally referred to as *flat-spectrum radio quasars* (FSRQs).

AGN radio jets exhibit a large range of apparent velocities, from mildly relativistic to highly relativistic motion. As in previous chapters, we use the notation of normalized velocity $\beta = v/c$ and the Lorentz factor $\gamma = (1 - \beta^2)^{-1/2}$, to quantify such velocities.

The important questions regarding central AGN jets are how such structures are formed, accelerated, and confined. Freely expanding jets are likely to have large opening angles, thus the observed narrow structures require confinement and collimation. Confinement relates to the external pressure that can be estimated from the observed synchrotron power, the value of β, and the power such jets are believed to transport.

Three general categories of jet models are usually considered. The first is a thermal pressure model of the jet. Such models assume two antiparallel channels that propagate adiabatically from the vicinity of the BH. The second involves the strong AGN radiation that can overcome gravity along certain directions and produce radiative pressure-driven jets. The third class are hydromagnetic jet models that use hydromagnetic stresses exerted by magnetized accretion disks. Such flows are centrifugally driven and magnetically confined. It is possible to show that under very general conditions, MHD winds will always be collimated asymptotically, even in relativistic flows.

Electromagnetic power can be extracted from a spinning BH immersed in a magnetic field via the Blandford and Znajek process. The magnetic field can be anchored in the accretion disk and can be generated through a dynamo in the disk. It can extract energy from the spinning BH, eject plasma in two opposite directions, and produce electromagnetic radiation. The extracted power is proportional to $B^2 M_{\rm BH}^2$ and, for a $10^8 M_\odot$ BH and magnetic field of 10^4 gauss, can reach about 10^{44} erg s^{-1}. The total energy supply of the BH, under these conditions, can last more than 100 Myr.

Regarding blazar γ-ray jets, there are two general ideas and a third possibility that combines the two. The *leptonic model* considers electrons and positrons as the emitting relativistic particle population. In the *hadronic model*, the emitting particles are mostly relativistic protons and electrons. An important historical reason that led to the suggestion of a hadronic jet in AGNs is the search for the origin of the ultra-high-energy cosmic rays. General energy budget considerations seem to show that the total luminosity of pure leptonic models falls short of the observed luminosity is several well-studied cases. Thus electron–positron pairs cannot be the sole source of the observed γ-ray radiation and some combinations of the two types were also considered. A major question in the models is the origin of the relativistic hadrons. For example, it is generally assumed that the origin of the relativistic particles is a magnetized blob that, in itself, moves at a relativistic speed along the jet direction.

Another way to test leptonic jet models is related to luminosity variations. Blazars of both types show highly variable γ-ray emission. Rapid variability requires a compact emitting region. Such a region should be opaque to pair production due to the interaction of a high-energy γ-ray photon with a lower-energy photon to produce an electronpositron pair (Chapter 2). The opacity to pair production is large, unless most of the photons are moving in the same direction, as in a jet.

The jet phenomenon can be extended to other wavelength bands. Deep X-ray imaging shows that a large number of sources with Doppler-boosted radio emission also show elongated X-ray structure in the direction of the jet. One such example was shown in Figure 6.5. A smaller number of sources show highly collimated radio and optical jets. Examples are the very low luminosity LINER M 87 and the very high luminosity type-I AGN, 3C 273.

7.9.2. *The electromagnetic signature of relativistic jets*

Relativistic motion and Doppler boosting

Many flat radio spectrum AGNs show nuclear clumps that vary, in location and luminosity, over time scales of months to years. Following the clumps' motion across the sky with instruments like the VLBI suggests superluminal motion with observed normalized velocity, β_{obs}, that exceeds unity. As illustrated in Figure 7.47, such velocities can be explained by a simple geometry.

Consider a two-clump radio source where one clump is stationary at point A and the second clump is moving from point B1 to point B2 over time t with an angle θ to the line of sight. The motion along the plane of the sky corresponds to a distance Δx (see diagram). Because of the motion toward the distant observer, the signal from the stationary source is delayed by $\Delta t = \beta t \cos \theta$, and the observed time between the two measurements is $t_{obs} = t(1 - \beta \cos \theta)$. The observed velocity will then be

$$\beta_{obs} = \frac{\beta \sin \theta}{1 - \beta \cos \theta}. \tag{7.41}$$

For example, if $\beta = 0.95$ and $\theta = 10°$, we get $\beta_{obs} = 2.56c$. The apparent expansion velocity is maximal when $\theta = \cos^{-1} \beta$, where $\beta_{obs} = \gamma \beta$.

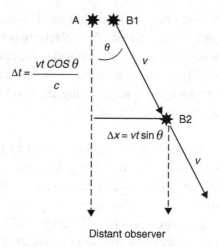

Figure 7.47. Superluminal motion of bright radio clumps. Clump A is stationary, and clump B is moving from point B1 to point B2 at velocity v. On the plane of the sky, the motion is along distance Δx, and the time of arrival of the signal from clump A is delayed by Δt.

Superluminal motion involves beaming of the emitting radiation in the direction of motion and Doppler boosting due to changes in flux density by relativistic time dilation. The effect can be calculated using relativistic radiative transfer, which will not be detailed here. For two optically thin, antiparallel radio jets, with angles θ and $\pi - \theta$ to the line of sight, the ratio of the observed fluxes of the approaching and receding jets, F_1/F_2, depends on β and the SED emitted by the jets (assumed to be identical in the rest frame of the jets). For $L_\nu \propto \nu^{-\alpha_R}$ radio sources, the observed ratio is

$$\frac{F_1}{F_2} = \left[\frac{1 + \cos\theta}{1 - \cos\theta}\right]^{2+\alpha_R}. \tag{7.42}$$

The ratio can be large enough to completely hide the radiation from the receding jet, even for mildly relativistic velocities. For the case considered earlier with $\theta = 10^0$, $\beta = 0.95$, and a source with $\alpha_R = 0.5$, we get $F_1/F_2 \simeq 5 \times 10^3$. It is not surprising that many compact AGN radio sources look like a one-sided jet.

Blazar and jet SEDs

Broad-energy-range spectra of blazars have been illustrated in Chapter 1. They are dominated by two broad peaks, one in the millimeter to NIR range and one in the X-ray to γ-ray band. There are noticeable distinctions between the broadband SED of BL-Lac objects and flat-spectrum radio-loud AGNs. The two former show lower-luminosity, higher-energy peak emission extending up to several GeV. The high energy peak of the latter is typically in the MeV range. The large difference in total luminosity between the groups is similar to the (isotropic) luminosity difference between FRI and FRII sources. However, blazars do not show the morphological radio differences that were used in the FR classification because they are seen with a small viewing angle to the direction of the jet. The differences between the two subgroups is evident from many observations: very high resolution sub-parsec-scale radio observations, poor spatial resolution γ-ray observations, the total luminosity, the different SEDs, and the differences in optical emission lines. It is tempting to speculate that a side view of the blazars that are BL-Lac γ-ray emitters will show a FRI radio sources, and a side view of a flat radio spectrum blazar will be seen as an FRII radio source.

A possible model that explains the different types of blazar SEDs combines the intrinsic properties of the jet and the nature of the inverse Compton (IC), high-energy bump. IC radiation can proceed in two different channels, via internal Compton and external Compton (referred to as EC) processes. IC internal is the case where the seed-scattered photons are the locally produced synchrotron photons. In this case, a lower-luminosity jet will result in a lower luminosity and hence a lower optical depth γ-ray source. Such synchrotron-self-Compton sources (SSC; Chapter 2) show a very high energy Compton peak with a power-law spectrum of the same slope as the synchrotron source. The more powerful jets are probably dominated by external seed photons (EC sources). Such jets are optically thick to the internally produced γ-ray photons. However, the external radiation field near the boosted core can be of high enough energy density to explain a very powerful IC bump. The dimension of the sub-parsec central radio source in such cases is similar to the typical dimension of the BLR. IC scattering of broad-line photons can produce a high-luminosity, high-energy peak that has higher luminosity but lower frequency compared with the high-energy peak in BL-Lac objects. Other potentially important sources of external photons are the central accretion disk itself and also disk radiation that is scattered by very hot, very low density gas filling the BLR. The cosmic background (CMB) radio radiation is yet another external source of soft photons, although this is more suitable for larger-scale IC sources, perhaps related to the radio lobes.

Some confirmation of the general features of such models is obtained from fitting the observations of well-observed blazars like PKS 1222+216. As shown in Figure 7.48, the accretion disk and torus unbeamed SEDs are dominating the energy range between the two jet-produced peaks and hence can be directly observed. The model that is shown in the right panel of the diagram produces synchrotron and EC peaks and can reasonably explain the entire SED of this blazar.

The direct observations of the accretion disk radiation in blazars' spectra allow an estimate of the isotropic luminosity, L_{bol}, of such highly beamed sources. The broad emissions together with the disk continuum allow an estimate of M_{BH}. This allows us to plot L/L_{Edd} versus the observed γ-ray luminosity, L_γ, for blazars. Such diagrams show a clear transition between low L/L_{Edd} BL-Lac objects and

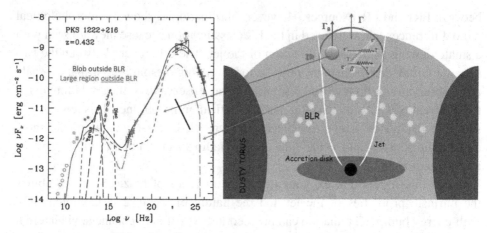

Figure 7.48. Model fit to the SED of the flat radio spectrum blazar PKS 1222+216 (from Stamerra et al., 2011; reproduced by permission of John Wiley & Sons Ltd.). Such sources have two emission peaks at low and high energies. The low-energy peak is the result of synchrotron emission. The high-energy peak is thought to be due to inverse Compton scattering of soft NIR–optical–UV photons, probably from the BLR and the accretion disk, by the relativistic electrons in the jet. The main model ingredients, with the location of the high-energy source, are shown in the right panel. The direct signature of the central torus and accretion disk, and the broad emission lines, are often seen in the spectrum of such sources (see left panel). (See color plate)

high L/L_{Edd} flat radio spectrum blazars at around $L/L_{Edd} = 0.01$. This is roughly the value of L/L_{Edd} that marks the transition between standard thin accretion disks and RIAFs (Chapter 4). This somewhat speculative idea connects the different types of blazars to different accretion modes of the AGN. It fits other ideas about the large range of properties of type-I and type-II radio-loud AGNs within the unification scheme presented in Chapter 6.

Such models are still evolving, and the preceding is not the only explanation of the different types of blazars and the origin of the γ-ray source. One open issue is related to the exact location of the blazar main emitting region, which can be outside or inside the BLR.

7.10. Multicomponent AGN models

Can all the preceding AGN components, BH, disk, torus, BLR, NLR, HIG, and so on, be combined into a general model where the various locations and the different dynamics are combined into a coherent picture? One such picture, which is partly based on observation and partly on speculation, is that of a multicomponent, multidimension wind that goes under the name of *AGN wind* model. Such winds are thought to originate from the vicinity of the central BH, perhaps from the surface

OUTFLOW PICTURE

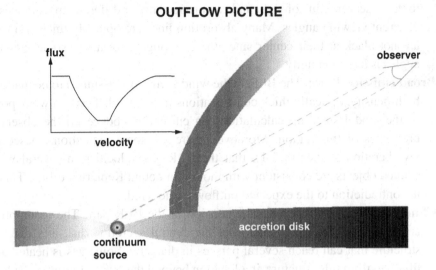

Figure 7.49. A schematic disk wind geometry showing a preferred line of sight and the resulting absorption line profile (courtesy of N. Arav).

of the accretion disk, and extend all the way to kpc distances, beyond the NLR. Most of the components discussed earlier are linked to this global picture, some with more justification than others. The main ingredients of the wind model are as follows:

Fast disk winds: This part of the wind is the one nearest to the central source and hence involves the fastest-moving material. According to some models, the wind can be lifted from the thin accretion disk along the magnetic field lines. Radiation and magnetic pressure provide the first "kick," and further momentum is given by the central radiation field at heights above the disks where the gas is exposed to this radiation. The original direction is therefore perpendicular to the disk surface. Later on, the flow is accelerated due to radiation pressure force, and the motion is basically radial. The general flow lines can change direction due to the combined effect of all forces and the net angular momentum of the original gas. The result is a global velocity field, which is rather different from a simple spherical outflow. It can also be associated with relativistic motions that might have been detected in several such cases (Chapter 9). A simple visualization of this kind of flow is shown in Figure 7.49.

Broad and narrow absorption lines: These are most naturally explained by the wind model. They are observed to be broad or narrow, depending on the location and the line of sight to the center. An important factor for disk winds is the amount of spectral shielding since very high velocity outflows require shielding at their base. The spectral appearance of such flows is very sensitive to the observers' line of sight and can produce rather different appearances at different viewing angles. Many absorption lines are optically thick, yet they are not black at their center since the covering factor along some viewing angles is less than unity.

Broad emission lines: The BLR in the wind scenario is assumed to be made of high-density, optically thick condensations in the wind. This is a weak point of the model since no calculation can currently produce all the observed properties of this region. Moreover, there are some indications, based on reverberation studies (§ 7.1), that the BLR gas velocities in several well-studied objects are consistent with motion in bound Keplerian orbits. This is in contradiction to the expected outflow in the wind.

Clumpy torus: This is a natural extension of the BLR part. The assumption is that the large-scale extension of the central accretion disk is a disklike structure that can reach several parsecs in diameter. Some gas is heated and lifted up from this structure at a location beyond the dust sublimation radius. A continuous flow of such clouds forms almost time-independent, clumpy, disklike structure. Simple estimates show that the mass of the gas lifted from the disk must exceed the amount that eventually reaches the central BH. Thus $M_{\mathrm{outflow}} > M_{\mathrm{accreted}}$.

Narrow emission lines: The general dynamics suggests that the outflow can reach distances that are as large as the NLR, and perhaps break into large condensations at these locations, forming the NLR. Additional acceleration at this location may be due to radiation pressure force on dust grains.

HIG outflow: This is, perhaps, the most successful aspect of the general wind model since outflowing, highly ionized X-ray gas, with a modest mass outflow rate, is indeed observed in many AGNs.

What are the strong points and shortcomings of the wind model? Disk winds, especially MHD winds from the inner part and thermal winds from the outer part, seem to be a general, perhaps natural consequence of many disk models. Moreover, acceleration by radiation pressure provides a natural way to accelerate such winds to large terminal velocities once the material is lifted from the disk surface. However, calculating the exact wind properties is beyond the capability of present-day disk models. In particular, the mass outflow rate is not known since it depends on the disk magnetic field, density structure, and various local instabilities. General considerations suggest that the mass outflow rate can exceed the accretion rate onto the black hole, but precise calculations are not available.

There are several open questions and contradicting observational and theoretical issues that need to be sorted out before such a model is completed. One such issue is the clumpiness of the gas. Several AGN observations suggest clumpy gas at different locations: the BLR, the torus, the NLR, and perhaps also the HIG. No disk wind model is capable of calculating the size and mass of such clumps.

Table 7.1. *AGN components: Location density and ionization parameter*

Component	Distance in r_g	Density	Ionization parameter
Accretion disk	~100	$\sim10^{15}$ cm^{-3}	$U_{\text{oxygen}} = 10^{-3}$–$10^{-1}$
BLR	10^4–10^5	$\sim10^{10}$ cm^{-3}	$U_{\text{hydrogen}} \sim 10^{-2}$
Torus	10^5–10^6	10^3–10^6 cm^{-3}	$U_{\text{oxygen}} \sim 10^{-2}$
HIG	$\sim10^6$	10^3–10^5 cm^{-3}	$U_{\text{oxygen}} \sim 10^{-2}$
NLR	10^7–10^8	10^3–10^5 cm^{-3}	$U_{\text{hydrogen}} \sim 10^{-2}$
Starburst	10^7–10^8	10^0–10^3 cm^{-3}	$U_{\text{hydrogen}} = 1$–10^{-2}

The mass outflow rate is another problematic issue. While the total mass in the BLR is small, this is not the case in the NLR. In fact, it is hard to imagine a situation where disk outflows can explain the total mass in an outflowing NLR. Moreover, many observations show that much of the NLR gas is not outflowing from the galaxy, thus disk wind cannot explain all the observed NLR properties. Regarding the HIG properties, the outflow picture is confirmed, but the observed velocities are too small to be consistent with a wind that is launched from the central disk. A different origin for the HIG gas must be considered, for example, evaporation from BLR clumps and/or the inner "walls" of the torus.

A major problem of the disk wind model is the prediction of a radial velocity field along most, or perhaps all, lines of sight. While this is consistent with BAL and HIG outflows, it seems to contradict at least some of the reverberation mapping results. As explained earlier (§ 7.1), two-dimensional reverberation mapping provides information about the entire phase space of the BLR gas. A very strong prediction of an outflowing BLR is that the blue wings of the broad-line profiles will show a shorter time lag compared with the red wing of the profile. This is not confirmed by detailed observations of several nearby AGNs.

To summarize all these possibilities, we show in Table 7.1 a list of all locations considered here with their known (or estimated) densities and the corresponding ionization parameters. The values in the table refer to the illuminated faces of the various components and assume no attenuation between the central source and the location in question. The dimensions are given in gravitational radii, r_g, and the ionization parameters in either U_{oxygen} for X-ray-emitting components or U_{hydrogen} for optical–UV–emitting components. For the conversion between the two, see Table 5.2.

7.11. Further reading

Physical conditions in the BLR: The general description is based on Davidson and Netzer (1979), Netzer (1990), Osterbrock and Ferland (2006), and Netzer (2008). The cloud model is described in Kaspi and Netzer (1999), and X-ray evidence for clouds can be found in Maiolino et al. (2010). The LOC model is described in Korista and Goad (2000, 2004). The discussion of the physics and observations of Fe II emission lines follows Wills et al. (1985) and Baldwin et al. (2004). Alternative excitation mechanisms for Fe II lines are discussed in Collin and Joly (2000). For gas metallicity in the BLR, see the review by Hamann and Ferland (1999) and more information in Shemmer et al. (2004) and Matsuoka et al. (2011). Very high density BLR gas is discussed in Negreta et al. (2012), and references therein. For the Baldwin relationship, see the original paper by Baldwin (1977) and numerous references in Netzer et al. (1992) and Xu et al. (2008). For disk orientation and line EW, see Netzer (1985) and Risaliti et al. (2011). For the origin of the BLR gas, see Elitzur and Ho (2009), Cao (2010), and references therein.

Gas kinematics and broad-line profiles: This topic is discussed in Netzer (1990), Robinson (1995), Quintilio and Viegas (1997), Bottorff and Ferland (2000), Netzer and Marziani (2010), and references therein. Recent references on the role of radiation pressure force are Marconi et al. (2008) and Netzer and Marziani (2010). The relation of line profiles and eigenvector 1 is discussed in numerous papers, starting from Boroson and Green (1992). For extensive review, see Sulentic et al. (2000, 2009). For double-peak emission lines, see Eracleous and Halpern (2003).

Bolometric correction factors: See Marconi et al. (2004) and Vasudevan and Fabian (2007).

Emission-line variability, RM method, transfer function, M_{BH}: The general discussion is based on Netzer (1990), Netzer and Peterson (1997), Peterson and Bentz (2006), and Peterson (2008). For RM experiments, see description and references in Kaspi et al. (2000, 2005, 2007) and Bentz et al. (2009). For BH mass determination, see Peterson (2008), Vestergaard and Peterson (2006), Netzer et al. (2007b), McLure and Dunlop (2004), and Shen et al. (2008). For the problematic use of the C IV λ1549 line, see Trakhtenbrot and Netzer (2012). For a different view of this issue, see Assef et al. (2011).

Ionization cones, coronal lines, and dusty NLR clouds: The discussion here and in Chapter 6 is based on reviews by Tadhunter (2008) and Meuller-Sanchez et al. (2011). Dusty NLR models and confinement by radiation pressure is discussed by Dopita et al. (2002). For emission-line widths and their correlation with M_{BH}, see Greene and Ho (2005) and many references in Dasyra et al. (2011). For luminosity estimates based on narrow emission lines, see Kauffmann and Heckman (2009), Netzer (2009), and references therein. For extended NLRs, see Fosbury (2006).

HIG and gas outflow: See reviews by Crenshaw et al. (2003) and additional references in Turner and Miller (2009), Kaastra (2008), Netzer et al. (2003), and Chelouche and Netzer (2005). BAL flows are described in numerous papers, for example, Trump et al. (2006), Arav et al. (2008), and Capellupo et al. (2011).

The central torus: For the general idea, see Antonucci (1993). A recent view of dusty torus models is given in Elitzur (2008). Fitting dusty torus SEDs is described in Fritz et al. (2006), Alonso-Herrero et al. (2011), Mor and Netzer (2012), and references therein. For AGNs without a torus, see Mor and Trakhtenbrot (2011). For alternative obscuration mechanisms, see Lawrence and Elvis (2010).

AGN megamasers: This section is based mainly on Herrnstein et al. (1999) and Kuo et al. (2011).

AGN accretion disks: For polarization measurements, see Kishimoto et al. (2008). The X-ray spectrum of AGN disks, and the relativistic iron lines, are reviewed in Reynolds and Nowak (2003), Dvorčiak (2004), Ross and Fabian (2005), and references therein. For an alternative explanation of broad X-ray features, see Miller et al. (2009). The analogy with BHBs is reviewed in McHardy (2010).

Radio, X-ray, and γ-ray jets: For the Blandfor–Znajek process, see the original paper in 1977. Detailed descriptions of AGN jets can be found in Robson (1996), Tavecchio et al. (2007), Tadhunter (2008), Worrall (2010), and Lister et al. (2011).

Disk wind and a multicomponent AGN model: For the general idea, see Elvis (2000). Several types of disk winds are discussed in Proga and Kallman (2004), Chelouche and Netzer (2005), Tombesi et al. (2010), and King (2010).

8

Host galaxies of AGNs

8.1. Observations of host galaxies

The earliest discovered quasars, in the 1960s, were so bright that their optical images showed no signs of the host galaxies. This resulted in the names *quasi-stellar objects* (QSOs) and *quasars* and caused a lot of confusion and even some unusual explanations and models. Interestingly, the much earlier discovery of the first Seyfert galaxies, by Seyfert in 1943, raised no such questions. The luminosity of the central sources in these galaxies was about 2 orders of magnitude below the luminosity of the first discovered quasars, and the galaxy was clearly seen in all cases.

The host galaxies of the earlier quasars were soon discovered by ground-based telescopes in sites with good seeing conditions. Faint nebulosities were discovered in all objects with redshift less than about 0.5, where conditions allowed such detection. The launch of the Hubble Space Telescope (HST), in 1990, resulted in superb resolution observations and the extension of such studies to redshifts 2 and 3. Some of the information was detailed enough to enable a systematic study of the morphology, color, and even stellar population of the hosts. Such information is now available for numerous low-redshift AGNs where the study is made easier because of the lower luminosity of low-z AGNs and the vast improvement in the performance of adaptive optics (AO) systems on giant ground-based telescopes. Similar observations of type-II AGNs, where the optical–UV radiation of the central source is completely obscured, allowed us to extend such studies to more sources and to higher redshift. Figure 8.1 shows several examples of the hosts of luminous type-I AGNs observed with the HST. Several such hosts show clear signs of distortion and interaction.

The picture that emerged after several decades of study suggests that the morphologies of most AGN host galaxies are similar and even indistinguishable from the morphologies of inactive galaxies at the same redshift. In particular, the color and stellar population of AGN hosts are very similar to those of inactive galaxies

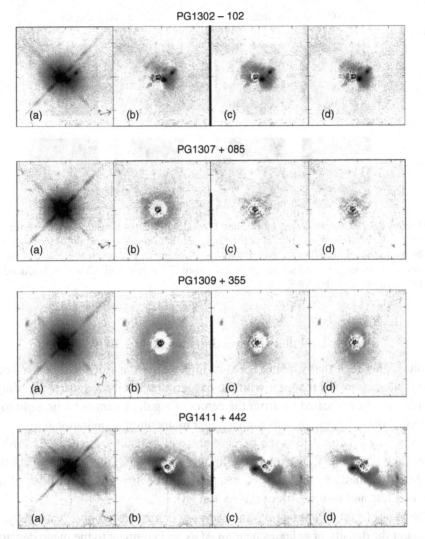

Figure 8.1. HST images of host galaxies of several type-I AGNs showing clear signs of interaction and disturbed morphology (from Veilleux et al., 2009; reproduced by permission of the AAS). In all objects, (a) is the raw image, and the other frames show the results of the removal of the PSF and various assumed Sérsic profiles with index n: (b) $n = 1$, (c) $n = 4$, (d) unconstrained Sérsic index.

of similar mass and morphology. Studies of the stellar mass distribution of AGN hosts at $z < 2$ show relatively few such galaxies with low stellar mass, below about $10^9 M_\odot$. Accurate information about higher-redshift hosts is difficult to obtain and more uncertain. Finally, there are indications that star formation (SF) in low-redshift AGN hosts is similar to that measured in inactive galaxies. All these issues are discussed here and also in Chapter 9.

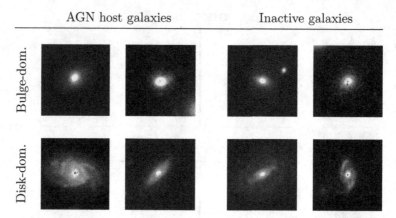

Figure 8.2. Example active and nonactive galaxy images arranged into different morphological classes. There is no obvious distinction in the morphology of AGN hosts and inactive galaxies. Black spots at the center of some of the galaxies are residuals from the point source removal (from Cisternas et al., 2011; reproduced by permission of the AAS).

8.2. Galaxy interaction and AGN activity

An interesting idea that was put forward in the early days of the field connected galaxy interaction and merging with the triggering of AGN activity. According to this idea, cold gas located far from the center of a galaxy cannot be brought to the vicinity of the BH because of its high angular momentum. Gravitational interaction with a nearby galaxy distorts the parent galaxy morphology, changes the orbits of the gas and stars, and allows the gas to find its way to the center. Close interactions that lead eventually to mergers were suggested to trigger short episodes of high accretion rate and fast growth of the central BH.

The idea of galaxy interaction and merging can be tested observationally. We can count the density of galaxies near an AGN and compare to the mean density in the field. We can also search for galaxy–galaxy encounters and look for distortion and deviations from smooth and ordered morphology. The early observations of a small number of high-luminosity AGNs revealed, indeed, cases similar to the ones shown in Figure 8.1. This was taken as an indication that BH accretion is related to galaxy interaction, yet it did not provide enough information about the frequency of such events.

The big advance in understanding AGN hosts is the result of systematic studies of larger AGN samples and a more consistent comparison with samples of inactive galaxies. High-spatial-resolution observations, mostly with the HST, are major tools in such studies. The picture that emerged applies to AGNs up to $z \simeq 2$. It seems that the vast majority of such hosts do not show indications of strong interaction, and there is no significant difference in the fraction of galaxies with distorted morphology in samples of active and inactive galaxies. The only exception may be the hosts of the most luminous AGNs that show more disturbed morphologies compared with similar mass galaxies with no AGN. Several examples are shown in Figure 8.2. They illustrate the big differences between the more common AGN hosts and the ones shown in Figure 8.1. This has profound implications for the evolution of galaxies with active BHs, especially those with the lower M_{BH} that are thought to have gathered most of their mass at later times (see Chapter 9). Thus, there must be alternative mechanisms to trigger fast BH growth and supply large amounts of gas to the center. These can be internal (so-called secular) processes and also minor interactions that do not much affect the global morphology.

The suggestion that most low-z AGN activity is not related to major mergers raises several other issues. It has been suggested that the effect of major mergers was washed out, and was not observed, because of the long time lag between mergers. However, there are observations that do not agree with this idea. For example, recent major mergers can be ruled out due to the high fraction of disk galaxies hosting AGNs. Such disks would not survive a major merger. In fact, elliptical galaxies that are thought to be the result of a merger of similar mass disk galaxies are the only clear evidence for a significant change of morphology of this type. These seem to be the minority of AGN hosts. It can also be argued that strongly interacting systems may hide their active BH such that the galaxies will not be classified as AGNs. This is not likely to be the case since there is no sign of X-ray emission from such systems, yet interacting AGN hosts are strong X-ray emitters at soft and hard X-ray energies. Thus, there is no escape from the conclusion that secular processes are more important in triggering AGN activity at all redshifts up to about 2.

The situation regarding the most massive active BHs at higher redshifts is not so clear. Here the study of the host morphology is more challenging, and the discovery of faint companions that can trigger such an activity is more difficult. There are well-documented cases in which extremely fast accretion onto very massive BHs are associated with galaxy–galaxy interaction. There are other cases of interaction that are associated with intense SF yet no detected AGN activity. At the extreme end of this phenomenon, at low and high redshifts, we find the ultraluminous infrared galaxies (ULIRGs) and the submillimeter galaxies (SMGs), respectively. Such systems are discussed in § 8.5.4.

8.3. Stellar mass and black hole mass

8.3.1. BH mass in nonactive galaxies

High-spatial-resolution observations of the central parts of several nearby galaxies clearly show deviations from the smooth luminosity profile expected from undisturbed stellar population. High-resolution spectroscopy of such regions shows evidence, based on velocity curves, of a large mass concentration in the center. The clearest cases are found in the centers of several elliptical and bulge-dominated nearby galaxies.

Modeling the stellar flux distribution and following the mass increase toward the center leads to the suggestion that in most, perhaps all, bulge-dominated systems, the central mass concentration is due to a massive black hole at the very center of the galaxy. These are nonactive "dormant" BHs that do not grow in size. These BHs must have had an active history with short or long episodes of fast accretion, resulting in a significant increase in their mass. The best-studied example of this type is the BH in the center of the Milky Way galaxy. Here $M_{BH} \simeq 4 \times 10^6 M_\odot$, and the measurement is based on the orbits of stars in the center. This BH may not qualify as "dormant" because there must be some accretion events as indicated by IR and X-ray flares from this region. The accretion rate is extremely low, which may suggest a RIAF.

Except for the BH in the galactic center, the most accurate BH mass measurements are obtained by probing regions that are within the sphere of influence of the BH. In elliptical and bulge-dominated galaxies, the radius of this sphere is

$$r_{BH,sph} = \frac{GM_{BH}}{\sigma_*^2} \simeq 10.7 \frac{M_{BH}}{10^8 M_\odot} \left[\frac{\sigma_*}{200 \, \text{km/s}} \right]^{-2} \text{pc}, \tag{8.1}$$

where σ_* is the stellar velocity dispersion in the bulge (for a large elliptical galaxy, $\sigma_* \sim 200 \, \text{km s}^{-1}$). Mass measurements of dormant BHs suggest that larger bulges host larger BHs. A well-known example is M 87, a large elliptical galaxy with a stellar mass of about $10^{12} \, M_\odot$. This galaxy contains a BH with $M_{BH} \sim 7 \times 10^9 M_\odot$. At the time of writing (2011), the "record holders" are several BHs in central galaxies of large galaxy clusters (so called cD galaxies) with masses that slightly exceed $10^{10} M_\odot$.

BHs are so common in bulge-dominated galaxies that the working hypothesis is that all such systems contain BHs. About 60 such systems, all at low redshift (which is essential to achieve the required spatial resolution), have been studied in great detail, and the observations are accurate enough to derive the BH mass with only a small uncertainty.

The measurements lead to several useful correlations between M_{BH} and other properties of bulge-dominated galaxies. The first is a correlation between M_{BH} and σ_*, the stellar velocity dispersion in the bulge (hereinafter the $M-\sigma^*$ relationship). The fit is usually expressed in the following way:

$$\log (M_{BH}/M_\odot) = a_{BH} + b_{BH} \log \left(\frac{\sigma_*}{200 \, \text{km s}^{-1}} \right). \tag{8.2}$$

The latest results (2011) based on some 60 sources are $a_{BH} \sim 8.38$ and $b_{BH} \simeq 4.53$ for early-type galaxies (ellipticals and S0s), $a_{BH} \sim 7.97$ and $b_{BH} \simeq 4.58$ for spiral galaxies, and $a_{BH} \sim 8.29$ and $b_{BH} \simeq 5.12$ for the entire sample. A somewhat smaller data set used to obtain such relationships is shown in Figure 8.3.

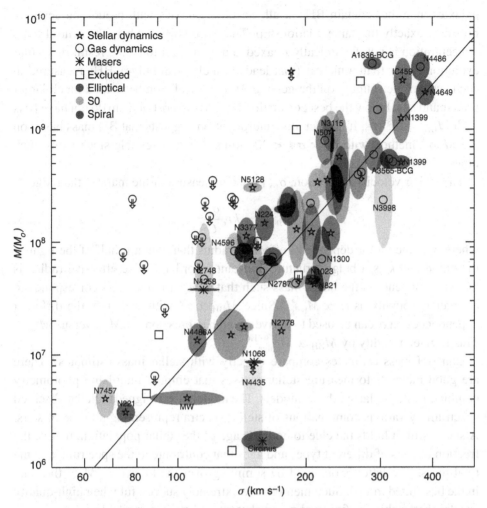

Figure 8.3. The $M–\sigma^*$ relationship for galaxies with dynamical measurements (from Gultekin et al., 2009; reproduced by permission of the AAS). The symbol indicates the method of BH mass measurement: stellar dynamical (pentagrams), gas dynamical (circles), masers (asterisks). The color of the error ellipse indicates the Hubble type of the host galaxy: elliptical (red), S0 (green), and spiral (blue). The line is the best fit relation to the full sample. (See color plate)

While the $M–\sigma^*$ relationship has been used to estimate BH mass in thousands of local galaxies, σ_* is not a reliable measure of the mass of the largest known galaxies. In particular, central galaxies in several large clusters (Coma and a few Abell clusters) are hosting the largest BHs in the local universe, some 2–4 times the mass of the central BH in the giant elliptical galaxy M 87, yet in those systems, $\sigma \sim 300$ km s^{-1}, very similar to the stellar velocity dispersion in M 87. The situation regarding disk-dominated galaxies, irregular galaxies, and galaxies with pseudo-bulges is also not so clear. First, many of those, especially the irregular galaxies, may not contain BHs at all. Second, galaxies with pseudo-bulges may not show exactly the same relationship. These are objects where the central mass concentration is not dynamically relaxed and is thought to be the result of secular evolution rather than a merger (that leads to a classical bulge). BHs measured in such systems lie roughly on the general M–σ^* correlation but with a much larger scatter and a bit below the best correlation line. Also, most of them do not have BHs with $M_{BH} > 10^8 M_\odot$. If correct, this interpretation suggests that BH mass based on the M–σ^* method with $60 < \sigma* < 200$ km s^{-1} is less reliable since many such objects have pseudo-bulges.

The stellar velocity dispersion, σ_*, is also a measure of the mass of the bulge,

$$M_{\text{bulge}} = k r_e \frac{\sigma_*^2}{G}, \tag{8.3}$$

where r_e is the (color-dependent) effective radius that contains half of the light of the bulge and k is a bulge structure constant of order 3. The effective radius is known to depend on the bulge mass such that a larger bulge mass corresponds to a larger r_e roughly as $r_e \propto M_{\text{bulge}}^{1/3}$. Since $M_{BH} \propto \sigma^\alpha$ with $\alpha = 4$–5, the different dependences on σ can be used to derive a rough expression for M_{BH} versus M_{bulge}. This is given roughly by $M_{BH} \propto M_{\text{bulge}}^{1.3-1.6}$.

Can BH mass estimates compare directly with stellar mass estimates? There are good methods to measure stellar masses that employ multiband photometry combined with stellar synthesis models. The models are designed to fit the observed spectrum by various combinations of stellar spectra representing all type of stars. A successful fit holds the clue to both the age of the stellar population, that is, the fraction of stars of different types and ages that contribute to the spectrum, and the total mass. The latter is obtained by summing over the known L/M of the stars in the best-fitted model. Such methods are extremely successful when high-quality spectra that enable the fitting of many absorption lines are available.

Broadband photometry can also provide adequate stellar mass estimates. In this case, the stellar mass is given by an expression that contains the absolute magnitudes in several bands, for example,

$$\log M_* = a_\lambda + b_\lambda (M_B - M_V) + \log (L_\lambda/L_{\lambda,\odot})\, M_\odot, \tag{8.4}$$

where a_λ and b_λ are constants that depend on the band (e.g., R or K) and M_B and M_V are the absolute magnitudes in the B- and V-bands. For example, if λ represents the K-band, then $a_K = -1.39$ and $b_K = 1.176$.

Applying the stellar mass measurement to bulge-dominated galaxies allows us to calculate the BH-to-bulge mass ratio in such systems. For example, a particular study of 60 galaxies shown in Figure 8.4 with dynamically measured M_{BH} and bulge stellar masses obtained from broadband photometry gives

$$\log \frac{M*}{M_{BH}} \simeq 8.1 - 0.64 \log M_{BH}. \tag{8.5}$$

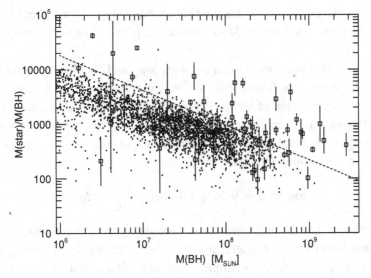

Figure 8.4. $M_{\mathrm{bulge}}/M_{\mathrm{BH}}$ and M_*/M_{BH} for two galaxy samples. Large symbols are $M_{\mathrm{bulge}}/M_{\mathrm{BH}}$ from the work of E. Sani and collaborators (courtesy of E. Sani) with two additional points corresponding to the largest known local BHs (as of 2011). Small points represent bulge-dominated ($B/T > 0.85$ in the r-band) red SDSS galaxies where M_* is obtained from SED fitting and M_{BH} from the M–σ^* relationship. The dashed line is a fit to the Sani et al. (2011) data.

This expression describes the observed trend well in the BH mass range of 10^7–$10^9 M_\odot$. It is not very reliable at the two extremes where there are only very few measurements. The conclusion is that for the smaller BHs with $M_{\mathrm{BH}} \sim 10^7 M_\odot$, the ratio is close to 4000, while in the case of the most massive BHs, with $M_{\mathrm{BH}} = 1$–$2 \times 10^{10} M_\odot$, the ratio is of order 100.

Figure 8.4 also shows calculated M_*/M_{BH} for about 4000 SDSS red galaxies that were chosen to have bulge-to-total (B/T) luminosity in the r-band of $B/T > 0.8$. These are bulge-dominated objects that are mostly early-type galaxies and hence are likely to produce reliable BH mass estimates based on the M–σ^* method. Stellar masses for these galaxies were computed from SED fitting using the SDSS photometry and are probably good estimates of the bulge mass. The sample shows a similar trend of M_*/M_{BH} with BH mass but with slightly different normalization (smaller M_*/M_{BH} for the same M_{BH}) compared with the sample of 60 local galaxies shown also in the diagram.

A major source of concern in deriving the masses of dormant BHs is the modeling of the light and mass ratio in the center of the galaxy, even in cases where precise velocity observations, very close to the BH, are available. Such modeling includes the BH gravitational field, the stellar mass distribution, and the contribution of dark matter to the gravitational potential. Accurate dark matter modeling is tricky and was neglected in earlier works of this type. This resulted in underestimation of several BH masses prior to 2009. Another complication related to applying the relationship to large samples is the possibility that a single slope does not provide adequate approximation to the M–σ^* relationship over 4 orders of magnitude in mass.

There is a second relationship that involves M_{BH} and the absolute luminosity of the bulge. The two were found to be strongly correlated, which is not surprising given the tight correlation of stellar mass and luminosity in galaxies of all types. The correlations can be expressed in different photometric bands, for example, V and K. Using the K-band has the advantage that it avoids, almost completely, the tricky issue of reddening. One such expression is

$$\log M_{BH} = 9.16 + 1.16 \log \frac{L_V}{10^{11} L_\odot} \, M_\odot. \tag{8.6}$$

The scatter and hence the uncertainty in the preceding mass–luminosity correlation is somewhat larger than the scatter in the M–σ^* relationship. However, in many cases, measuring magnitudes is simpler than measuring stellar velocities.

A third method that is useful for nearby elliptical galaxies involves the fitting of the stellar light distribution of the bulge in a chosen photometric band. Elliptical galaxies are assumed to be dynamically relaxed systems. As such, their stellar light distribution, except for the very center, is very well fitted by a Sérsic function,

$$I(r) = I(0) \exp\left[-b_n \left(\frac{r}{r_e} \right)^{1/n} \right], \tag{8.7}$$

where $I(0)$ is the central intensity, n is the shape parameter, and b_n is a function that is chosen such that r_e encloses half the total luminosity. The particular choice of $n = 1$ reduces to an exponential profile, while the case of $n = 4$ results in the so-called de Vaucouleurs light profile.

Fitting the light distribution of nearby elliptical galaxies shows that, except for the inner few percentage of their radius, they are very well fitted with Sérsic profiles. However, there are clear deviations very close to the center that divide this population into two subgroups. In one group, core ellipticals, the light profile falls below the Sérsic function, and in the other group, the coreless ellipticals, the intensity is clearly above the Sérsic profile. This indicates lack of stars and hence lack of stellar light in core ellipticals compared to the other group.

A plausible interpretation of the two distinct groups is that elliptical galaxies, being the result of a merger, evolve differently in their cores depending on the amount of cold gas in the system. In a case of a dissipative ("wet") merger, where at least one of the merging systems contains cold gas, stars continue to form in the resulting single galaxy. This is the case of a coreless elliptical. In dissipationless ("dry") mergers, the two galaxies contain mostly stars, and much of the action in the core, during the final stage of the merger, is due to the interaction of the two BHs that form, in the later stages, a binary BH (Chapter 9). The merging of the two BHs is a long process involving motion and oscillations of the two. This can result in the evacuation of the central core from stars and leads to core ellipticals. Indeed, the measured size of the core is of the order of the BH sphere of influence (Equation 8.1).

Detailed surface photometry shows that the light deficit near the center of core ellipticals, relative to the Sérsic profile, is strongly correlated with the mass of the BH,

$$M_{V,\text{def}} \simeq -18.05 - 2.63 \log \left(\frac{M_{\text{BH}}}{10^9 \, M_\odot} \right), \qquad (8.8)$$

where $M_{V,\text{def}}$ is the light deficit in V absolute magnitude relative to the Sérsic profile. This relationship provides a third method for measuring BH mass in bulge systems. Unfortunately, the method can only be applied to nearby systems because it requires very high spatial resolution observations.

Can BH mass estimates based on RM be directly compared to those based on the σ_* and the M_V methods? As of 2011, such a comparison can be made for about 30 type-I AGNs where σ_* has been measured in the host. There is also a similar number of cases with reliable V or K magnitudes of the bulge. The latter measurements are tricky and less accurate since they require a careful separation of the AGN and the stellar light. Figure 8.5 shows the comparison of the RM-based estimates of M_{BH} with the results of the $M-\sigma^*$ method. As evident from the diagram, the agreement is good with a mean scatter of about 0.3. dex. There are several complications and some claims that the method is good for some but not all AGNs. Nevertheless, the overall agreement is good, confirming the validity of the RM-based method. The result can be used to estimate the value of the mass normalization factor in the RM-based method.

Can the relationship between the mass of the bulge and the mass of the BH be extended to include the entire stellar mass, M_*? We note that in bulge-dominated systems, the observed stellar velocities (the absorption line widths) reflect the star motion in the bulge. These are the cases that show the cleanest $M-\sigma^*$ relationship. In galaxies with massive disks, and in bulgeless systems, the observed stellar velocity can be dominated by the rotation of the disk. As far as we know, there is no correlation between this velocity and the BH mass. In fact, it is not at all clear that all such systems contain central BHs. Thus, estimating BH mass from $\sigma*$ in some galaxies, especially lower-mass systems with large disks, can be highly uncertain.

The accuracy of the BH mass estimate depends on the relative masses of the bulge and the disk in the galaxy. In nearby galaxies, a good separation of the two can be obtained from accurate surface photometry. At larger redshift, this is difficult or, sometimes, impossible. In many cases with $z > 0.1$, the galaxy morphology is unclear. In this case, the only information about the relative mass of the bulge and the disk is obtained from broadband colors of the entire galaxy. Since bulges contain older stars and are, on average, redder than disks, the relative mass in the

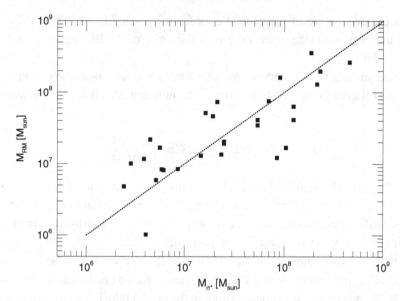

Figure 8.5. A comparison of two methods of measuring M_{BH} in 30 low-luminosity AGNs. The virial method mass is based on the combination of $R_{BLR}(H\beta)$ and FWHM(Hβ) (vertical axis) and The M–σ^* method is based on the width of the stellar absorption lines. The line is the 1:1 ratio.

two components affects the overall color of the galaxy. This leads us to the more general issue of bimodality and the classification into red and blue galaxies that is discussed in § 8.4.

8.3.2. BH mass and emission-line width

The location of the NLR, the places where most AGN narrow emission lines are emitted, is way outside the sphere of influence of the BH. Such lines are thought to be emitted in clouds whose kinematics are determined by two major factors: the gravitational potential of the host galaxy and radiation pressure force due to the strong AGN radiation. In bulge-dominated systems, the stellar velocity dispersion is the best indicator of the gravitational potential and hence the velocity dispersion of the emission lines is expected to be very similar to $\sigma *$. The additional component due to radiation pressure is more difficult to estimate. Absorption by the ionized gas is not expected to be important, but radiation pressure on dust grains can add significantly to the outward velocity component. Indeed, as explained in Chapters 7 and 9, fast outflows of NLR gas are seen in several well-studied AGNs.

The expected correlation of emission-line width and M_{BH} can be tested obser-vationally in type-I AGNs whose BH masses were determined by RM. Such a comparison was carried out for the [O III] λ5007 line, where a strong correlation was indeed found. The scatter in the correlation is rather large, and the usefulness

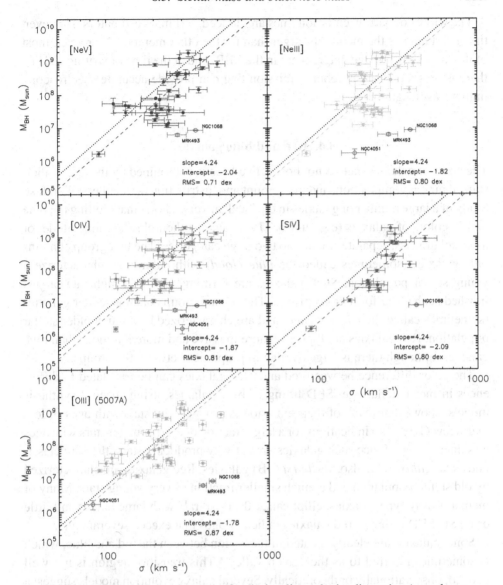

Figure 8.6. BH mass as a function of emission-line velocity dispersion (note that σ in this diagram refers to the observed emission lines and not the stellar absorption lines). A clear correlation is evident in all lines, suggesting that line width measurements can be used to estimate BH mass (from Dasyra et al., 2011; reproduced by permission of the AAS).

of the method is still to be demonstrated. In particular, it seems that the line width depends also on L/L_{Edd}, which makes it a less accurate mass indicator. Several NIR and MIR lines are potentially more important because they reflect a larger range of levels of ionization and because reddening, which may affect the observed profiles, is much smaller. Such a comparison is shown in Figure 8.6.

The widths of narrow emission lines are, indeed, BH mass estimators. However, the uncertainty of the method is larger than for the other methods. Moreover, most of the lines shown in the diagram are in the MIR, and at the time of writing (2011), there is no ground or spaceborne mission that can provide accurate spectroscopy in this wavelength band.

8.4. Red and blue galaxies

The color of galaxies that do not host active BHs is determined by the age of their stars, their gas composition, and the amount and distribution of the interstellar dust. Study of a large number of galaxies in the local universe shows that plotting the blue or UV colors of galaxies (e.g., u–r or UV–r) against their absolute magnitude, or total stellar mass, separates them into two large categories. The first group contains blue galaxies (sometimes called the *blue cloud*) with colors that characterize a young stellar population. Such galaxies are forming stars at a high rate or just finished such a star formation episode. The second group contains redder galaxies (sometimes called the *red sequence*) and are characterized by a much older stellar population. Red galaxies are typically more massive and more luminous than blue galaxies, although there is large overlap in properties between the groups.

The color difference between red and blue galaxies can be translated to differences in mean stellar age. SED fitting of blue galaxies, using spectral synthesis models, shows that much of their emission is produced by stars with ages of less than a few Gyrs, with indications for a large fraction of even younger stars with ages less than 10^8 yrs. Since such galaxies are actively producing stars, they are classified as *SF galaxies* or, also, *starburst* (SB) galaxies. Red galaxies are characterized by old stellar population and a much smaller fraction of very young stars. Many of them are early-type galaxies, ellipticals, S0s, or spirals with large bulges and little or no SF. SED fitting of red galaxies indicates ages that exceed several Gyrs.

Some galaxies are clearly located in the region between the red and blue, which is sometimes referred to as the "green valley." This transition region is not well defined, observationally or theoretically. Several galaxy evolution models suggest a transition in time between an early stage of a star-forming system into a later stage with little or no star formation. The evolution includes several phases of intense SF when a large amount of cold gas is converted into stars. Blue SF galaxies happen to be in this early stage, while "red and dead" galaxies have already finished this phase of their evolution. In this scenario, the green valley is a transition region in color, in time, in gas mass, and in stellar mass. This scenario predicts that the stellar mass of red galaxies is, on average, larger than the stellar mass of blue galaxies.

Galaxy evolutionary scenarios assume that the initial SF is triggered by external conditions. Such conditions may be the result of a merger between galaxies or due

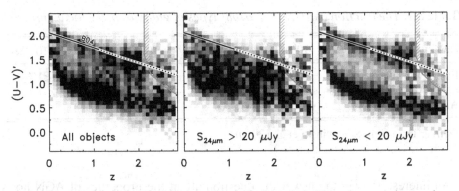

Figure 8.7. Blue and red galaxies as determined by their UV color at various red-shifts. The right panel shows reddening-corrected UV colors. The separation into two groups becomes clearer, suggesting that many green valley sources are highly reddened blue galaxies (from Brammer et al., 2009; reproduced by permission of the AAS).

to gas accretion from the halo followed by secular processes inside the galaxy. Merger-triggered star formation is indeed seen in interacting systems, especially in more luminous systems like local ULIRGs. This represents the minority of SF galaxies. Regarding secular evolution, this seems to be the case in most SF galaxies. It is evident in many barred galaxies in the local universe and is thought to be related to bar instability (see § 9.4). The separation into groups is not so obvious at high redshift. For example, a large number of $z = 2$–3 galaxies are forming stars at a very high rate, yet they are not associated with any merger. The morphology of such galaxies, which are clearly at an early stage of their evolution, is not regular, and the mechanism that triggers their secular SF is not fully understood. One possibility is that the SF is triggered by cold or hot gas which is falling on the galaxies from the halo they occupy (hot filaments, cold streams, etc.).

A galaxy may look blue or red for several reasons. Young stars are hot and blue, and a SF galaxy contains many. However, the color of a galaxy depends on the *relative* number of young stars, that is, the mass of young stars compared with the total stellar mass (this is equivalent to the specific star formation described later). Thus, many galaxies that form stars at a high rate are not as blue as other SF galaxies because they are more massive and contain also a large fraction of old stars. Such large-mass SF galaxies can occupy the green valley. The presence of dust can entirely change the color of a galaxy. The amount of dust and obscuration increases with the SFR, and the most luminous SF galaxies are red because of extinction. Many SF galaxies in the green valley are known to contain a large amount of dust and hence show more extinction and reddening than other SF galaxies. In this case, the red color is not an intrinsic property of the stars in the galaxy. Figure 8.7 shows that this may indeed be the case in a large sample of $z \leq 2.5$ galaxies.

Table 8.1. *The fraction of AGNs in various types of galaxies in the local universe*

Galaxy type/M_{BH}	10^5–$10^6 M_\odot$	10^6–$10^7 M_\odot$	10^7–$10^8 M_\odot$	10^8–$10^{10} M_\odot$
Early-type galaxies	>3%	~2%	~1%	<1%
Late-type galaxies	<2%	~4%	~5%	~1%
All galaxies	<2%	~4%	~2%	~1%

Source: Data from Schawinski et al. (2010).

An interesting way to answer the question about the properties of AGN hosts compared with other galaxies is to use the results of the ambitious classification project "Galaxy Zoo," where many thousands of SDSS galaxies are classified by amateur astronomers. The project is providing valuable answers to questions such as the fraction of active BHs in galaxies with different morphologies. Figure 8.8, which is based on such a study, shows the fraction of AGN hosts with different mass BH, in the general population, in early-type galaxies and in late-type galaxies. The fractions, which are also summarized in Table 8.1, range from less than 1 percent in early-type galaxies with large M_{BH} to more than 4 percent in low-M_{BH} late-type galaxies. This information can be used to test evolutionary scenarios where BH growth and duty cycles are changing together with the changes in the morphology of the host galaxy.

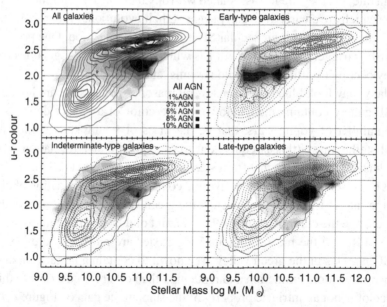

Figure 8.8. The fraction of AGNs with different M_{BH} in host galaxies of different types (from Schawinski et al., 2010; reproduced by permission of the AAS).

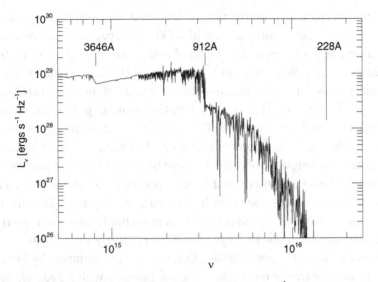

Figure 8.9. Model spectrum of a continuous $10\ M_\odot\,\mathrm{yr}^{-1}$ starburst, with solar metallicity gas and a Kroupa IMF, 10^8 yrs after the beginning of the burst. The hydrogen Lyman (912 Å), He$^+$ Lyman, and hydrogen Balmer (3646 Å) edges are marked.

8.5. Star-forming galaxies

8.5.1. Observed properties of SF galaxies

SF galaxies are easily recognized by their color and by their spectra. Their overall color is typical of young stars, which suggests that a good way to detect them is by using a two-band photometry: a blue or UV band (e.g., U or one of the GALEX UV bands) and a red band (e.g., r or R). High-spatial-resolution observations of SF galaxies show numerous SF regions (luminous HII regions) distributed across the galaxy. Such regions are detected around the galactic center and all over the disk in disk-dominated systems in the local universe. The structure of many high-redshift systems is not a simple bulge-disk morphology, but individual SF regions are clearly detected. Another clear indication of intense SF activity is interstellar dust. SF regions contain a large amount of cold gas and dust. Many of these regions are heavily extincted, which, as explained, can change their color considerably.

Spectral indicators of SF are found at several wavelength bands. The optical–UV spectrum is dominated by young stars. This is demonstrated in Figure 8.9, which shows a theoretical synthetic spectrum of a continuous $10\ M_\odot\,\mathrm{yr}^{-1}$ starburst, with solar metallicity gas, 10^8 yrs after the beginning of the burst. The Balmer and Lyman edges, at 3646 and 912 Å, are clearly seen in the spectrum, and there is practically no radiation beyond the He$^+$ edge at 4 Ryd. The model shows the calculated stellar spectrum, but real systems are, in many cases, heavily obscured by dust with a much redder looking spectrum.

Additional SF indicators in the optical spectrum are strong absorption Balmer lines and a weak spectral jump at around 4000 Å. A spectral break at around this wavelength is a feature seen in the spectra of many stars and galaxies. It reflects the typical temperature in the atmosphere of the stars and is more conspicuous in colder stellar atmospheres. In individual stars, the magnitude of the break is related to the spectral type. The later the type is, the larger the break. In galaxies, the magnitude of the break is an indication of the stellar population and the mean stellar age. The so-called $D_n 4000$ index is a number that gives the change in the measured stellar flux shortward and longword of 4000 Å. Since the observed break is a combination of spectral features in many stars of different spectral types, this index can be used to determine the fraction of young, early-type stars. As explained later, the $D_n 4000$ method for measuring specific SFR (SSFR) is particularly useful in type-II AGNs.

Young stars are born out of large clouds of relatively cold mostly molecular gas. The spectrum of the more massive O-type stars is dominated by high-energy photons, beyond the Lyman edge. This radiation interacts with the gas, photoionizes it, and produces a reach emission-line spectrum typical of HII regions (Chapter 5). Strong Balmer emission lines are hence a clear indication of a SF region, provided the hot stars are the only source of Lyman continuum radiation (i.e., there is no nearby AGN). Much of the radiation emitted by the gas and stars is absorbed by the dust and reemitted at much longer wavelengths. This explains why the IR part of the spectrum provides, in many cases, the cleanest signature of SF activity. Most of the thermal emission of the dust is emitted at around 40–100 μm.

Several IR spectral features are associated with SF regions. Some of the easier ones to detect are silicate absorption and/or emission features, water absorption lines, and polycyclic aromatic hydrocarbon emission features (PAHs). The latter are a family of planar molecules of carbon atoms arranged in a honeycombed structure of six-membered rings surrounded by hydrogen atoms. The molecules are probably produced in circumstellar envelopes, very much like dust grains. There are various types of such molecules, and the ones most common in SF regions contain between 20 and 50 carbon atoms. Typical dimensions (radii) of those molecules are 5–10 Å, considerably smaller than most dust grains. The molecular emission features include several bands with characteristic widths that are much broader than the width of a single atomic emission line. Such emission is very pronounced in the NIR–MIR spectrum of SF galaxies, with the most intense features centered at wavelengths around 3.3, 6.2, 7.7, 8.6, 11.3, and 12.7 μm.

PAH emission features have been studied in galactic HII regions, in SF galaxies, in the laboratory, and in the MIR spectrum of many AGNs. It is thought that their excitation is due to soft, $h\nu < 13.6$ eV, radiation typical of photodissociation regions (PDRs). Such conditions are common in SF regions and in the vicinity of O and B stars. PAHs are very sensitive to the presence of high-energy radiation. The molecules are easily destroyed by absorption of a single EUV/X-ray photon. Such photons are not produced in SF regions where even the youngest stars emit

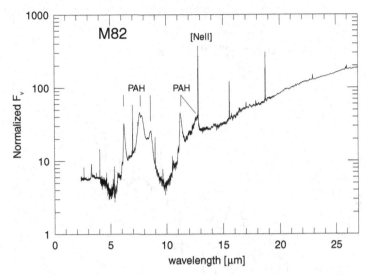

Figure 8.10. The NIR–MIR spectrum of the star-forming galaxy M 82 showing the strong PAH emission features. The [NeII] line at 12.8 μm is, in many cases, the strongest IR line in the spectrum of SF galaxies (courtesy of D. Lutz).

extremely little radiation at energies beyond 4 Ryd, the ionization edge of He^+. Conversely, most AGNs produce much of their radiation at higher energies, all the way to the hard X-ray band. Because of this, PAH emission is considered to be a sensitive probe of "pure" SF regions with little or no nearby AGN activity. Figure 8.10 shows the MIR spectrum of M 82, a well-known low-redshift SF galaxy with strong PAH emission features.

The luminosity of several of the PAH features is strongly correlated with the FIR luminosity. The measurements of the PAHs can be tricky since other dust-related absorption features, for example, silicate absorption at around 9.7 μm, are also present. A couple of examples are shown in Figure 8.11, where a sample of nearby ULIRGs was divided in two according to the strength (optical depth) of the silicate feature. Many ULIRGs contain a significant contribution due to a "buried" AGN (Chapter 9), but in the case under study, this has little affect on the MIR SED.

8.5.2. Highlights of SF theory

Star formation is the result of the gravitational collapse of gas in a dark matter halo (Chapter 9) or the increasing cold gas density in an individual galaxy. The phenomenon can occur on a large range of scales, from less than a pc to more than a kpc. Its details are determined by various physical conditions and different physical processes: gas temperature and pressure, magnetic pressure, radiative cooling, angular momentum, various types of torques, energy injection by earlier generations of stars, gas turbulence, and more. Observations show that there are

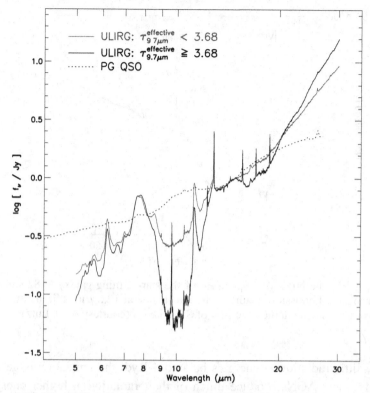

Figure 8.11. The MIR spectrum of low-redshift ULIRGs showing the large range of properties, in particular, the large variation in the optical depth of the 9.7 μm silicate absorption feature. The diagram shows two subgroups of ULIRGs and compares their spectra with a mean high-luminosity AGN (QSO) spectrum (courtesy of D. Lutz).

two distinct modes of SF: a quiescent mode in HII regions in disk galaxies and a starburst mode in starburst galaxies. The time scales associated with the two modes can differ by 2 orders of magnitude.

SF is usually described as a two-stage process: the collapse of the ISM to form giant molecular clouds, and the fragmentation and collapse of these clouds to form stars. The collapse and fragmentation are explained more in § 9.4. The overall efficiency of the SF process is small. For example, in galactic scale SF, only a few percent of the gas are converted to stars in 10^8 yrs.

There are several tight correlations connecting almost all SF objects. In particular, there is a strong correlation between the SFR per unit area, Σ_{SFR}, and the surface density of the gas, Σ_{gas}. Fitting large samples of SF galaxies reveals two types of expressions that fit the data equally well. The first is the Schmidt–Kennicutt law.

$$\Sigma_{SFR} = K \Sigma_{gas}^N,$$

(8.9)

where $N \simeq 1.4$, Σ_{gas} is measured in $M_\odot$ pc^{-2} and Σ_{SFR} in $M_\odot$ yr^{-1} kpc^{-2}. The constant K is different for disk galaxies that are forming stars and for starburst systems like ULIRGs (see § 8.5.6). For the first group, $K \sim 1.5 \times 10^{-4}$, and for the second, the number is ~ 4 times larger. The index N is close to what is expected from local gravitation collapse. The second relationship is related more to the global properties of the system. It is given by

$$\Sigma_{SFR} = \epsilon \frac{\Sigma_{gas}}{\tau_{dyn}}, \tag{8.10}$$

where $\epsilon \simeq 10^{-2}$ and $\tau_{dyn} = R/V$ is the dynamical time of the system with R its typical size and V the typical, rotational of free-fall velocity. For SF disk galaxies, R is the galactic scale and $\tau_{dyn} =$ few $\times 10^8$ yr. For starburst systems like ULIRGs and SMGs, the size is much smaller, and τ_{dyn} is much shorter, of order 30–50 Myr. Given the value of ϵ, we see that the SF time scale is about 100 times longer than τ_{dyn}.

8.5.3. Estimating star formation rates

Star formation rate (SFR) is the amount of gas converted to stars per unit time. The standard unit to measure SFR is solar mass per year ($M_\odot$ yr^{-1}). The SFR can be measured in small regions in the galaxy, like a single molecular cloud, in large galaxies, and in the entire universe. The SFR at a given redshift gives an essential information about the evolution of a certain galaxy or the entire galaxy population at that time. Following the SFR in time is equivalent to following the stellar mass assembly in the universe.

There are several empirical methods to estimate the SFR from ultraviolet, optical, and infrared observations. All methods are based on the standard theory of star formation and employ different types of emission associated with such events.

Basic star formation theory can be used to predict the number of stars of mass m, $N(m)$, that are formed from an initial cold gas cloud under various conditions. This is usually referred to as the *initial mass function* (IMF). Observations suggest that the functional form of the IMF over a limited stellar mass is given by a simple power law,

$$\frac{dN(m)}{dm} = am^{-\alpha}, \tag{8.11}$$

where the slope α is of order 2–2.5 for stars with $m > 0.5 M_\odot$. The special case of $\alpha = 2.35$ for all stars between 0.1 and 100 $M_\odot$ was considered for many years to be a good approximation for young stellar clusters. This IMF is referred to as the "Salpeter IMF." Detailed analysis of many field and cluster stars suggests that this IMF overpredicts the number of low-mass stars below about $0.5 M_\odot$. A more realistic shape starts at the high-mass end and rises to a peak at around 1 $M_\odot$ with the previously mentioned slope. The function becomes flat over a small range in mass and decreases toward the small mass end. This gives a top-heavy (larger fraction of high-mass stars) IMF compared with the Salpeter IMF. Such IMFs are thought to describe better regions undergoing intense star formation. There are also suggestions that the IMF is sharply truncated at the very high mass end. A commonly used functional form is the so-called Kroupa IMF, which is similar to the Salpeter IMF for masses above $0.5 M_\odot$ but has a much flatter slope at lower masses.

Star-forming events can proceed in various ways. A *starburst* (SB) is a term usually used for a short-duration event that transforms a large fraction of a gas cloud into stars. In such an event, the most massive stars evolve very rapidly and explode as supernovae (SNs) after a few million years. This results in an intense, short-duration energy release by the young cluster. Later stages, referred to as *decaying starburst* or *poststarburst*, are dominated by the emission of the longer-lived stars over a period of tens of Myrs with a slowly decreasing luminosity. A *continuous SB event* refers to the case of a large gas supply that results in SF over a much longer period of time, 10^8 years and even longer. Here, again, the first massive stars undergo SN explosion within a very short time, but more such stars are formed at later times, extending this phase of the burst.

An equivalent way to characterize SF events is to use the star formation luminosity, L_{SF}, which gives the mean bolometric luminosity released by the burst at different times. The conversion factor (the SF radiative efficiency η_{SF}) can be estimated by noting that the primary source of radiation is nuclear energy with its typical efficiency of about 1 percent. This should be combined with the fraction of gas undergoing nuclear fusion during the more luminous phase of the burst. For a long-duration burst, this fraction is about 0.05–0.1 over the first 10^8 years. Combining the two, we get $\eta_{\mathrm{SF}} \approx 5 \times 10^{-4}$–$10^{-3}$. This gives

$$L_{\mathrm{SF}} \simeq 10^{10} \left[\frac{\eta_{\mathrm{SF}}}{7 \times 10^{-4}} \right] [\mathrm{SFR}] \, L_\odot, \qquad (8.12)$$

where [SFR] is measured in units of $M_\odot \, \mathrm{yr}^{-1}$. The numbers are not precise as various SF events differ in their time evolution, their IMF, and so on.

The early phases of a short-duration burst are dominated by the most massive early-type stars, mostly O and B stars. These short-lived stars produce much of their energy in the UV part of the spectrum. A small fraction of this energy is released at wavelengths above 912 Å, and most of it at the shorter-wavelength Lyman continuum, where the photons are capable of ionizing the surrounding gas, which results in recombination line emission. At later stages, the relative flux in the $\lambda > 912$ Å continuum increases and the ionizing continuum flux decreases as the youngest stars end their life in SN explosions. Both the $\lambda > 912$ Å continuum photons that are directly observed and the Lyman continuum photons that can be indirectly measured from emission-line observations can serve as SFR indicators if the stage of the burst is known.

For a short-duration SB (a *delta function SB*), the emission-line (e.g., Hα) intensities rise on the shortest time scale, about 10 Myr. The time scale for a considerable change of the stellar UV is 3–5 times longer, and the time scale for dust emission (see later) is the longest, of order 10^8 yrs. In the same way, in the poststarburst phase, the Hα line luminosity (and other recombination lines) decays first, the UV stellar luminosity decays second, and the FIR luminosity decays over a much longer time. For a continuous burst, the Hα luminosity rises first and reaches a constant value after about 10 Myr, the UV luminosity reaches a constant value after 30–60 Myr, and the FIR luminosity reaches a high value after a somewhat longer time but keeps rising, slowly, as the number of older stars that are capable of heating the surrounding dust increases over very long time scales.

The various methods to estimate SFR that are listed subsequently were calculated for *a continuous SB* 10^8 *years after its start*. They are based on observations of the primary sources of radiation (the young stars), recombination lines, forbidden emission lines, dust emission in the IR, X-ray emission by low-mass and high-mass X-ray binaries, and radio emission in SF regions. Unless otherwise specified, the conversion factor used throughout, which is appropriate for a typical (but not extreme) top-heavy IMF, is $L_{\rm SF} = 10^{10}[{\rm SFR}]\,L_\odot$.

Estimating SFR from hydrogen recombination lines: The strongest recombination line, Lyα, is heavily extincted and cannot be used to estimate, reliably, the Lyman continuum luminosity. Hydrogen Balmer lines are more useful since the dust opacity in the optical band is smaller and since the line photons are not scattered many times by the neutral hydrogen gas, like the Lyα photons, which increase the probability of being absorbed by dust grains. Out of the Balmer lines, the best line to use is Hα, which is the strongest and the one least affected by extinction. In addition, the Hα absorption line in the stellar continuum is relatively weak and does not affect much the observed emission-line intensity.

Basic recombination theory suggests that, under case B conditions, the number of Hα photons emitted by the gas is about half the number of the absorbed ionizing Lyman continuum photons. Thus, the unattenuated Hα line intensity is a direct measure of the photon flux in the ionizing continuum, provided the covering factor (Chapter 5) is close to unity (a good approximation for stars that are located well inside molecular clouds) and provided all the Lyman continuum is absorbed by the gas and not by the dust. For example, in the case of ionization by an early-type star with a surface effective temperature of 40,000 K, the Hα luminosity is about 1.7 percent of the total star luminosity and ~ 4 percent of its Lyman continuum luminosity. The total Lyman continuum luminosity, and the bolometric luminosity of a SB, is obtained by assuming a certain IMF that gives the fraction of various mass stars in the cluster during the various stages of the event. Thus, Hα observations can be converted to L_{SF} and hence to SFR.

An expression that is often used in combination with the Kroupa IMF assumes heating and ionization by stars that are less than 30 Myrs old. Under these assumptions,

$$[\text{SFR}] = 5.5 \times 10^{-42} L(H_\alpha) \, M_\odot \, \text{yr}^{-1}. \tag{8.13}$$

For Salpeter IMF, the constant 5.5 is replaced by 7.9. L_{SF} can be computed from the same $L(H\alpha)$. For example, in a case of a Kroupa IMF,

$$L_{\text{SF}} \simeq 220 L(H\alpha)_{\text{corr}} \, \text{ergs s}^{-1}, \tag{8.14}$$

where $L(H\alpha)_{\text{corr}}$ is the *unattenuated* line luminosity.

Reddening corrections must be applied before using Equation 8.13. This can be a tricky issue because it requires some knowledge of the total dust mass in the galaxy, the dust distribution, and the exact location of the HII regions producing the observed Hα. In general, higher SFR is associated with regions of higher gas surface density and hence higher dust column density. Thus, galaxies with higher SFR are those with more attenuated line and UV continuum radiation. In low-mass, low-SFR galaxies, the standard extinction correction based on the known (case B) ratio of $I(H\alpha)/I(H\beta)$ is a reasonable approximation to the unattenuated $L(H\alpha)$. These are also the cases where the FIR indicators listed later fail to give the correct SFR because much of the emitted stellar and line flux is not absorbed by dust. The other extreme are sources with high SFR, like LIRGs and ULIRGs (see later), where most of the line flux is absorbed by dust and deriving the unattenuated $L(H\alpha)$ is most uncertain.

A possible way to bridge the gap between low- and high-SFR systems, with very different amounts of dust obscuration, is to construct empirical relations that combine the *attenuated* line flux, $L(H\alpha)_{\text{obs}}$, with a fraction of the IR luminosity. Such expressions can be written as

$$L(H\alpha)_{\text{corr}} = L(H\alpha)_{\text{obs}} + a_\lambda L(\text{IR}), \tag{8.15}$$

where a_λ is a wavelength-dependent correction factor measured for the IR band in question. For example, if $L(\text{IR}) = L(24 \, \mu\text{m})$, then $a_\lambda \simeq 0.02$. In the important case where $L(\text{IR})$ represents the total 8–1000 μm emission, $a_\lambda \simeq 0.0024$. Such expressions have been obtained and verified over the range of $L(H\alpha)_{\text{corr}} = 10^{39}$–$10^{43}$ erg s^{-1}.

The derived values of a_λ are based on observations of entire galaxies, or at least a large part of a galaxy that contains many individual HII regions. In such systems, a large fraction of the IR emission is due to dust that is heated by stars that are much older than the young stars in HII regions (the ionizing lifetimes of O-stars are less than 5 Myrs). The ionizing radiation of the entire galaxy at any given time, and hence the Hα luminosity, is determined by the young O-stars, but the dust emission includes the reprocessed radiation of the much older stars whose fraction depends on the duration of the burst. Thus, the ratio of bolometric luminosity to Hα luminosity increases with time. This change is reflected in the exact value of the parameter a_λ. For example, the constant 0.02 used previously for $L(24\ \mu\text{m})$ of an entire galaxy should be changed to about 0.03 for individual HII regions.

Finally, in all the preceding, we can replace Hα by different hydrogen Balmer and Paschen lines, in particular, replacing $L(\text{H}\alpha)$ by $2.85L(\text{H}\beta)$. This is important when trying to estimate SFRs in $z > 0.4$ sources from ground-based spectroscopy, where the Hα line is shifted into the unobserved NIR.

Estimating SFR from forbidden emission lines: Other emission lines are important coolants in SF regions, and hence their luminosity is proportional to L_{SF}. For example, the [O II] $\lambda3727$ doublet is thought to emit a roughly constant fraction of L_{SF} such that, for a Kroupa IMF,

$$L_{\text{SF}} \simeq 218L([\text{OII}]\lambda3727)\ \text{ergs s}^{-1}, \tag{8.16}$$

where, again, the line intensity to use is after correcting for reddening (not a trivial issue, given the short wavelength of this line). This relation is less accurate than the one based on the Balmer emission lines because the nebular physics involved in the production of this line depends on conditions that are not always known, such as gas density and composition.

Estimating SFR from ultraviolet spectroscopy: Young stars produce some of their radiation in the $\lambda > 912$ Å part of the spectrum. For low-redshift sources, this wavelength band is accessible to various space experiments and can be used to estimate the SFR. At high redshift, the rest-frame UV is observable by ground-based telescopes and can provide a useful way to estimate the SFR.

As shown in Figure 8.9, the UV spectrum is basically flat in L_ν between 1200 and 2500 Å. Thus, an expression of the form

$$L_{\text{SF}} = 3.4 \times 10^{44} \left[\frac{L_\nu}{10^{29}} \right]\ \text{ergs s}^{-1}, \tag{8.17}$$

which is appropriate for a Kroupa IMF, can be used to estimate the SFR. Reddening correction must be applied to the UV observations before calculating L_{SF}. Alternatively, the *observed* UV flux can be used in combination with an estimate of the fraction of FIR emission that is due to reradiation by dust.

The $L_\nu(\text{UV})$ SFR calibrator is different from the one based on Hα in two important ways. First, the observed UV continuum represents the star formation history over almost 10^8 yrs, while $L(\text{H}\alpha)$ reflects more the luminosity of SF regions with ages of at most few $\times 10^7$ yrs. Various methods have been developed to calibrate the UV indicator versus the $L(\text{H}\alpha)$ indicator and to obtain consistent values for the inferred SFRs. Second, the reddening correction is different because the stars are mixed with the dust in a way different than the Hα-emitting gas. In fact, the UV-based estimate of the SFR is the one most sensitive to extinction. As already mentioned, the effect is extreme at very large SFRs to the extent that such estimates, on their own, are highly uncertain. For example, at $[\text{SFR}] = 1\ M_\odot/\text{yr}$, only about 10 percent of the (rest-frame) 1500 Å flux can escape the SB region, and at $[\text{SFR}] = 10$ $M_\odot \text{yr}^{-1}$, the fraction is of order 3 percent.

Estimating SFR from the $D_n 4000$ index: This index, which uses the magnitude of the 4000 Å jump to obtain SSFR, was described in the previous section. Combined with estimates of the total stellar mass, it gives reliable estimates of the SFR in SF galaxies. The estimates of SFRs in type-II AGN hosts are more uncertain. The method is not without fault. In particular, it fails badly in massive early-type galaxies with $D_n 4000 > 1.8$.

Estimating SFR from fitting the stellar spectrum: The most accurate method to obtain the SFR from optical–UV spectroscopy is to use a combination of modeled stellar spectra to fit the extincted spectrum of an entire galaxy or a single SF region. The combination of at least one UV band with several optical bands and many stellar absorption lines enables us to find the right mix of stellar types as well as the extinction. This is an elegant way to avoid the use of a simple expression like Equation 8.17, which is extremely sensitive to extinction. It can also be used to estimate SFRs in different types of bursts, not only the continuous burst used here. Such methods are occasionally combined with Hα measurement, and photoionization modeling, to improve SFR estimates.

Estimating SFR from infrared observations: The IR spectrum provides the best SFR indicators in those cases where much of the stellar continuum is absorbed by dust. Several IR indicators are commonly used. They are based on the 8 μm, 24 μm, and FIR continuum and the luminosity of PAH features.

The case of a complete absorption of the young star radiation by the cold dust is particularly simple since in this case,

$$L_{\text{SF}} \simeq L(\text{IR})(1 - \eta_{\text{old}}), \tag{8.18}$$

where $L(\text{IR})$ is the IR luminosity, which is normally taken to be the integral over the 1–1000 μm or the 8–1000 μm range. The factor η_{old} takes into account dust heating by old stars. Its typical value is 0.1–0.3.

Several approximations allow the conversion of a single band photometry to L_{SF}, for example,

$$\log L_{SF} \simeq 1.16 + 0.92 \log L(70\,\mu m) \tag{8.19}$$

and

$$\log L_{SF} \simeq 1.5 + 0.9 \log L(160\,\mu m), \tag{8.20}$$

where L in the chosen band refers to λL_λ in units of $L_\odot$. In many cases of interest, the cold dust temperature is in the range 30–60 K, and most of the flux is emitted between rest wavelengths of about 30 and 150 μm. In this case, $L_{SF} \simeq 2\nu L_\nu(60\,\mu m)$.

Similar considerations provide the calibration factors for the 8 and the 24 μm continua. These are further removed in wavelength from the peak dust emission, and as a result, the calibration is less accurate. For example, for a Kroupa IMF, the expressions

$$[\text{SFR}] \simeq 2.5 \times 10^{-43} L(24\,\mu m)\, M_\odot\, \text{yr}^{-1} \tag{8.21}$$

and

$$[\text{SFR}] \simeq 1.5 \times 10^{-43} L(8\,\mu m)\, M_\odot\, \text{yr}^{-1} \tag{8.22}$$

give reasonable estimates of the SFR in galaxies where $L(24\,\mu m)$ and $L(8\,\mu m)$ are of order 10^{44} erg s^{-1}. Much beyond this luminosity, the dust in the galaxy becomes hotter and transparent at MIR wavelengths, and the conversion factor itself depends on luminosity. Much below this value, the dust content in the galaxy is low enough to allow much of the stellar radiation to escape without being absorbed by the dust. In this case, expressions like Equation 8.21 underestimate the total SFR.

A complication that must be taken into account is that much of the energy around 8 μm is due to several strong PAH features around this wavelength. The exact conversion factor depends on the band width of the instrument in question (e.g., Spitzer) and the redshift of the source. In fact, the luminosity of these features in galaxies that are on the SF sequence (see later) is, by itself, a good SFR indicator. A useful expression that can be used for such galaxies is based on the combined strengths of the 7.7 μm and 8.6 μm PAH features,

$$L_{SF} = f_{PAH} L(\text{PAH}\, 7.7 + 8.6), \tag{8.23}$$

where $f_{PAH} \sim 12$. In general, f_{PAH} depends on the gas metallicity and the distance from the center of the SF sequence in the SFR versus M_* plane (see § 8.5.5). For powerful starburst systems like LIRGs and ULIRGs, the distance is large (these sources are found above the SF sequence), and f_{PAH} is larger too.

Estimating SFR from X-ray observations: SBs of all types result also in a population of X-ray-emitting point sources, mostly low-mass X-ray binaries (LMXBs) and high-mass X-ray binaries (HMXBs). The X-ray emission of such sources is known to depend on the SFR. For low to moderate SFR, one can use

$$[\text{SFR}] \simeq 10^{-39} L(2\text{--}10\,\text{keV})\, M_\odot\,\text{yr}^{-1}. \qquad (8.24)$$

For very high rates, the conversion factor is some 2–4 times smaller.

Estimating SFR from radio observations: Nearly all of the radio emission from FIR-emitting galaxies at wavelengths longer than a few centimeters is due to synchrotron radiation from relativistic electrons and free–free emission from ionized hydrogen. The relativistic electrons are thought to be accelerated in supernovae remnants, the results of the explosion of stars with $M > 8 M_\odot$ – mostly type-II and type-Ib supernovae. These are the same stars that cause most of the ionization of the gas. Thus, star formation and radio emission are related, and we expect a strong correlation between radio continuum and FIR emission. Indeed, galaxies of all luminosities show a strong correlation between radio emission at around 1.4 GHz and IR emission at 8, 24, and 60 μm. An expression that is often used in galaxies with intermediate to high SFRs is

$$[\text{SFR}] \simeq 4 \times 10^{-38} L(1.4\,\text{GHz})\, M_\odot\,\text{yr}^{-1}, \qquad (8.25)$$

where $L(1.4\,\text{GHz})$ is λL_λ in units of erg s^{-1}.

There are several known discrepancies between the various methods that are mostly related to three fundamental issues: the time evolution of the burst, the shape of the IMF, and the extinction correction. Regarding the time evolution, the confusion between continuous and decaying bursts can result in very different estimates based on the short time-scale evolution of $L(\text{H}\alpha)$ and the much longer time-scale evolution of $L(\text{FIR})$. Also, neglecting the contribution of old stars to the heating of the dust can introduce a factor of up to 2 uncertainty on SFR estimates based on $L(\text{FIR})$ in systems with low SFR. As for the extinction, older stars seem to be physically removed from the location of the ionized gas and thus suffer less extinction. They are also mixed with the dusty gas and hence should not be corrected for extinction by using a simple dust screen model. Several methods have been developed to take all this into account, in particular, the use of different extinction curves to correct for the reddening for young stars and recombination lines.

The calculation of SFRs in AGN hosts must be treated with care since several of the preceding considerations may not apply anymore. For example, the observed Hα luminosity can be dominated by AGN ionizing photons, and much of the 8 and 24 μm continuum emission can be due to AGN-heated dust in the central torus. Even more problematic is the use of the radio emission SF indicator in radio-loud

AGNs. As shown later, the FIR radiation at wavelengths longer than about 40 μm is still a reliable SF measure in such cases.

8.5.4. Various types of SF galaxies

As already alluded to, the term *SF* is general and is used to describe all star-forming events. The term *SB* is occasionally used to describe short-duration SF events. A SB event can be observed in a certain molecular cloud in the galaxy or over an entire galaxy following a merger. In red galaxies, the SFR can be very small, below one $M_\odot$ yr^{-1}. In the Milky Way galaxy (which is not a SF galaxy), it is in the range 2–4 $M_\odot$ yr^{-1}. Some high-redshift galaxies show intense SF activity that can reach 1000 $M_\odot$ yr^{-1} and even higher.

Various names are being used to classify the subgroups among SF galaxies. They are usually related to the IR luminosity of the system since much of the SF-produced radiation is emitted in this part of the spectrum. The term *IR luminosity* is not very well defined. In many, but not all, papers, it refers to the integrated luminosity over the 8–1000 μm range. The classification of SF galaxies is based on their SFR, on L_{SF}, and on $L(IR)$.

Luminous infrared galaxies, LIRGs: These are galaxies with SFR in the range 10–100 $M_\odot$ yr^{-1}, which corresponds to a SF-produced luminosity of at least 10^{11} $L_\odot$.

Ultraluminous infrared galaxies, ULIRGs: These are larger, more luminous systems where the star formation rate is in excess of 100 $M_\odot$ yr^{-1}, corresponding to $L_{SF} > 10^{12}$ $L_\odot$. Most low-redshift ULIRGs are found in interacting systems, and it is thought that their intense SB activity is related to the merger event. A significant fraction of these galaxies also show indications for AGN activity. A large number of galaxies at $z \sim 2$ show SF activity with a rate which is similar to the one observed in local ULIRGs. These galaxies do not show signs of interaction. In this case, we are witnessing a steady SF activity over a period of many Myrs, similar to the steady SF in local galaxies with much lower SFR.

Submillimeter galaxies, SMGs: This group is at the top of the SF scale, with a SFR that exceeds 1000 $M_\odot$ yr^{-1}. The term *hyperluminous infrared galaxies* (HyLIRGs) is occasionally used to describe such systems. Most SMGs are found at redshifts 2–3. Their SFR is so large that a relatively short episode of star formation, about 10^8 yrs, can produce a total stellar mass similar to the mass of a large galaxy.

The standard classification detailed here is quite successful in describing the properties of the various groups of SF galaxies in the local universe and the transition from one type to the next as a function of SFR. The classification can also be generalized to include other properties, for example, the IR–SED or, equivalently, the dust temperature or range of temperatures. The classification starts to break down when we follow the different groups in redshift. ULIRGs at redshift ~ 2 resemble LIRGs in the local universe in terms of their SED and global morphology. Their typical dust temperature is lower and they are larger and not as compact as the local ULIRGs. Many such systems do not show any sign of interaction and are producing most of their new stars in large "blobs" of gas, typical of secularly evolving systems. The same is true for the SMGs that, at high redshift, resemble more the lower-redshift ULIRGs. These differences indicate strong evolution of the population of SF galaxies are correlated with the gas fraction in the galaxy, mostly molecular gas, which is higher at higher redshift. A more complete description of SF galaxies must include several parameters, such as gas content, SFR, and level of interaction, as a function of cosmic time. In particular, caution must be exercised when attempting to determine the SFR of a certain galaxy at high redshift based on known relationships in the local universe.

8.5.5. *The star formation sequence*

One way to visualize the properties and the evolution of SF galaxies is to draw the SFR versus M_* for all galaxies at a given redshift. Such diagrams show two distinct features. The first is a broad strip containing almost all SF galaxies where the lower-stellar-mass galaxies are also the ones with the lower SFR. All blue galaxies and many green-valley galaxies are concentrated in this band, which is sometimes referred to as the *main sequence*. Here we refer to it as the *SF sequence*. Since galaxy colors depend on both the SFR and the stellar mass, the bluest galaxies are at the bottom of the band, where the mass is the lowest. The redder SF galaxies (which are *not* red galaxies) are at the top because of their large M_*. The second region of the diagram is occupied by large-mass red galaxies with very low SFR. These galaxies are located well below the SF sequence. An example of these groups at redshift ~ 2 is shown in Figure 8.12.

The SFR for a given stellar mass increases with redshift. Combining with the observations of SF sequences at various redshifts leads to an expression of the form

$$[\text{SFR}] = a(z) \left[\frac{M_*}{10^{11} M_\odot} \right]^b , \tag{8.26}$$

where $b = 0.8 \pm 0.2$, with indications for a weak redshift dependence. The normalization of this relationship is given by the redshift-dependent term $a(z)$, expressed in units of $M_\odot \, \text{yr}^{-1}$. The observations show that $a(z)$ increases with redshift up to about $z = 2.5$ and then remains constant. A reasonable estimate over this range is $a(z) = 7(1 + z)^{2.5}$. The variation with redshift can be explained by the larger gas fraction of galaxies at earlier cosmological times.

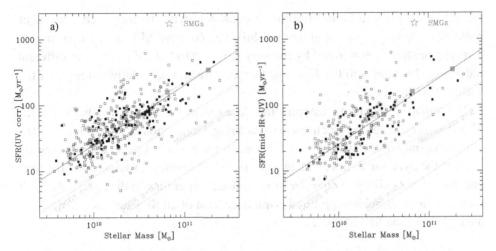

Figure 8.12. SFRs for $z = 2$ galaxies (points with a solid line) measured by two different methods as indicated on the vertical axes (from Daddi et al., 2007; reproduced by permission of the AAS) illustrating the SF sequence, which is the crowded band going from the bottom left to the top right of the diagram. The best-fit SF sequences for $z = 0.1$ and $z = 1$ galaxies are also shown with different colors. The theoretical prediction for $z = 2$ galaxies obtained from the millennium simulation is shown as a dashed line.

As noted earlier, the term *specific star formation rate* (SSFR) is used to describe the SFR per unit stellar mass. Equation 8.26 shows that for a given redshift, the SSFR is a slowly *decreasing* function of the stellar mass. The slope in this case is roughly -0.2 ± 0.2, but numbers as small as -0.5 have been suggested. Here, again, the normalization factor increases with redshift up to about $z = 2.5$.

8.5.6. The various modes of star formation

Analysis of galaxy samples at all redshifts, and SFR measurements especially at FIR wavelengths, for example, by Herschel, illustrates the two distinct modes of stellar mass growth (see the various values of the constant K in Equation 8.9). The first is a steady growth in disklike galaxies, which defines a tight correlation between SFR and stellar mass: the SF sequence. Most ($\sim 90\%$) SF galaxies are situated in this well-defined band. A small fraction of SF galaxies grow in a different, more intense starburst mode. They have larger SFRs for a given M_* and are situated above the SF sequence. These systems are thought to be the result of mergers between galaxies.

The common interpretation of the steady growth in disklike galaxies is that such systems contain many HII regions where SF is taking place. For a given redshift, the larger the galaxy is, the larger is the number of HII regions and hence the total SFR. Measurement of the molecular gas content in galaxies on the SF sequence at different redshifts suggest redshift-independent SF efficiency. The upward shift of the entire sequence at higher redshift is consistent with the increased fraction of molecular gas in galaxies at higher redshift (i.e., constant SFR per unit gas mass). This explanation is confirmed by the very similar SED of SF galaxies at different redshifts. This relates to the FIR dust temperature and the relative strength of the various PAH features.

The situation regarding mergers like low-redshift ULIRGs and high-redshift SMGs is very different. Such events can bring large amounts of cold gas to the common center of the system, which increases the gas density and column density, reduces the dynamical time, and changes the SF efficiency. In such cases, the SFR per unit gas mass is up to an order of magnitude larger than in disk galaxies. Such starburst galaxies, which represent a small fraction of all SF galaxies, are the ones located above the SF sequence.

8.5.7. *The star formation history of the universe*

Much of the interest in SF galaxies is related to their evolution and the global star formation history in the universe. This is closely related to the evolution and the growth of massive BHs in galactic centers, which is discussed in Chapter 9. The emerging picture is of a redshift-dependent SF history where the peak activity, that is, the largest SFR per unit co-moving volume (SFR density), is at around $z \sim 1.5$, corresponding to an age of the universe of about 4 Gyrs. At lower redshifts and later times, there is a noticeable decrease in the SFR density. Thus, the present SFR is much below the rate near the peak.

The separation into two modes of SF allows us to estimate the relative contribution of the different modes to the general growth of stellar mass in galaxies. It seems that the steady mass growth in disklike galaxies is the main contributer to the SFR density at all redshifts of up to at least 2. The merger-driven SF is a significant contributer to the cosmic SFR density only when considering systems with SFR $> 1000\ M_\odot\ yr^{-1}$.

At redshifts above about 3, the measurements are more uncertain. They suggest a slow decline in SFR toward very high redshifts with, yet, an undetermined slope. There were some attempts to estimate the cosmic SFR at redshifts as high as 8 or 9, but many of the observations are questionable. A diagram showing the trend of SF with redshift is given in Figure 8.13.

A challenging task, closely related to the cosmic SFR history, is to determine the metallicity evolution of SF galaxies. Methods for obtaining the metal abundance in ionized gas have been discussed in Chapter 5. They involve the comparison of various emission-line intensities mostly in the optical part of the spectrum. Such methods require several combinations of emission lines at different wavelengths. These methods are based on the dependence of the gas temperature on metallicity and give reliable results for abundances that are half solar or less. The analysis of higher metallicity gas requires the combination of several lines and photoionization

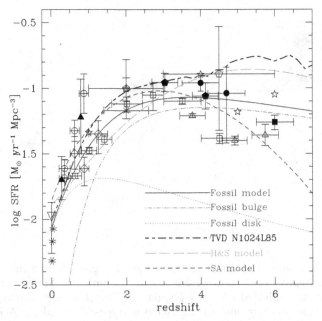

Figure 8.13. The SF history of the universe showing models (as marked) and observations (from Nagamine et al., 2006; reproduced by permission of the AAS).

modeling. Almost all methods are based on optical emission lines, and the studies at high redshift are most challenging since they require NIR spectroscopy of faint sources. Nevertheless, there are some interesting results based on studies of a handful of sources. One example, shown in Figure 8.14, presents fits to several

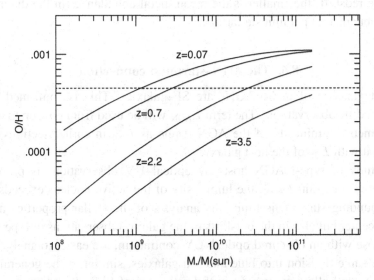

Figure 8.14. The metallicity indicator, N(O)/N(H), as a function of the total stellar mass at various redshifts, as marked. The horizontal dashed line marks the solar N(O)/N(H) (based on data presented in Maiolino et al., 2008).

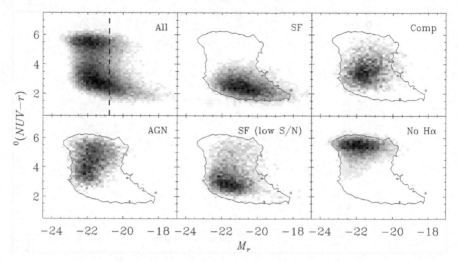

Figure 8.15. Classification of AGN hosts and other galaxies in a color absolute magnitude diagram. The maps were obtained by combining SDSS spectroscopy and photometry with GALEX NUV photometry. The classification into groups is based on diagnostic diagrams, and the AGNs are type-II low-redshift AGNs. Note that many AGNs hosts are located in the green valley between the red and blue galaxies (from Salim et al., 2007; reproduced by permission of the AAS).

measurements of N(O)/N(H) in SF galaxies of various masses at various redshifts. It suggests that, like the local systems, more massive galaxies are associated with higher metallicity even at high redshift. However, for a given galaxy mass, the larger the redshift, the smaller is the mean metal abundance (in the diagram, the oxygen abundance) per unit stellar mass.

8.6. The AGN–starburst connection

The host galaxies of many AGNs are SF galaxies. This is confirmed by UV, optical, and IR observations. The term L_{AGN} will be used from now on to describe the bolometric luminosity of the AGN (same as L_{bol} in earlier sections) and to compare it with L_{SF} of the host galaxy.

The study of type-I AGN hosts by optical–UV observations is problematic since, above a certain L_{AGN}, the luminosity of the active nucleus exceeds that of the surrounding stars. This limits the analysis of the stellar properties not only near the center but also in the outskirts of the galaxy. Observations of type-II AGN hosts, those with an obscured optical–UV continuum, are easier to analyze. They show the same division into blue and red galaxies, similar to the general galaxy population, and allow more meaningful estimates of SFRs based on optical–UV calibrators.

AGN hosts cover the entire range from blue to red galaxies. A diagram that illustrates this point is shown in Figure 8.15. There seems to be a notable concentration of AGN hosts in the green valley, the region between blue and red galaxies in the color versus stellar mass or color versus absolute magnitude diagrams. This may be interpreted as an indication that many of these hosts are in an intermediate transition stage between blue SF galaxies and "red and dead" galaxies. Such ideas go nicely with the suggestion that the AGN phase in the life of a galaxy is seen mostly toward the end of the intense SF phase and, perhaps, even after it (note, however, the earlier comments about the excessive reddening of green valley sources). Interestingly, there are many high-mass SF galaxies in the same region of the diagram, in agreement with the similarity between AGN hosts and similar mass SF galaxies.

The fraction of AGN among low-mass blue galaxies is much smaller than among the more massive objects. As for very high mass red galaxies, it is well known that some high-luminosity AGNs are associated with giant red galaxies. Notable examples are hosts of powerful radio-loud AGNs. The fraction of such sources in the high-ionization (i.e., not including LINERs) AGN population is about 10 percent. However, exact statistics regarding the properties of the hosts of luminous radio-loud AGNs is lacking, especially at redshifts larger than about 0.3.

How can we separate the SF-produced radiation from the AGN-produced radiation in type-I AGNs, especially in high-redshift sources where the host galaxy is barely resolved? For example, how can we separate the two when the UV continuum is dominated by the AGN radiation and the narrow $H\alpha$ line includes contributions from AGN-excited gas? More generally, out of the various methods that are used to measure SFR in nonactive galaxies, which ones can be applied to AGN hosts?

It seems that the most reliable way to measure SFR and L_{SF} is to use FIR observations. Several studies based on Spitzer and Herschel observations show that the FIR continuum in intermediate- and high-luminosity type-I AGN is almost entirely due to SF activity. This conclusion is based on two fundamental observations. The first is the comparison of several MIR PAH emission features with $L(FIR)$ in AGNs and in ULIRGs. As explained, PAH features are, perhaps, the cleanest SF indicators. They are very strong in SF galaxies (Figure 8.10) and are believed to be suppressed by the hard AGN radiation. Spitzer MIR spectroscopy clearly shows the presence of intense PAHs in a number of luminous type-I AGNs, like the ones shown in Figure 8.16. The strong PAH emission seen in both AGN spectra, near 7.7 μm, resembles similar features seen in many LIRGs and ULIRGs.

The more complete statistics is obtained from observations of a large number of AGNs that are used for comparing the FIR luminosity that is known from IRAS, Spitzer, and Herschel surveys, with the luminosity of the PAH features measured from ISO and Spitzer spectroscopy. Such a comparison is shown in Figure 8.17 for a sample of PG-QSOs and nearby ULIRGs. The diagram reveals a strong

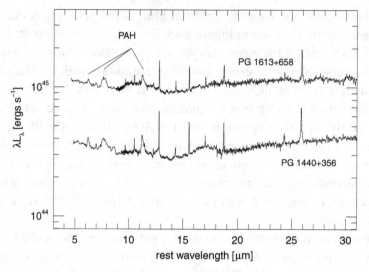

Figure 8.16. The MIR spectra of two AGNs, PG 1613+658 ($z = 0.129$) and PG 1440+356 ($z = 0.079$), showing strong PAH features in the spectrum. These are similar to the features observed in the spectrum of M 82 (Figure 8.10) and other SF galaxies. They are thought to be due to intense SF activity in the host galaxies of these objects in regions shielded from the hard AGN radiation field.

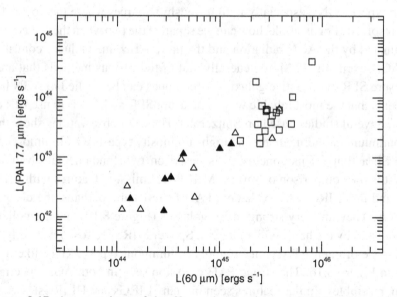

Figure 8.17. A comparison of the 60 μm luminosity and L(PAH 7.7 μm) in a sample of luminous AGNs and ULIRGs. Open squares are low-redshift ULIRGs. Triangles are PG-QSOs where open symbols are upper limits and closed symbols are detections.

correlation of $L(7.7\,\mu\text{m})$ and $L(60\,\mu\text{m})$, which is used here as a proxy for $L(\text{FIR})$. If the observed PAH features are due entirely to SF, then this must also be the case for the observed FIR emission. This would mean that basically all the FIR emission in those AGNs is due to SF in their hosts. Further support is obtained from the realization that lower-luminosity SF galaxies that do not show any indication of AGN activity (unlike ULIRGs, which are suspected to have some AGN contribution to the bolometric luminosity) follow a similar trend.

The second supportive evidence is the great similarity in the FIR colors of SF galaxies with no indication of an AGN and AGN hosts of similar FIR luminosity. The two IRAS FIR bands, and the six Herschel FIR bands, can be used to characterize the broadband SED over a large luminosity and redshift range. The similarity is hard to explain if the main source of the heating of the 30–60 K dust is due to a central point source in one case and a large number of obscured HII regions in another.

The findings about the stellar origin of the FIR emission in AGN hosts can be used as a basis for measuring L_{SF} and SFR at all redshifts. As will be shown in Chapter 9, the comparison of L_{AGN} and L_{SF} can provide a powerful tool to compare the growth rate of the BH and the stellar mass in the same galaxy and in the general galaxy population.

As already alluded to, two other indicators can be used to compare L_{SF} and L_{AGN} in type-II AGNs at optical and UV wavelengths: UV photometry and the $D_n 4000$ method. The most accurate UV-based determination of SFR is based on multicolor spectral synthesis to determine the properties of the (extincted) stellar population using, among other things, UV measurements provided by GALEX. Thousands of such measurements are available for the SDSS sources. As for the $D_n 4000$ index, this is available for many thousand type-II AGN observed by the SDSS.

How many AGNs are located on the SF sequence or above it? The answer to this question for high-ionization AGNs at low redshift is no less than 2/3. The fraction at higher redshift is more difficult to assess, but the indications that we have suggest that it is larger, perhaps 80 or even 90 percent, depending on the redshift.

The combination of all methods – measuring $L(\text{FIR})$ from IRAS, Herschel, and Spitzer photometry; UV imaging and spectroscopy of type-II AGN, the calibrated $D_n 4000$ method and the $L(\text{PAH})$ method – can be used to correlate L_{AGN} and L_{SF} for a large number of AGNs over a large luminosity and redshift range. One such comparison is shown in Figure 8.18. In this example, the relationship is plotted only for *AGN-dominated* systems, that is, those AGNs with $L_{\text{AGN}} > L_{\text{SF}}$. It shows a very strong dependence of the two luminosities, $L_{\text{SF}} \propto L_{\text{AGN}}^{\alpha}$, with $\alpha \simeq 0.7$. Extrapolation to low luminosity indicates that $L_{\text{SF}} \simeq L_{\text{AGN}}$ at around 10^{43} erg s^{-1}. This luminosity corresponds to very low SFR, below 0.3 $M_\odot$ yr^{-1}. At the other extreme of very high luminosities, where $L_{\text{AGN}} \sim 10^{48}$ erg s^{-1}, the correlation shows that $L_{\text{AGN}} \sim 25\, L_{\text{SF}}$.

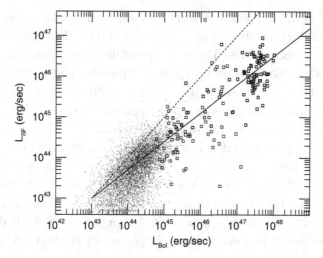

Figure 8.18. A comparison of L_{SF} versus L_{bol} for several AGN samples in the redshift range 0–3. The diagram is adapted from Netzer (2009) and includes only AGN-dominated systems where $L_{bol} > L_{SF}$. The solid line is drawn by hand and goes through the points with a logarithmic slope of 0.7. The dashed line shows the 1:1 relationship.

The comparison of L_{AGN} and L_{SF} looks different when all AGNs, including those with $L_{SF} > L_{AGN}$, are included. Such a case is shown in Figure 8.19, where Herschel observations of a large number of X-ray-selected AGNs are depicted. In this case, there are many more sources to the left of the correlation curves, and the different colors and lines show that each redshift corresponds to a roughly horizontal line of constant L_{SF}. Such lines can be interpreted as marking short intervals of time during which L_{AGN} increases following the onset of a SF event with constant SFR.

The relationship in Figure 8.18 is intriguing. It suggests that for AGN-dominated sources, the amounts of cold star-forming gas and gas that is accreted onto the central BH are related. This is surprising given that SF regions are known to occupy very large regions, several kpc in scale, while the BH sphere of influence is of order several pc and the accretion itself occurs in a region that is a small fraction of a parsec. Indeed, some nearby low-luminosity AGN show intense SF activity in their inner ~ 10 pc, yet this is not necessarily true for the entire population and is also unlikely in sources near the top of the diagram, where L_{SF} indicates SFR $> 1000 \, M_\odot$/yr. An interesting, not entirely confirmed suggestion is that most of the objects on the correlation line are in merging systems, and the reason for this trend, at least in high-redshift sources, is related to the conditions prevailing during the more luminous phases of the merger.

The explanations of the various relationships between L_{AGN} and L_{SF} described here must be oversimplified for various reasons. First, they take no account of red

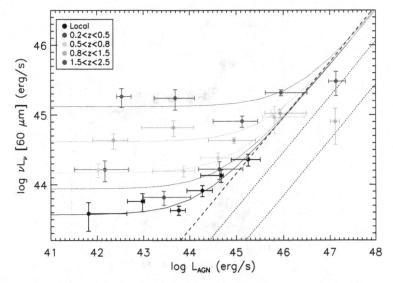

Figure 8.19. Same as Figure 8.18, except that all AGNs, including those where L_{SF} exceeds L_{bol}, are included. The points represent X-ray-selected AGNs whose SF luminosity was obtained by Herschel. The different redshift groups are marked with different colors, and the horizontal lines illustrate specific evolutionary scenarios leading to the correlation line on the right (adapted from Rosario et al., 2012). (See color plate)

and dead AGN hosts. As explained, the fraction of these among high-ionization AGN at high redshift is still an open question. Second, there is no account of redshift evolution, while this is obvious from the mere fact that the great majority of the objects at the high L_{AGN} end are high-redshift sources. A more natural functional form, which is likely to emerge from a more complete comparison, may look like

$$L_{SF} \propto L_{AGN}^{\alpha_1}(1 + z)^{\alpha_2}, \tag{8.27}$$

where α_1, α_2 are still to be determined. Again, at the high L_{SF} end, many of the sources shown here are probably in interacting systems, while at $z \leq 1$, most AGN are in noninteracting hosts, and the feeding of the BH is due to secular processes.

The preceding relationships, the close connection between the mass of the central BH and the bulge mass, and several other indicators, all point to a correlated evolution of galaxies and the massive BHs in their centers. The next chapter examines these questions from the viewpoint of the evolution of the entire AGN population.

8.7. A combined AGN–galaxy SED

Present-day observations provide photometric information over a very wide wavelength range, from hard X-ray energies to the FIR. For luminous type-I AGN, the

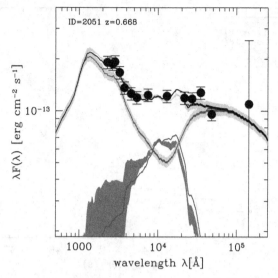

Figure 8.20. The combined AGN–galaxy SED. The observations (solid points) are modeled with a combination of a typical AGN (top light color line) and a host galaxy (bottom line with a peak around 1 μm). For very luminous AGNs, the galaxy contribution is negligible. For intermediate-luminosity AGNs, like the one shown here, the host galaxy can contribute significantly, especially around 1 μm, where the AGN SED has a local minimum (courtesy of Angela Bongiorno).

entire range, except for the FIR part, is dominated by the AGN. In type-II sources, much of the light is due to the host galaxy. There are many type-I sources where the galaxy contribution in the optical–NIR band is large enough to affect the shape of the SED. In such objects, the broad emission lines can provide an estimate of the BH mass and the galaxy light an estimate of the stellar mass. This is a direct way to compare the evolution of the BH and its host over a large redshift range.

Figure 8.20 shows a fit of the combined AGN–host spectrum in an intermediate-luminosity type-I source. In this case, the galaxy contribution is large enough to dominate the emission at wavelengths around 1 μm, where there is a dip in the AGN spectrum (Figure 1.5). This provides a measure of both the SFR and M_* in the host. In more luminous AGN, this is impossible to do because of the more dominant AGN contribution in this part of the spectrum.

8.8. Further reading

Observations of host galaxies: See review by Veilleux (2008). For evidence of undisturbed morphology, see Cisternas et al. (2011).

Scaling relationships between M_{BH} and various host properties: See reviews in Kormendy and Richstone (1995), Peterson (2008), and Ho (2008) and discovery papers by Ferrarese and Merritt (2000) and Gebhart et al. (2000). For recent measurements of the $M-\sigma^*$ relationship and the $M_{BH}-M_*$ correlation, see Gultekin et al. (2009), Sani et al. (2011), and McConnell et al. (2011).

BHs in core ellipticals: See discussion and examples in Kormendy et al. (2009).

Red and blue galaxies: There are hundreds of papers discussing these issues. See Blanton et al. (2003) and Brammer et al. (2009), and references therein. For AGN hosts, see Kauffmann et al. (2003) and Kewley et al. (2006).

Statistics of AGNs in different host galaxies: See Schawinski et al. (2010) for a comprehensive study based on the Galaxy Zoo and references to older work. See Kauffmann et al. (2005) for the special case of local radio-loud AGN.

Basic SF theory and observations and estimates of SFR: See the review by Kennicutt (1998) and some refinements in Hirashita et al. (2003), Cardiel et al. (2003), Kennicutt et al. (2009), and references therein. Veilleux (2008) is a comprehensive review of AGN hosts with emphasis on starburst galaxies like ULIRGs.

The SF sequence (main sequence): See review and references in Daddi et al. (2007) and Dutton et al. (2010). For various modes of star formation, see Daddi et al. (2010) and Wuyts et al. (2011).

The AGN–starburst connection: Of the very large number of papers published every year on various aspects of this topic, the following contain multiwavelength evidence and many references to other publications: Salim et al. (2007), Veilleux (2008), Netzer et al. (2007a), Sani et al. (2010), Schweitzer et al. (2006), Wild et al. (2010), and Trichas et al. (2009).

9

Formation and evolution of AGNs

9.1. The redshift sequence of AGNs

AGN are now detected with ground-based telescopes all the way to $z \sim 7$. This is done by observing thousands of sources discovered by large surveys like SDSS and 2DF. Arranging the objects according to their redshift, like in Figure 9.1, clearly shows the redshift progression of absorption by intergalactic neutral hydrogen starting at an observed wavelength of $(1 + z)1215$ Å. At $z = 7$, this is close to the long wavelength limit of ground-based spectroscopy. Detailed study of such sources, and measurement of their mass and accretion rate, can be used to follow BH evolution through cosmic time and to compare it with the evolution of dark matter halos and galaxies. The main tools that are used for this study are the luminosity and mass functions of AGNs. Before investigating these functions, we review, briefly, some aspects of galaxy evolution.

9.2. Highlights of galaxy evolution

9.2.1. Hierarchical structure formation

The most successful cosmological model of today is the Λ cold dark matter (ΛCDM) model with its three ingredients: dark energy, dark matter, and baryonic matter. Support for this model comes from measurements of the acceleration of the universe, from observations of the cosmic microwave background (CMB), from the abundance of the light elements, and from several other observations. The success of the model in explaining the observed temperature fluctuations at the recombination era, about 380,000 years after the Big Bang, when the radiation and baryon fluids stopped interacting with each other, is perhaps its main strength.

The theoretical framework of the ΛCDM model is based on an early stage of inflation that made the universe flat. Inflation also generated a spectrum of adiabatic density fluctuations that grew, at later stages, by gravitational instabilities to form the structure observed today. The power spectrum of the fluctuations

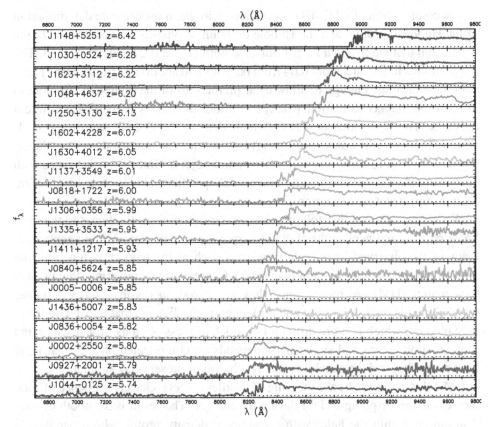

Figure 9.1. The redshift sequence of AGNs: Spectra of various redshift SDSS AGNs showing intergalactic absorption at progressing rest wavelength (from Fan et al., 2006; reproduced by permission of ARAA).

has been measured on smaller and smaller scales starting with the COBE satellite and continuing with WMAP and several successful balloon experiments and (soon to be announced) PLANCK. The most noticeable features that are accurately reproduced by the model are the three Doppler peaks due to acoustic oscillations in the photon–baryon fluid at the last scattering surface. The CMB temperature fluctuations provide the most precise constraints on the basic cosmological parameters.

The ΛCDM model is also supported by other observations based on the imprint of acoustic oscillations on the distribution of local galaxies. Such studies are based on the distribution of galaxies in several large local surveys such as SDSS, 2DF, and COSMOS (see Chapter 1). These are used to measure the power spectrum of the primordial density fluctuations that lead to the formation of structure in the universe. The model predictions are also supported by observations of weak lensing and the Lyman absorption forest.

The combination of all CMB observations with the most accurate determination of the Hubble constant, and the Hubble diagram of type-Ia supernovae, led to the conclusion that we live in a $\Omega \simeq 1$ universe with $\Omega_M \simeq 0.27$ and $\Omega_\Lambda \simeq 0.73$. The contributions to Ω_M are about 0.04 from baryons and about 0.23 from CDM. These values are accurate enough to allow clear predictions of the growth of structure in the universe, galaxy formation, galaxy evolution, and galaxy properties at different redshifts. The basic idea is that for small density fluctuations with $\delta\rho/\rho \ll 1$, perturbations on all scales grow as $1/(1 + z)$ after recombination and collapse, when $\delta\rho/\rho \sim 1$, to form gravitationally bound objects. Since $\delta\rho/\rho$ increases with decreasing mass scale, the smaller objects collapse first. This leads to hierarchical (bottom-up) structure formation.

N-body simulations that follow structure formation from the linear to the non-linear regime, and neglect the effect of baryons, are quite successful in explaining many of the observed large-scale properties such as the halo mass function (the number density of halos per unit halo mass). Cosmological predictions that involve dark matter (DM) are robust since they are based on dissipationless collapse, which depends only on gravity and dark energy. Predictions that depend on baryonic physics, which is more difficult to model, are less obvious. In particular, the accumulation of gas and stellar mass depends on complicated physics that is, at least partly, beyond the capability and resolution of present-day simulations.

DM halos assemble by the merging of smaller objects. One consequence of such mergers is the stripping of DM during the merger. The result is a smooth density distribution within the halo, with a canonical density profile whose parameters depend on the halo's mass and the redshift. Galaxies form within these structures due to radiative cooling of the hot baryons in the halo. Additional processes that must be considered include feedback (see later) and cold baryons flowing onto the newly formed galaxies.

Figure 9.2 shows the result of a set of simulations that follow the merging of dark matter halos. This simulation starts from a certain mass halo in the present universe ($10^{13} M_\odot$) and follows one of its various possible histories by going backward in time. Obviously there are several other paths that can lead to the same present mass halo, each with its own galaxy formation and evolution history.

9.2.2. Formation and evolution of galaxies

Heating and cooling of baryonic matter in DM halos

Galaxy formation follows the collapse and merging of DM halos. This leads to the heating of the dark and baryonic matter due to the conversion of gravitational potential energy into kinetic energy. For the collisionless DM, this process results in the increase of internal energy. For the baryons, it results in the conversion of

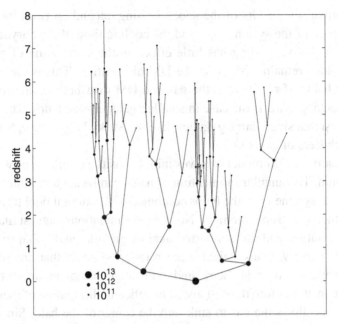

Figure 9.2. The merger tree: The merging history of a medium-size halo, with mass of $10^{13}\ M_\odot$ at $z = 0$. Larger and larger dark matter halos assemble as the age of the universe increases (courtesy of Irina Dvorkin).

gravitational potential energy into thermal energy in shocks through the supersonic collision of clumps and streams of baryons.

One way to estimate the characteristic temperature involved in the process is to calculate the gas temperature, assuming *no radiation cooling*. This gives the virial temperature of the halo,

$$T_{\text{vir}} \sim \frac{1}{2} \frac{m_p}{k} V_{\text{vir}}^2\ K, \qquad (9.1)$$

where m_p is the mass of the proton and V_{vir} is a characteristic velocity of the halo defined by its mass M_H and size r_{vir}.

$$V_{\text{vir}} = \sqrt{\frac{GM_H}{r_{\text{vir}}}}. \qquad (9.2)$$

The size r_{vir} depends on the virial density $\rho_{\text{vir}} \propto (1 + z)^3$. Using this definition, we find that

$$T_{\text{vir}} \simeq 6 \times 10^5 \left[\frac{M_H}{10^{12}\ M_\odot}\right]^{2/3} (1 + z)\ K. \qquad (9.3)$$

The hot baryons cool through various atomic and molecular two-body processes (Chapter 5). While the initial velocity distributions of the DM and the baryons are very similar, the results of the shock heating depend on two time scales: the dynamical time of the system, t_{dyn}, and the cooling time of the baryons, t_{cool}. For $t_{cool} \gg t_{dyn}$, radiative cooling has little effect on the distribution of gas after the shocking, which remains similar to the DM distribution. This is the case in large cluster-size halos. If $t_{cool} \sim t_{dyn}$, the gas and DM distributions are very different after the shocking. This is the case in smaller, galactic-size halos. These considerations suggest that single large galaxies, like the Milky Way galaxy, formed earlier than large clusters of galaxies.

The standard ΛCDM model involves initial density perturbations with no angular momentum. The angular momentum of halos can be acquired due to departure from spherical symmetry of the halos combined with external tidal torques that are coupled to their quadruple moment. Numerical simulations suggest that DM halos are weakly rotating and are supported against gravitational collapse by random motions. Conversely, many galaxies are rotating systems that are supported by rotational motions. This must be related to the angular momentum of the halo and the baryonic matter before the collapse. The efficient dissipation of thermal energy by the baryons allows the gas to sink into the center of the halo. Since the initial specific angular momentum of the gas particles is conserved, the collapse of the cooled baryonic matter stops when the newly formed object becomes rotationally supported. Simulations suggest that the process results in a large ($\sim$10 kp) central disk galaxy and several smaller satellite galaxies. At later times, the smaller satellite galaxies sink to the center by dynamical friction and merge with the central galaxy.

Gas heating in shocks, followed by cooling and sinking to the center of the halo, is not the only possibility. *Cold accretion* via cold streams (cold gas accretion along filaments) have also been suggested to bring cold gas to the center without heating it to the virial temperature. Regardless of the exact nature of the process, the increase of gas density and column density in the central object results in SF in the newly formed disk. This leads to decrease in the gas-to-star ratio and to various feedback-related processes (see later).

Mergers and galaxy morphology

Observations of the Hubble morphological sequence, combined with the theory of structure formation, suggest that galaxy mergers are important in transforming late-type disk galaxies into the very common early-type, bulge galaxies observed in the local universe. The basic idea is that merging happens in a common halo following halo merger, where each of the progenitor halos contains at least one disk galaxy. Dynamical friction in the common halo leads to the decay of the orbit and the formation of a new central galaxy. The time scale of the merger depends on the size and density of the halo and is, in general, shorter at higher redshifts, when the density is larger. A typical time scale for the merger of large, similar mass galaxies ("similar" in this context meaning mass ratio smaller than 3) is about 10–20 percent of the age of the universe at the time of the merger.

The tidal forces operating during mergers of large similar-size galaxies lead to the destruction of both disks. Gas is pulled from the disks and sinks to the center of the system. This results in rapid SF (a starburst). Most of the gas is used up in the process, and the final galaxy is a gas-poor system with ellipsoidal appearance, similar to elliptical galaxies. Such spheroids can grow new disks through later mergers. The merger of galaxies with nonsimilar masses is different. Much of the effect in such mergers is a thickening of the disk and not its destruction. The name *dry merger* is often used to describe the merging of two gas-poor systems.

Isolated spiral and bar spiral galaxies can also evolve via *secular evolution*. These processes are very important for the understanding of BH feeding and are described in § 9.4.

Luminosity function, mass function, and downsizing of galaxies

Hierarchical structure formation suggests a bottom-up evolution where small structures lead to larger structures. This seems to be in good agreement with the observations of large, halo-size systems. However, observations of galaxies at high redshift suggest that the baryonic matter grows in an antihierarchical fashion. In particular, the most massive galaxies of today grew almost to their observed size at redshifts larger than 3, while the smaller galaxies accumulated most of their stellar mass at much later times.

A useful way to follow galaxy evolution observationally is to construct luminosity functions (LFs) and stellar mass functions (MFs) of all galaxies at a given redshift. The LF, $\Phi(L, z)$, is defined such that $dN = \Phi(L, z)dVdL$ is the number of galaxies per unit comoving volume, dV, at redshift z in the luminosity interval $L, L + dL$. The use of comoving volume, which is obtained from integrating the volume between $r(z)$ and $r(z + \Delta z)$ using the assumed cosmology, enables a straightforward comparison of the same population of sources at all redshift. Thus, $\Phi(L, z)$ provides a concise way to describe galaxy evolution at different cosmological times.

Galaxy count and photometric observations show that, in many cases, the shape of the LF is given by a so-called Schechter function,

$$\Phi(L, z)dL = \Phi(L_*(z)) \left[\frac{L}{L_*(z)} \right]^{-\gamma} \exp(-L/L_*(z))d\left(\frac{L}{L_*} \right), \qquad (9.4)$$

where $L_*(z)$ is a characteristic luminosity at redshift z in the photometric band in question. The change of $\Phi(L, z)$ with redshift can be described by a "pure luminosity evolution,"

$$\Phi(L, z) = \Phi(L/L_*(z), 0), \qquad (9.5)$$

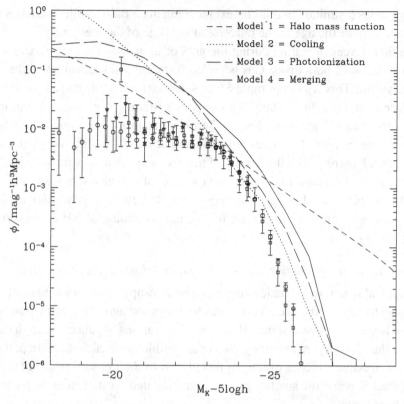

Figure 9.3. Computed (continuous curves) and observed (individual points) K-band $z < 0.1$ galaxy LFs (from Benson et al., 2003; reproduced by permission of the AAS). Model 1 (dashed line) shows the result of converting the dark matter halo mass function into a galaxy luminosity function by assuming a fixed mass-to-light ratio chosen to match the knee of the luminosity function. Model 2 (dotted line) shows the result when no feedback, photoionization suppression, galaxy merging, or conduction are included. Models 3 and 4 (long dashed and solid lines, respectively) show the effects of adding photoionization and then galaxy merging.

a "pure density evolution,"

$$\Phi(L, z) = N(z)\Phi(L, 0), \qquad (9.6)$$

or a combined "density and luminosity evolution,"

$$\Phi(L, z) = N(z)\Phi(L/L_*(z), 0). \qquad (9.7)$$

Redshift-dependent LFs can be compared with simple theoretical predictions that include only hot baryon cooling, as shown in Figure 9.3. The comparison shows that the agreement between model and observations is good for a narrow range of luminosities around L_*. However, the calculated number of the most massive galaxies and the least massive galaxies deviates substantially from the

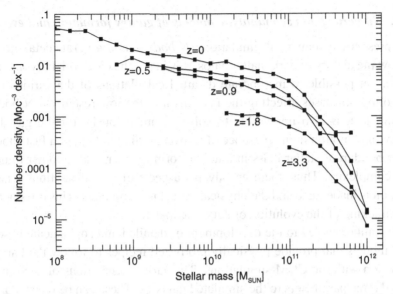

Figure 9.4. Stellar MFs at different redshifts (constructed from data provided in Pérez-González et al., 2008).

observations. Moreover, there is a significant discrepancy between the prediction that atomic cooling should lead to the condensation of about 80 percent of all available baryons into gas and stars in galaxies and the observations that show that this fraction is less than 10 percent at $z = 0$.

A possible solution is to invoke a "feedback" process, that is, a mechanism that inhibits gas cooling and SF at late stages of galaxy formation. This can be due to stars ("stellar feedback") or to the active central BH ("AGN feedback"). For example, it is thought that stellar feedback, in particular, the photoionization of the baryonic gas in the halo, can slow down the formation of small-mass galaxies in DM halos, thus explaining the lack of objects at the low-luminosity end of the LF. This, however, cannot explain the lack of very large galaxies predicted by the simple models, and other feedback processes must be invoked. Such processes are explained in more detail in § 9.8.3.

The next step is the construction of redshift-dependent stellar MFs. Such functions are obtained from the available LFs by using known L/M relationships for different types of stars. As explained in Chapter 8, good estimates of the fractional number of stars of different types in a certain galaxy can be obtained by fitting the observed SED of the galaxy with a combination of stellar spectra, assuming a certain IMF. Stellar MFs constructed in this way, for example the ones shown in Figure 9.4, suggest that at low redshift, the MF evolves more strongly at low mass, that is, the larger galaxies grew first and the lower-mass ones are still growing significantly today. This antihierarchical growth is referred to as *downsizing*.

9.2.3. *Numerical and semianalytic models of galaxy formation and evolution*

While present-day numerical simulations (*N-body simulations*) are detailed enough to follow the history of dark matter halos, this in not the case with baryonic matter. It is not yet possible to incorporate detailed calculations of the various heating and cooling processes affecting the baryons into the low-resolution N-body DM simulations. It is also impossible to solve, simultaneously, for the gas density, pressure, and motion in the presence of a strong radiation field or a fast shock. SF, and the buildup of stellar mass in a nonuniform manner, adds another dimension to the complexity. Thus, there are always aspects of the calculations where the resolution is inadequate and the physical model is incomplete. This does not allow proper tracking of the evolution of baryonic matter.

Such limitations led to the development of detailed and sophisticated "semianalytic" models that provide physically motivated recipes to follow the baryons in conjunction with the CDM simulations. The earlier generations of such methods assigned three parameters to the simulated galaxies. These can be the stellar mass, the cold gas mass, and the hot gas mass. The BH mass can be used as a fourth parameter. The recipes can then be used to describe the physical processes in the galaxy. They can also be used to determine the SFR given a certain amount of cold gas, to fix the fraction of thermal energy that is returned to the gas following a SN explosion, or the fraction of AGN-emitted radiation that interacts with the gas and the stars, thus inhibiting further SFR (feedback).

The combination of N-body simulations with semianalytic models allows us to test a large range of physical scenarios by comparing the results with real observations. This can later be used to improve the recipes used by the models. One advantage of this approach is that the models can easily be varied to test the influence of a specific assumption on the results. An important example is the calculation of the SEDs of the simulated galaxies that are related to the various SF histories, the assumed IMF, the size of the galaxy, and its age. Another important example is the attempt to predict the chemical composition of various types of galaxies at various ages. All these predictions can be compared directly with observations. Unfortunately, this is all only a substitution for a full description of all time-dependent physical processes.

The most advanced numerical calculations include a more realistic attempt to follow the hydrodynamics of the gas in an expanding universe. Two methods that are commonly used are based on the smooth particle hydrodynamics (SPH) technique and on grid-based Eulerian schemes. The idea is to track both the dark matter particles, which respond only to the gravitational force, and the baryonic particles that can also feel pressure and dissipate thermal energy through cooling. Such codes can follow the gas density, temperature, and dynamics on a much smaller scale, alleviating some of the need for the semianalytical approximation. The need for recipes still exists, but they can be applied to "individual particles" (as defined by the simulation), or clumps, and not to an entire galaxy. For example, given the density and total amount of cold gas computed by such codes, the assumed SFR is no longer a property of an entire galaxy.

9.2.4. The role of supermassive BHs in galaxy evolution

The properties of accreting supermassive BHs are clearly correlated with the properties of their host galaxies. The strongest correlations are observed in early-type galaxies with relaxed bulges. Here M_{BH} is correlated with the stellar velocity dispersion $\sigma*$, with the mass of the bulge of the host, and perhaps also with the total stellar mass (Chapter 8). This suggests that supermassive BHs are affected by and probably influence the evolution of their hosts. This parallel evolution is usually referred to as *coevolution* of BHs and their hosts. The understanding of this coevolution is a major goal of numerous observational and theoretical studies. In particular, most of today's large-scale cosmological simulations include, as an integral part, the additional ingredient of BH formation, merger, and growth.

An important clue to understanding the parallel evolution is the mass ratio of the BH and stellar (or bulge) mass of the host. As shown in Chapter 8, this ratio is known in the local universe, where it seems to depend on M_{BH}, being smaller in galaxies that host the most massive BHs (Figure 8.4). Unfortunately, similar studies at high redshift do not seem to reach a clear conclusion about this ratio for several reasons. The only reliable measurements of BH mass at high redshift is in type-I AGNs, where we can use the virial, RM-based mass estimate. However, in such sources, the very luminous nonstellar continuum masks the much fainter stellar light, making stellar mass estimates most uncertain. Much better estimates of M_* are available in galaxies where the BH is dormant or in type-II AGNs. Unfortunately, there is no way to obtain reliable estimates of M_{BH} in such objects at high redshift. Most important, there is no confirmation that the $M-\sigma^*$ relationship obtained at low redshift applies also at high redshift because of the uncertain evolutions of M_* and M_{BH}. There are various ways to try to overcome this limitation, so far with limited success. The little evidence we have suggests that for a given galaxy, M_*/M_{BH} was larger in the past than it is today.

The remainder of this chapter is devoted to the understanding of BH formation, evolution, and growth that will lead, eventually, to a better understanding of the coevolution of galaxies and the supermassive BHs at their centers.

9.3. Seed BHs

Observations of high-redshift AGNs, at $z \sim 6$, show that some BHs with $M_{BH} \sim 10^9 M_\odot$ already exist at these redshifts, when the age of the universe was only about

1 Gyr. Observational limitations prevent the measurements of smaller BHs at such redshifts. However, it is likely that the number of BHs with much smaller mass far exceeds the number of the most massive objects. This means that massive BHs must have been common in the early universe immediately after the first stars were born. Going back in time, we observe the properties of these first supermassive BHs, those referred to as "seed BHs" with mass M_{seed}, that represent the first stage of growth of the most massive BHs observed today. It is thought that such objects existed in the very early universe, as early as $z = 10$ or perhaps even at $z = 20$.

There are at least three possibilities regarding the physical processes that can lead to the formation of seed BHs.

Population III stars and stellar remnant BHs: The first galaxies started their growth when the universe was less than 0.5 Gyr old, and it is suggested that the first BHs can be traced to about the same epoch. The theory of structure formation in a universe dominated by CDM was described earlier. It involves the collapse of local perturbations on larger and larger scales. This bottom-up hierarchal growth takes place when the growth of the initial density perturbations in the universe, which started before recombination, enter the nonlinear regime. The first galaxies form from the baryons in the first halos. Initially, the baryons are shock heated to the virial temperature. Their cooling leads to the formation of the first stars and also the first class of seed BHs.

The first population III (POPIII) stars were formed from zero-metallicity gas. It is thought that this process took place in halos (or "minihalos") of about $10^6 \, M_\odot$ at $z = 20$–40. There are many open questions as to the formation, evolution, and death of massive POPIII stars. The cooling of such stars must be dominated by atomic or molecular hydrogen, and calculations show that they must be very massive objects with masses well over $100 \, M_\odot$. The lifetime of such stars is very short, a few Myr. Low-metallicity stars of $M_* \sim 100 \, M_\odot$ are thought to form BH directly, with masses that are roughly half their initial mass. At somewhat larger mass, $\sim 200 \, M_\odot$, pair production becomes important and leads to SN explosion without a dense remnant. At still higher mass, $250 \, M_\odot$ and heavier, conditions are again different, and BHs with about half the initial mass are the end result of the stellar evolution. Thus, remnant BHs with mass $M_{BH} \sim 100 \, M_\odot$ are predicted in some models but are definitely not inevitable. Such remnants can initiate the later growth of supermassive BHs.

Direct collapse and stellar dynamics processes: The second class of seed BHs is speculated to result from the direct collapse of a very massive cloud of dense gas. Such metal-poor clouds are thought to exist in the centers of protogalaxies. Unlike metal-enriched gas clouds that cool efficiently and

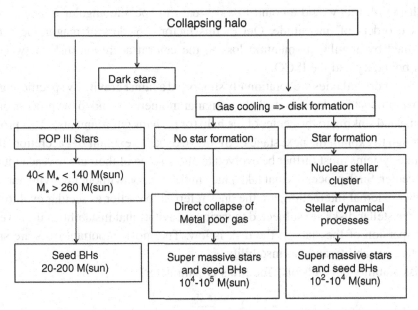

Figure 9.5. Various pathways to the formation of supermassive seed BHs in the earliest galaxies (adapted from Volonteri, 2010).

hence fragment, very low metallicity gas clouds retain their original masses and temperatures. An additional important ingredient is the cloud's angular momentum, which must be small to allow collapse. The evolution of such objects involves various types of collapse at various stages and perhaps disk and bar formation. It also involves the formation of supermassive stars at the final stage. Such stars evolve and produce BHs that continue to accrete from the surrounding gas. The evolution ends with the formation of seed BHs with masses in the range $10^4 - 10^5 \, M_\odot$.

BH formation via dynamical stellar processes: The third suggested scenario involves processes in dense stellar systems that were formed at the earlier stages of the collapse, before the earliest BHs. Such processes among stars can lead to the formation of seed BHs with typical masses of 1000–2000 $M_\odot$.

Figure 9.5 illustrates the three possibilities as well as the next stages in the growth of such BHs.

9.4. Feeding the monster

9.4.1. Angular momentum disk instability and secular evolution

A major gap in our understanding of AGN physics is the process by which far-away particles in the galaxy lose almost all of their angular momentum and get close enough to the central BH to be accreted via the central accretion disk. As explained in Chapter 4, this would amount to a decrease in specific angular momentum by about 6 orders of magnitude. Out of this factor, 2 orders of magnitude can be explained by angular momentum loss in the central accretion disk, between its outer boundary and the ISCO.

Newly born galaxies are rotationally supported against gravity by specific angular momentum that is related to the halo's angular momentum and to the process of gas cooling and sinking to the center of the halo (or perhaps gas filaments). This process is enough to explain the next stage of SF in the disk. To explain BH accretion, these disklike systems must further be evolved to allow some of their gas to reach into the very center. Such processes that take place in the galaxy, without involving mergers or other external interaction, are the ones referred to earlier as secular evolution.

Large stellar disks are subjected to various gravitational instabilities if the velocity dispersion of the stars in the disk is low. The most important ones are spiral structure instability and bar instability.

Disk stability depends on "Toomre's Q parameter"

$$Q = \frac{\kappa \sigma_{gas}}{\pi G \Sigma_{gas}}, \tag{9.8}$$

where

$$\kappa = \left[R \frac{d\Omega^2}{dR} + 4\Omega^2 \right]^{1/2} \tag{9.9}$$

is the epicyclic frequency, σ_{gas} is the velocity dispersion, and Σ_{gas} is the gas surface density (note that the denominator is the vertical component of the self-gravity of the gas; see Equation 4.47). The parameter Q determines the stability of a thin rotating gas disk. If $Q < 1$, the disk is locally unstable and will fragment and form stars. If $Q > 1$, the disk is stable against axisymmetric perturbations. A similar analysis for a stellar disk results in a similar Q parameter, where σ_* and Σ_* replace σ_{gas} and Σ_{gas}.

Spiral structure causes stellar velocity dispersion to increase. This results in an increase in Q and the stability of the disk. The spiral structure persists as long as gas inflow and new star formation take place. Stellar bars are observed in the majority of local spiral galaxies and also at redshift 1 and beyond. Such bars drive the dynamical evolution of the galaxy by exerting gravitational torques that redistribute mass and angular momentum. It is thought that this can cause gas inflow into the inner few hundred parsecs of the galaxy. Such processes may involve bars within bars (sometimes referred to as "primary" and "secondary" bars), where the smaller (nuclear) bar can help bring the gas even closer to the center. The final stage of a bar-driven evolution may be a bulge or a pseudo-bulge.

There is ample evidence that bars can drive SF, and their presence can help explain circumnuclear SF regions. There is also evidence that many local AGN hosts are bar-dominated systems. However, there is no direct link between the presence of a bar and the feeding of the central BH. In fact, the amount of gas within the inner few hundred parsecs is several orders of magnitude larger than required to feed the BH at the observed rate. Thus, the small fraction of this gas reaching the BH may or may not be related to the inner bar.

Numerical simulations confirm all these processes. They show that instabilities can result in gas accretion, and loss of angular momentum, all the way to about 100 pc from the center. The exact distance is not known, and some models suggest an even smaller number, of about 10 pc. However, currently there is no clear understanding of the way the gas can cross the final 100 pc (or perhaps 10 pc) barrier. One idea involves dynamical friction operating on massive gas clumps at small radii. Others invoke tidal disruption of gas clumps by the central BH, resulting in a thick gaseous disk at several parsecs from the center.

9.4.2. Merger-driven BH accretion

Major mergers provide a simpler way to get rid of the initial angular momentum of the gas in the colliding systems. In such an event, two spiral galaxies merge to form a spheroid. Most of the angular momentum is lost in the event, allowing the gas to reach very close to the center. It is thought that much of the angular momentum in the early stages of the merger is lost through hydrodynamical torque (shocks) and gravitational torque due to large-scale bars. The later stages may involve rapidly varying gravitational torque, dynamical friction, and perhaps SF feedback (see § 9.8.3).

The final stages of a merger, beyond the 100 pc barrier, are unclear, much like the final stages of secular evolution. They may involve runaway gravitational instabilities, tidal disruption of clumps, viscous torque, and (much closer in, on a parsecs scale) interaction with hydrodynamical wind from the disk. Most of these are only names for processes that are waiting to be more accurately calculated. Small mergers may not drive gas directly to the center but can affect the gravitational potential at the center, resulting with an evolution that is, perhaps, not too different from the secular evolution discussed earlier.

Table 9.1 shows the angular momentum ladder that must be climbed to explain BH accretion in systems with large initial specific angular momentum.

9.4.3. Nuclear star clusters

Nuclear star clusters are common in many local AGNs. Their typical dimensions are 10–100 pc, and their total mass is comparable to, or can even exceed the mass of, the central BH. Observations show that many such clusters are relaxed systems with stellar velocities that are dominated by random motion and little angular momentum. While the exact mechanism for removing the initial angular

Table 9.1. *The angular momentum ladder for disklike galaxies showing the*
required reduction in specific angular momentum necessary to bring cold gas
from the outskirts of the galaxy to the central accretion disk

Distance (pc)	Specific angular momentum ($cm^2 s^{-1}$)	Physical process
5000	few $\times 10^{29}$	Gravitational torques from large-scale bar
500	few $\times 10^{28}$	Gravitational torques from nested nuclear bar(s)
10	$\sim 10^{26}$	Self-gravitational instabilities Tidal disruption of clumps
1		Hydromagnetic disk wind
10^{-5}–$10^{-1} M_8$	10^{24}–$10^{25.5} M_8$	Viscous torques in accretion disk

Note: Adapted from Jogee (2006).

momentum during the formation of such clusters is not known, their existence
raises the interesting possibility that stellar winds from evolved stars in the cluster
may be an important source of gas to the central BH.

Stellar winds are common during the final stages of evolution of almost all mass
stars. In early-type stars, such winds can be very fast, with velocities that exceed
the escape velocity of the cluster and even the entire galaxy. For late-type stars,
the wind velocity and the mass outflow rate are much smaller, enabling the BH's
gravitational attraction to direct some of this gas into the center. Simple estimates
show that the amounts of stellar-ejected material and BH-accreted material are
comparable, suggesting that such gas may explain a large part of the accretion
onto small and intermediate mass BHs. While this is an interesting idea that can be
related to low-accretion-rate sources such as LINERs, it cannot explain the much
faster accretion onto more massive BHs in very luminous AGNs.

9.5. The growth of supermassive BHs

9.5.1. General considerations

The growth rate of supermassive BHs is determined by their mass and by mass
supply from the galaxy. There are several different modes of growth and various
ways to estimate the rate and the associate growth time. We consider first BH
growth via radiation efficient processes, where all the released gravitational energy
is converted to electromagnetic radiation with an efficiency η that depends only on
the spin of the BH (Chapter 4). We then include growth via other routes like direct
mergers and RIAFs (Chapter 4). In all the following, we keep to the notation that
$\dot{M}$ is the mass accretion rate through the disk or any spherical configuration. In
radiation-efficient systems, the growth rate of the BH is $(1 - \eta)\dot{M}$.

The simplest case is linear growth at a constant accretion rate,

$$\dot{M} = \alpha_{BH}, \tag{9.10}$$

where α_{BH} is a constant related to the bolometric luminosity of the source,

$$\alpha_{BH} = \frac{L_{bol}}{\eta c^2}. \tag{9.11}$$

The trivial solution is

$$M_{BH} = M_{seed} + \frac{1 - \eta}{\eta} \frac{L}{c^2} (t - t_0). \tag{9.12}$$

where M_{seed} is the mass of the BH at the start of the growth process.

The second case is exponential growth at a constant normalized accretion rate,

$$\dot{M} = \beta M, \tag{9.13}$$

which gives

$$M = M_{seed} \exp(t / t_{cg}). \tag{9.14}$$

The continuous growth time, t_{cg}, is

$$t_{cg} = \frac{1}{\beta} = \frac{t_{Edd}}{1 - \eta} \frac{1}{L / L_{Edd}} \ln \frac{M_{BH}}{M_{seed}} \text{ yr}, \tag{9.15}$$

where $t_{Edd} \simeq 4 \times 10^8 \eta$ is the Eddington time of Equation 3.18 and the growth time is

$$t_{grow} = \frac{t_{cg}}{f_{act}(M, t)}. \tag{9.16}$$

In this equation, $f_{act}(M, t)$ is the mass and redshift-dependent fraction of time when the BH is active. Its value is between 0 and 1, and it is often referred to as the *duty cycle* of the BH. For $\eta = 0.1$, the typical growth time via thin-disk accretion (the e-folding time) is roughly 4×10^7 yr for the fastest-growing holes with $L / L_{Edd} = 1$.

Both accretion scenarios discussed so far are rather simplistic, and there is no reason why a linear growth phase, or an exponential growth phase, should proceed at a constant rate for a long period of time. In particular, the relatively short duty cycles that are necessary to explain the growth of most BHs (see later) suggest several growth episodes that are separated by long periods when the BH is quiescence. This may be due to the regulation of accretion through the central disk, the episodic external mass supply to the center, and more. There is no reason why such separate episodes should have exactly the same values η, L_{bol}, or L / L_{Edd}.

Finally, most of the growth of a certain observed active BH can be due to accretion onto earlier, smaller BHs that merged to form this one.

It is interesting to consider the possibility that, over long periods of time, the BH growth is related to the accumulation of stellar mass in the bulge of the host galaxy. This would suggest an expression of the form

$$\dot{M} = \alpha_{SF(z)}[\text{SFR}], \tag{9.17}$$

where $\alpha_{SF(z)}$ may depend on the environment, the redshift, the mass, or the evolutionary state of the host galaxy. While the basic theory of SF and its relationship to BH growth is still missing many details, it is quite likely that an empirical relation between the SFR and $\dot{M}$ will provide the value of α_{SF}. One such example is given in § 9.8.

Most theoretical and numerical studies of BH growth adopt one of the preceding scenarios (usually the exponential growth one) and proceed to compute the distribution of BH mass and accretion rate at various redshifts. Such calculations must take into account the various processes that can initiate, enhance, or reduce the rate of BH growth. The parameters of the calculations are the mean L/L_{Edd} (or better $\dot{M}/\dot{M}_{Edd}$) during the growth period, the initial mass distribution of seed BHs, the rate and the type of galactic mergers, BH coalescence, and secular evolution in galaxies. Other important considerations are the amount of dust and obscuration associated with the merger events and the correlation, if any, between SF and AGN activity.

9.5.2. *Merging galaxies*

Galaxy merging were described in general terms earlier in this chapter and are known to be common at high redshift. They are thought to be responsible for galaxy growth and morphology, especially in cluster environments. Present-day numerical simulations are quite successful in predicting merger rates at different times and different-size dark matter halos. Such calculations can be tested observationally by studying the AGN population at different redshifts.

The merging of two large galaxies that contain two supermassive BHs is a violent event. The merger duration, from first encounter to the final relaxed single object stage, is very long and can last between 10 and 20 percent of the age of the universe at the time of the merger. Most of this time is spent when the galaxies are far apart. The final stages are short and more violent. The stellar orbits are disrupted, cold gas is compressed and funneled to the common center of mass, gas is compressed, new massive stars are formed and explode, and the morphology of the interacting galaxies undergo large variations. The final phase also involves the interaction of the two BHs. Details of the BH merger part are given later and a graphic illustration of a simulation of a merger between two similar-size large galaxies is show in Figure 9.6.

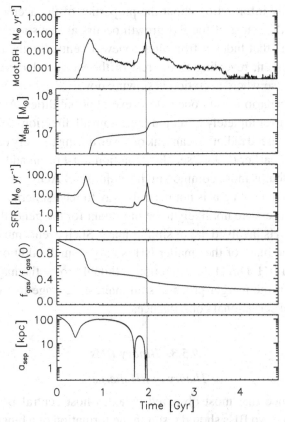

Figure 9.6. Various stages in the evolution of a large merger. From top to bottom:
BH accretion rate, global star formation rate, normalized galaxy gas fraction, BH
mass (MBH), and BH separation prior to the final merger. Vertical lines indicate
the time of BH merger. Note the delayed BH accretion compared with the SFR
(courtesy of Laura Blecha, adapted from Blecha et al., 2011).

Galaxy mergers are observed at all redshifts with a rate that is decreasing with
time $z = 2$. Almost all large mergers are associated with enhanced SF. Nearby
ULIRGs are merging systems with very high SFR and large amounts of dust. There
are indications that the BHs in some of these systems are active yet highly obscured
and hidden from direct view. SMGs represent another subgroup of mergers at high
redshift with even higher SFR.

Simulating the final stage of a large merger is a difficult task since the computa-
tional resolution is limited to about 10–100 pc, a dimension that is larger than the
sphere of influence of many BHs (or the binary BHs). There are various suggestions that gas accretion starts when the BH is completely obscured and the central
region is opaque to optical and even IR wavelengths. The AGN radiation in such
cases heats the nearby dust, and the hot dust emission is absorbed by further away
colder dust. The end result is FIR emission, which, besides the total luminosity,

retains little if any information about the properties of the accretion event. Some models suggest that most of the BH growth occurs at this stage, when the BH is inside a "cocoon" that hides it from direct view. Eventually, the radiation pierces the obscuring screen, revealing to the rest of the world a large-mass BH that is already at the phase where accretion is slowing down.

The cocoon scenario is only one out of several possibilities. According to other models, there is no completely obscured (i.e., from all direction) AGN, and some of the accretion-produced radiation can leak out even during the very early stages of the merger. As noted earlier (§ 9.4), secular evolution must be an additional important stage of growth that is more common in the majority of AGNs at least up to $z = 2$. Thus, merger-driven cold gas is not the only way to feed supermassive BHs.

Direct BH mergers are not likely to be important for the general growth of most BHs unless the BHs involved are of similar mass. Such events must be rare, and in most mergers, the mass of the smaller BH is only a small addition to the mass of the final, merged BH. Direct gas accretion is thought to be the major cause of BH growth. However, the merger process is an interesting stage in galactic evolution with important observational consequences.

9.5.3. Binary BHs

Theory of BH mergers

Given the evidence that most observed galaxies host central BHs, most galaxy mergers involving two BHs should result in the formation of a binary BH. Depending on the conditions and state of the merger, such binaries can "stall" for a very long time or can merge on a time scale that is much shorter than the galaxy merger time scale. The first scenario is thought to be typical in the merger of spheroidal, gas-poor galaxies. In such cases, the dynamical friction is very small, and a stable orbit, with a large separation of the two BHs, is the more likely outcome. Such systems are also the ones least likely to be directly detected since the lack of gas in the system prevents an efficient accretion onto the holes.

The situation is different in gas-rich galaxy mergers. The evolution of such systems can be described as a general three-step process. As the galaxies approach each other, the supermassive black holes sink toward the center of the newly formed galaxy via dynamical friction involving the mass in the two centers. This leads to the formation of a binary BH. The second stage involves the decay of the binary orbit due to interaction with stars on intersecting orbits. Such stars are ejected at velocities comparable to the binary's orbital velocity, carrying with them energy and angular momentum. This stage can also involve interaction with a gaseous disk if such a structure is formed in the center. The interaction can lead to further decay of the binary orbit. Numerical calculations indicate that in absence of stars or gas, the process can stal for a very long time, 1–2 Gyr, with a typical separation of the two BHs of about 1 pc. This problem, known as "the last pc problem," is not fully solved.

The third and final stage starts when the binary's separation is small enough and the emission of gravitational waves is efficient enough to carry away the remaining energy and angular momentum, which leads to the coalescence of the two BHs. The time scale for the merger in the final, gravitational wave-dominated stage is

$$t_{\text{merge}}(a) = 5.8 \times 10^6 \left[\frac{a}{0.01 \text{ pc}} \right]^4 \left[\frac{10^8 M_\odot}{m_1} \right]^3 \frac{m_1^2}{m_2(m_1 + m_2)} \text{ yr}, \quad (9.18)$$

where m_1 and m_2 are the BH masses in units of $M_\odot$ and a is their separation in parsecs.

Numerical simulations that follow this general scenario suggest that the time scale for BH merger is much shorter than the time scale of the galaxy merger. Such BHs are also the ones most likely to be observed as AGNs because of the gas-rich environment and the likelihood of forming one or two accretion disks. Given this scenario, the active phase of the binary system, before final merger, is short lived, of order 10^7 yrs.

The detection of the final stages of a merger of two BHs, which is associated with the emission of gravitational waves, is the main goal of several future gravitational waves experiments such as LISA. It is thought that this stage may also result in gravitational recoil of the merged BH if the system has any asymmetry. The asymmetry can be due to unequal masses, unequal spins, and different spin orientations. In such cases, asymmetric radiation of gravitational waves results in a net linear momentum flux from the merged BH and its recoil in the opposite direction.

Progress in accurate GR calculations enables us to calculate the resulting momentum and velocity of merged BHs under a variety of conditions. The distribution of recoil velocities ranges from very small (10–20 km s^{-1}) to very high (4000 km s^{-1}) velocities, much larger than the escape velocity from the system. In moderate-velocity cases, of few $\times$ 100 km s^{-1}, the BH can oscillate in the galaxy for a very long time, constantly changing its location. Such oscillating BHs can drag their accretion disk and broad-emission-line region along with them, which has interesting observational consequences. The recoil velocity depends on BH mass ratio, the BH spin ratio, and the relative orientation of the spins. For BHs with aligned or misaligned spins that are not in the plane of the binary orbit, the velocities are of order 100 km s^{-1}. For spins in the plane of the orbit, especially in opposite directions, the velocities are much larger and can reach several thousand kilometers per second. Several such scenarios are shown in a schematic way in Figure 9.7.

A major challenge of numerical and semianalytic models is to estimate the distribution of BH spin orientations in the final stages of gas-rich mergers. The BH recoil velocity, and hence its location in the galaxy, depends on these assumptions. Some calculations assume random spin orientation. Others relate the orientation to

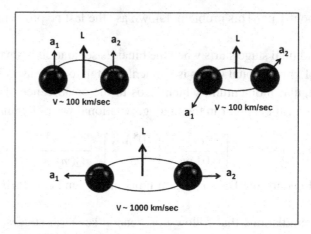

Figure 9.7. Possible final stages of a binary BH system and the expected recoil velocities of the merged BH. The BH spins are a_1 and a_2, and the orbital angular momentum is **L**.

the morphology and direction of approach of the merging galaxies. The presence of massive gaseous disks, stars, and SF regions in the center of the system are some of the more important factors that can affect the orientation of the spins and hence the recoil velocity of the BH.

BH recoil and escape from their galaxies must have been more important in the past because of the smaller galaxy masses and smaller escape velocities from such systems. In the very early stages, the escape velocities are of order 10–20 km s^{-1}. General BH evolution models must take this into account when following halo and galaxy mergers in the early universe. In particular, the largest BHs of today might have been associated with the most massive halos in the early universe. Some models suggest that high-velocity BH recoil can affect the correlation between BH and galaxy evolution and may result in an increased scatter in the M–σ^* relationship. In some of these models, BH oscillation around the galactic center may last up to a Gyr and even longer.

Observations of BH mergers and recoil

There are several well-documented cases of two active BHs in a merging system, for example, the type-II/ULIRG NGC 6240, where the two active BHs are separated by about 1.4 kpc. Other documented cases show even large, several kpc separation typical of early stages of a merger. Observational attempts to discover binary BHs in the later stages of the merger, with BH separation of a few parsecs, are still not conclusive. The best and most convincing case, so far, is a double radio source discovered in the elliptical galaxy 0402 + 379. The Very Long Baseline Array (VLBA) observations reveal two compact, variable, flat-spectrum AGNs with a projected separation of 7.3 pc. There are two velocity systems, and the combined mass of the two BHs is estimated to be about $1.5 \times 10^8 M_\odot$. The two nuclei appear stationary, while the jets emanating from the weaker of the two nuclei appear to move out and terminate in bright hot spots. Given the measured separation, the rotation period is roughly 1.5×10^5 yr. The gravitation wave frequency expected from the system is about 10^{-13} Hz, far below the expected LISA detection limit. The final merging time, assuming no dynamical friction and interaction with stars, exceeds the Hubble time by many orders of magnitude.

An additional interesting case that is still under debate is the Blazar OJ 287. In this case, there are observations of repeated outbursts every 11.86 yr that can be interpreted as related to the binary period. According to one suggestion, the secondary BH crosses the accretion disk of the primary BH at intervals of 11.86 yr, which results in an optical "flare." Assuming a real binary with both masses close to $10^8 M_\odot$, and neglecting interaction with the gas and stars in the system, the predicted merging time in this case is about 10^7 yr.

Finally, we note that large spectroscopic surveys, like the SDSS, show a small number of cases where the BLR and the NLR have very different velocities. In some cases there is a hint, from the emission-line profiles, of a double NLR and/or a double BLR. While superposition of two different AGNs must be considered, there is no indication for such a geometry in several well-studied cases. This raises the possibility that we are observing BHs, their accretion disks, and their BLRs that are moving at high speed relative to the center of the host galaxy, in agreement with the predictions of the BH recoil scenario. One such example is shown in Figure 9.8, where the broad $H\alpha$ and $H\beta$ lines are shifted by about 3500 km s^{-1} relative to the narrow [O III] $\lambda 5007$ line. The fractional number of cases with such a large velocity shift is very small, but cases with smaller shifts are more numerous. Moreover, there are alternative explanations to the large apparent wavelength shift like a double-horn emission line, where one side of the profile is enhanced relative to the other (see Figure 7.4). However, BH recoil is an interesting yet unproven hypothesis to such cases.

9.5.4. Various accretion modes of active black holes

BH growth by accretion can be divided into two general categories: radiatively efficient accretion and radiatively inefficient accretion. The first is referred to as *quasar mode*, or *AGN mode*, and the second as *radio mode*. The reasons for these names are explained later.

Gas accretion onto massive BHs is accompanied by the conversion of gravitational potential energy into other types of energy. In the radiation-efficient accretion mode, for example, through a thin accretion disk, the accretion rate is high ($L/L_{Edd} \geq 0.01$) and a fraction η of the rest mass of the accreted gas is converted to electromagnetic radiation. This process can be followed in time in individual active BHs and in the entire AGN population. For individual sources, we describe the process by using expressions like the ones derived earlier for t_{grow}.

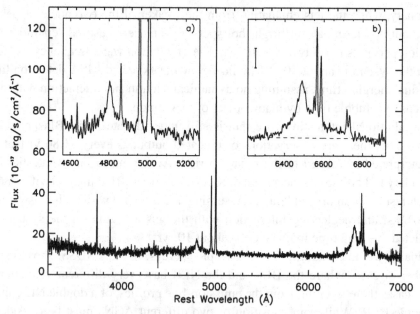

Figure 9.8. The high-luminosity AGN SDSS J510241+345631 is a possible candidate for a recoiling BH (from Shields et al., 2009; reproduced by permission of the AAS). The inserts show the expanded Hα and Hβ regions. In this case the broad Hβ line is shifted by about 3500 km s^{-1} relative to the narrow [O III] λ5007 line, which is thought to represent the systemic velocity of the host galaxy. This can be interpreted as a high-velocity BH that is moving at high speed through the galaxy, dragging along with it the accretion disk and the BLR. It can also be the blue horn of a double-peak emission line like the ones discussed in Chapter 7.

For population-based studies, we need a radiation census of all AGNs at all redshifts. Methods for using the two approaches are described in § 9.6 and § 9.7.

As discussed in Chapter 4, radiation-inefficient accretion flows (RIAFs) are characterized by lower accretion rates with mass-to-radiation conversion efficiencies much smaller than the smallest allowed η. The low efficiency is due to the fact that much of the accreted gas is advected to the center without radiating its gravitational energy. In such a case, the factor η must be replaced with a modified radiation conversion factor, ϵ_{rad}, which, in itself, depends on the accretion rate. In addition, a large fraction of the gravitational energy in such cases can be converted to winds and gas outflow from the vicinity of the BH. All such sources are recognized by their very small value of L/L_{Edd}. The number of sources in the local universe (mostly LINERs) that belong to this category far exceeds the number of the high L/L_{Edd} AGNs. The situation at higher redshifts is unclear since it is very difficult to observe low L/L_{Edd} sources at high redshift. Such considerations must be taken into account when following BH growth through time, and there are several ways to combine the information obtained from high L/L_{Edd} (radiatively efficient) sources with that obtained from low L/L_{Edd} (RIAFs) sources.

What is the cumulative growth history of *all* BHs? Answering this question requires information about efficient as well as inefficients accretors. We need to answer three basic questions:

1. In what state, averaged over cosmic time, do BHs radiate most of their energy?
2. In what state, radiatively efficient or radiatively inefficient, do BHs spend most of their time?
3. In which of the two states do BHs gain most of their mass?

The answer to question 1 is the simplest since the radiatively efficient state produces much more energy. It requires a "counting photon" exercise in the local universe, where all BHs up to a very low L/L_{Edd} are detected, and at high redshift, where very few, if any, low L/L_{Edd} are known. The answer to question 2 is also simple because it involves the classification of AGNs into various L/L_{Edd} classes and the direct comparison of the two groups. Such a comparison leads to a conclusion, which is based mainly on the fact that in the local universe, most AGNs are LINERs, that a larger fraction of the growth time is spent in the radiatively inefficient state. The answer to question 3 is more complicated. The time spent in the inefficient accretion state is much longer than the time spent in the efficient state, but the large numbers of slow accretors can compensate for the slower growth rate. Thus, the answer requires accurate counting of AGNs of all luminosities and all accretion rates at all redshifts – a formidable task that is limited by the sensitivity and resolution of present-day instruments.

A BH mass census, and radiation census of inefficient accretors, requires accurate estimates of the mass accretion rate, in particular, an estimate of the power released by such RIAFs in forms that are different from electromagnetic radiation. Several empirical relationships that couple radio and X-ray emission of such sources with the mass of the BH can be used for this purpose, in particular, the one known as the fundamental plane of BHs.

9.5.5. *The fundamental plane and the kinetic power of AGN outflows*

As discussed in § 7.8.4, there are interesting similarities between galactic black hole binary systems (BHBs) and massive BHs and accretion disks in AGNs. In particular, one can identify the "soft high" state of BHBs with the high accretion rate, radiatively efficient state in AGN. The similarities are manifested in the bend frequency (T_B) of the X-ray PSD and are most noticeable in very high normalized accretion rate systems such as NLS1s.

There is an additional similarity between the two populations at the "hard low" state that corresponds to very low normalized accretion rate, below about 0.01.

It goes under the title of the fundamental plane of BHs and is an observed relationship between the BH mass, the luminosity of the core radio jet (measured usually at 5 GHz), and the hard X-ray (2–10 keV) luminosity. The plane contains many low-accretion-rate sources, with L/L_{Edd} below a critical value of about 0.01, in galactic BHs and in AGNs. The relationship can be written as

$$\log L_R = a_X \log L_{2-10} + a_M \log M_{\mathrm{BH}} + a_K, \qquad (9.19)$$

where L_R is in units of watts, L_{2-10} in erg s^{-1}, and M_{BH} in $M_\odot$. The three constants are determined by fits to the observations and are roughly $a_X = 0.62$, $a_M = 0.55$, and $a_K = 8.6$.

The physical origin of the relationship is not fully understood. The idea is that jet formation in such systems is related to the BH mass, the accretion rate in the disk, and the BH spin. The jet power, due to synchrotron emission, is scale invariant, and the same mechanisms operate on all scales, from galactic BHs to AGNs. The X-ray luminosity is related to the radio luminosity through the accretion rate. All these relationships are observed in RIAFs but not in radiatively efficient accretion flows.

An important property of the RIAFs in question is that in such cases, most dissipative magnetic energy in the disk corona is converted to bulk kinetic energy of outflowing gas. This idea can be tested by measuring the kinetic energy associated with radio jets in low L/L_{Edd} AGNs. In many such sources, there is a low-power radio jet that coincides with the location of the BH. In some cases, the AGN is found in a cluster of galaxies, and the radio jet is associated with an observed "X-ray bubble" around the host galaxy. The energy required to create such a bubble can be calculated if the density and the temperature of the medium are known. Observations of such systems enable us to write an expression that associates the kinetic energy provided by the radio jet, E_K, with the radio luminosity of the source, L_R:

$$\frac{dE_K}{dt} = P_0 \left[\frac{L_R}{L_0} \right]^{0.8} \text{ erg s}^{-1}, \qquad (9.20)$$

where $P_0 \simeq 1.6 \times 10^{43}$ erg s^{-1} and $L_0 = 4 \times 10^{38}$ erg s^{-1}. For example, a low-accretion-rate AGN with $L_{2-10} = 10^{42}$ erg s^{-1} and $M_{\mathrm{BH}} = 10^8\ M_\odot$ is radiating, according to this relationship, $\sim 10^{39}$ erg s^{-1} at 5 GHz and produces a kinetic power of about 3.4×10^{43} erg s^{-1}. For this source, $E_K/L_{\mathrm{Edd}} \simeq 0.002$, about twice the observed L/L_{Edd}.

The improved understanding of RIAFs of this type provides better estimates of the effective radiation conversion factor ϵ_r which, when combined with the observed L/L_{Edd}, gives an estimate of $\dot{M}$ for each source. The combination with the census of low L/L_{Edd} sources leads to an estimate of the overall contribution of such events to the total accumulation of BH mass in the universe. Given today's numbers, and

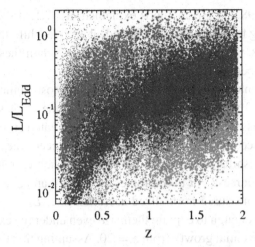

Figure 9.9. $L/L_{\rm Edd}$ versus BH mass and redshift for SDSS AGNs with $z < 2$ based on Hβ-based and Mg II $\lambda 2798$-based mass determinations. The different colors represent different mass groups: red, objects with $M_{\rm BH} \sim 10^7 M_\odot$; blue, objects with $M_{\rm BH} \sim 10^8 M_\odot$; and green, objects with $M_{\rm BH} \sim 10^9 M_\odot$. The smaller BHs are the fastest accretors at all redshifts, and $L/L_{\rm Edd}$ increases with redshift in all mass groups (adapted from Trakhtenbrot and Netzer, 2012). (See color plate)

the limitations associated with the observations of low $L/L_{\rm Edd}$ sources at high redshift, it seems that the answer to question 3 is that low-accretion-rate AGNs do not contribute significantly to the total BH mass in the universe.

9.6. Mass and luminosity evolution of individual AGNs

Can any of the growth scenarios be tested? Can BH growth be measured, and what are the duty cycles for BHs of different sizes at different redshifts? Observations of large AGN samples are starting to answer such questions.

The only reliable method to estimate BH mass at high redshift is the RM-based virial method described in Chapter 7. This method can only be applied to type-I AGNs. Moreover, not all broad emission lines provide adequate virial velocity estimates. A notable example is the C IV $\lambda 1549$ line that was discussed in § 7.1 and seems to provide unreliable mass estimates. BH masses in many thousands of AGNs have been measured with the virial method up to a redshift of about 2. However, none of the two more reliable lines, Hβ and Mg II $\lambda 2798$, can be observed in the visible range beyond $z \simeq 2$, and J, H, and K spectroscopy must be used instead. This limits the measurements to relatively narrow redshift bands (e.g., $z \sim 2.3$ and $z \sim 3.3$ for observing Hβ in the H- and K-bands, respectively) and hence a small number of sources.

A diagram showing the dependence of $L/L_{\rm Edd}$ on redshift for the entire SDSS type-I AGN sample (as of 2011) is shown in Figure 9.9. The masses were obtained with the Mg II λ2798 and Hβ methods explained in § 7.1. It seems that all active BHs were accreting faster at earlier times and, at each redshift, the smaller BHs are the faster accretors. Any evolutionary model must explain these trends observed for the active members of the BH population.

Explaining the masses and growth rates of the most luminous AGNs, those observed at $z = 2$–3 are particularly challenging. Some of these sources contain active BHs with masses that exceed $10^{10}\, M_\odot$. The calculated growth time, assuming $\eta = 0.1$ and large seed BHs, with $M_{\mathrm{seed}} = 10^4\, M_\odot$, exceeds the age of the universe at the observed redshifts unless $L/L_{\mathrm{Edd}} > 0.5$. Are such extreme accretion rates common in these sources? For two samples, at $z \simeq 2.4$ and $z \simeq 3.3$, the answer is definitely no. The measured normalized accretion rate, L/L_{Edd}, in most of these sources is not large enough to explain their size even under the extreme assumption of continuous exponential growth from $z = 20$. Assuming that the estimated mass-to-energy conversion efficiency is correct, and the measured M_{BH} accurate, how can we explain their very large masses given the age of the universe at those redshifts?

It seems that the most massive BHs observed at $z = 2$–3 must have grown even more rapidly at earlier epochs, and one should look for such fast growth episodes at higher redshift. A study of active BHs at $z \simeq 4.8$ seems to support this idea. The masses and L/L_{Edd} for these objects have been obtained from H-band spectroscopy, using the broad Mg II λ2798 line and the continuum next to it. The masses of the most massive BHs at this redshift are about $10^9\, M_\odot$, and the median L/L_{Edd} about 0.6. This allows enough time to grow to their observed mass since redshift 10 or 20 from seed BHs (§ 9.3). Assuming continuous ($f_{\mathrm{act}} = 1$) exponential growth with $\eta = 0.1$ from redshift 20, one can work out the required M_{seed} for each one of these sources.

Figure 9.10 shows some observations of the most luminous $z \simeq 4.8$ AGNs. The diagram indicates that about half the sources could have started their growth from seed BHs with masses that are consistent with being remnants of the first stars. The others require larger M_{seed}, in the range 10^3–$10^5\, M_\odot$. At the time of writing (2011), there is also a small group of sources at $z \sim 6.5$ whose BH masses have been determined. The number of such sources is too small to make an unequivocal statement, but the overall picture is consistent with what was found at redshift ~ 4.8.

The various studies of high-redshift AGNs can be combined to illustrate the time-dependent evolution of M_{BH}. This is shown in Figure 9.11. The diagram shows that active BH mass peaks at around $z = 2$ and that the largest BHs became quiescent at that epoch. The most massive active BHs of later times have smaller masses. At the time of writing, there is no systematic study of low-mass BHs, $M_{\mathrm{BH}} < 10^8\, M_\odot$, at redshifts larger than about 3 due to observational limitations.

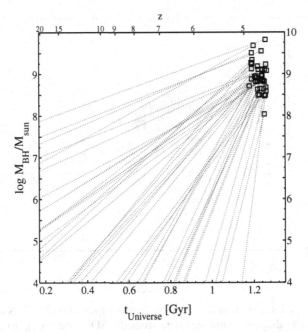

Figure 9.10. Suggested growth scenario for the most massive BHs at $z \simeq 4.8$ (adapted from Trakhtenbrot et al., 2011). The derived masses of 40 BHs are shown on the graph, and their evolution as a function of redshift, assuming continuous exponential growth with $\eta = 0.1$, is shown in dotted lines. The intersection with the vertical mass axis gives the required M_{seed} to reach the observed BH mass at $z = 4.8$, assuming continuous growth from $z = 20$.

As explained later, additional information about such objects can be obtained by analyzing the redshift-dependent luminosity and mass functions of all AGNs.

9.7. Mass and luminosity evolution of the AGN population

9.7.1. The luminosity function of AGNs

The earlier examples illustrated some aspects of BH evolution that were deduced by measuring the masses and accretion rates of active BHs. The more challenging task is to expand the study to the entire BH population, which requires the inclusion of dormant BHs.

As in galaxies, a powerful tool for following AGN evolution through time is to study their redshift-dependent LF, $\Phi(L, z)$ (§ 9.2.2). $\Phi(L, z)$ provides a concise way to describe the population properties at different cosmological times, which allows us to write the number of sources in a given luminosity and redshift bin as

$$N(L, z) = \int_{L}^{L+\Delta L} \int_{z}^{z+\Delta z} \Phi(L', z') \frac{dV}{dz'} dz' dL'. \qquad (9.21)$$

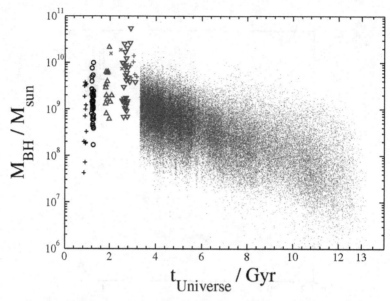

Figure 9.11. AGN evolution as traced by observations of active BHs all the way to $z \sim 6.5$. The diagram combines data from the SDSS survey up to $z \simeq 2$ with mass measurements from several dedicated NIR spectroscopy projects at higher redshifts (adapted from Trakhtenbrot and Netzer 2012).

The aim is to combine such LFs with BH mass estimates and to construct the MF of active and dormant BHs.

As alluded to in § 9.2.2, we can imagine LFs that evolve in redshift by changes in luminosity only, in space density only, or as a combination of both. A "pure luminosity evolution" means that the shape of the LF is retained in time but $L(z) = f_z L(z + \Delta z)$, with $f_z < 1$ for all luminosities. In the same way, "pure density evolution" would mean a constant-shape LF with a number density per unit volume that is going down in time. The "luminosity-dependent density evolution" is a combination of the two, where both the luminosity and the space density are decreasing with time. Schematics of the three possibilities are shown in Figure 9.12.

LFs can be constructed in various wavelength bands. A practical limitation is the incompleteness of most single-band surveys. This is related to the fact that counting sources over the redshift interval z, $z + \Delta z$ cannot fully take into account sources of luminosity L that fall below the flux limit of the sample over part of this redshift interval. A special statistical method (the V / V_{max} method) was developed to correct for these effects.

Optical luminosity functions of type-I AGNs, at very high redshift, have been constructed from large AGN samples. X-ray LFs for both type-I and type-II sources are known, to a reasonable accuracy, but to lower redshifts, $z \approx 3$. Several examples are shown in Figures 9.13 and 9.14.

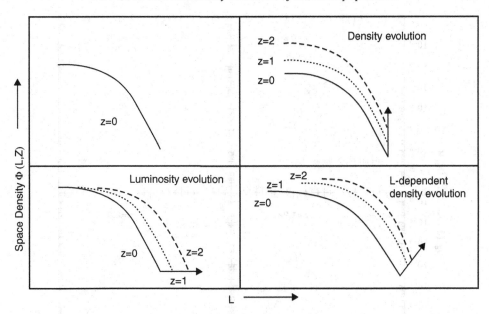

Figure 9.12. Various schematic luminosity functions for AGNs showing how such objects can vary in luminosity and space density through cosmic time.

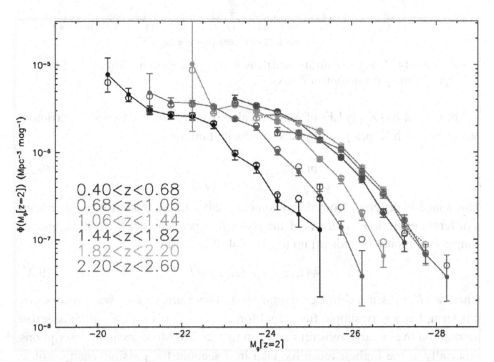

Figure 9.13. Optical luminosity functions at different redshifts, as marked (from Croom et al., 2009; reproduced by permission of John Wiley & Sons Ltd.). (See color plate)

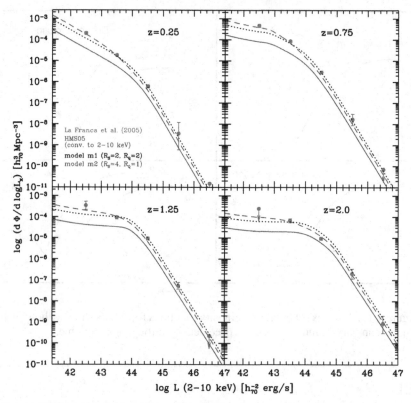

Figure 9.14. X-ray luminosity functions at different z (from Gilli et al., 2007; reproduced by permission of A&A).

The optical and X-ray LFs of AGNs, at all redshifts, can be described by a double power law with slopes γ_1 and γ_2 and a fiducial luminosity L^*:

$$\Phi(L) = \frac{A}{\left(\frac{L}{L^*}\right)^{\gamma_1} + \left(\frac{L}{L^*}\right)^{\gamma_2}}. \tag{9.22}$$

This shape is different from the canonical galaxy LF (Equation 9.4), suggesting a different evolution. Analysis of the redshift dependence of $\Phi(L, z)$ suggests a simple evolution with redshift up to $z = 3$–4 of the form

$$\Phi(L, z) = e_v(L, z)\Phi(L), \tag{9.23}$$

where $e_v(L, z)$ is a redshift evolution term. There are various ways to describe this term, but a very simple functional form, $e_v(L, z) = (1 + z)^p$, gives adequate description over a large redshift range, up to $z \simeq 2.5$. More accurate descriptions, especially at the highest redshifts, take into account the possible changes of γ_1 and γ_2 with redshift and additional multiplication terms in $e_v(L, z)$. An example of the various parameters that define the 2–10 keV X-ray luminosity function at $z < 0.5$ is $A = 5.04 \times 10^{-6} \text{ Mpc}^{-3}$, $\gamma_1 = 0.86$, $\gamma_2 = 2.6$, $L^* = 10^{43.9} \text{ erg s}^{-1}$, and $p = 4.2$.

Various other LFs have been constructed over the years, for example, radio LFs, IR LFs, and even [O III] λ5007 LFs. The most useful LFs are those involving the bolometric luminosity, $\Phi(L_{bol}, z)$. This LF gives a complete account of all AGN emission at all times and is the one most directly connected with BH evolution. This LF is not directly observable, and its construction requires the use of single-band LFs combined with bolometric correction factors like the ones discussed in Chapter 7.

9.7.2. The cosmological X-ray background

X-ray LFs, for example, $\Phi(L_{2-10\,keV}, z)$, play important roles in the reconstruction of the general population properties. The reason for this is that X-ray samples are more complete and uniform and, at high enough energy, less affected by obscuration. Another advantage is that X-ray LFs enable us to include type-II AGNs that are the majority of the AGN population at low redshift and a significant fraction of the population at high redshift. The exceptions are the Compton thick sources that are missing from most X-ray samples. This requires the introduction of obscuration correction factors that define the missing fraction as a function of L and z. A typical X-ray LF looks like $[1 + f_{obscure}(L_X, z)]\Phi(L_X, z)$, where $\Phi(L_X, z)$ is the observed LF and $f_{obscure}(L_X, z)$ is the correction factor that accounts for the missing sources. Another drawback is the low sensitivity of present-day X-ray instruments, which limits the level of completeness at $z \sim 2.5$. To better understand the X-ray LF of AGNs, and the values of $f_{obscure}$ at different redshifts, we must consider the general properties of the cosmological X-ray background.

The X-ray background (XRB) was discovered in the early 1960s. Its global SED is different from that of a single AGN showing a peak at around 30 keV. It is thought that the peak is due to the emission of a large number of AGNs, obscured and unobscured, at different redshifts. The most sensitive X-ray surveys clearly show that most of the X-ray background is due to the X-ray emission of individual AGNs with a small contribution at hard X-rays due to clusters of galaxies and at very low energy due to SF galaxies. In fact, some 90 percent of the XRB is resolved at energies of 2–10 keV.

The reconstruction of the XRB is based on the combination of X-ray LFs of various groups of AGNs. It depends on the distribution of obscuring columns and the shape of the X-ray SED. Fortunately, the typical (unobscured) X-ray SED is simple and made of a combination of a power law with an energy slope of ~ 0.9, a high-energy cut-off, and a reflection component (see Chapter 7). Obviously, there is no difference in this respect between type-I and type-II sources. A more tricky issue is the column density distribution, which is a function of X-ray luminosity and perhaps also the redshift (as usual, the two are difficult to disentangle at high redshift because of the flux limits of the observed samples).

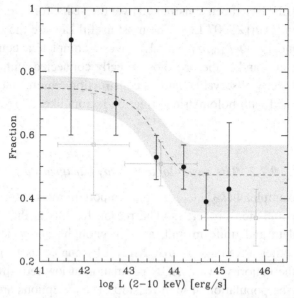

Figure 9.15. The fraction of AGNs with obscuring column $>10^{22}$ cm^{-2} as a function of the 2–10 keV luminosity. The data from two surveys are shown by different types of points, and the best-fit model and its uncertainty are shown by a dashed line and a shaded area (from Gilli et al., 2007; reproduced by permission of A&A). Similar studies of sources with column of $>10^{24}$ cm^{-2} indicate a large fraction, of about 0.4, at the low-luminosity end but a very fast drop toward higher luminosity.

Deep surveys can unveil the column density distribution, as shown, for example, in Figure 9.15.

The addition of several groups of AGNs with various assumptions based on modeling of the different X-ray LFs (Figure 9.16) gives a reasonable fit to the XRB over a large energy range. There is a well-known difficulty to fit the background near its peak, at ~ 30 keV, which seems to be related to the number of Compton thick sources, which are the most difficult to find and characterize. Much of the uncertainty in the modeling is to estimate the scattering efficiency of the central X-ray radiation in individual source. The mean scattering efficiency is thought to be about 0.01–0.03 but is not well constrained by the observations. This leaves a large uncertainty, especially near the peak of the XRB.

The general issue of Compton thick AGNs is of fundamental importance for the understanding of BH evolution. Some models assume that the early growth phase of AGNs, in particular, those with extremely massive BHs at high redshift, take place inside a "cocoon" – a completely opaque system that can only be seen at very long wavelength, due to emission by AGN-heated dust. Such objects are almost impossible to detect with present-day instruments, and their contribution to the XRB must be small but not necessarily negligible. This idea is the basis of several

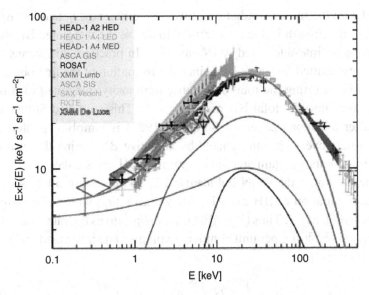

Figure 9.16. The XRB and its modeling by various X-ray LFs (from Gilli et al., 2007; reproduced by permission of A&A). The different surveys are marked on the top left and are shown by various points. The best-fitting model is a combination of four colored lines: red, unobscured AGNs; blue, obscured Compton thin AGNs; magenta, the preceding two groups plus galaxy clusters; black, Compton thick AGNs. (See color plate)

attempts to search for AGN-heated dust at IR energies and to correlate it with the X-ray luminosity. A very small number of sources fit this description and show no X-ray emission even in the deepest X-ray surveys (e.g., the 4Ms Chandra survey). A possible way to infer the X-ray properties of such sources is to stack many of them using the IR coordinates. Such stacking reveals, in some cases, a clear X-ray signal. A major uncertainty in such studies is the confusion with high-redshift SF galaxies that are also bright at FIR energies with almost no detected hard X-ray emission.

9.7.3. The total BH mass in the universe

A crucial step in building the mass function of all BHs is to construct MFs for active BHs and look for the relative normalization of the two populations, that is, the ratio of active and dormant BHs at a given time. An important quantity that provides such a normalization is the integrated BH mass at $z = 0$. This number is obtained from a census of all local BHs and can be compared with the total stellar mass, and the total dark matter mass, to constrain galaxy and BH evolution models.

A straightforward way to calculate the total BH mass in the universe is to use the fact that large BHs do not lose energy or mass and can only grow in time. This can be combined with the mass-to-radiation conversion efficiency, η, and the assumption that most of the growth is due to accretion, to derive the integrated BH mass from the volume and time-integrated AGN emission. In practice, this means summing all the energy emitted by active BHs since the formation of the first objects at high redshift and converting the total bolometric luminosity integrated over the entire cosmic history into the total BH mass at $z = 0$. This so-called Soltan argument (named after a famous paper by Soltan in 1982) is simply a photon counting exercise, where every photon radiated by an active BH during the history of the universe represents a certain amount of grams of BH mass today.

Assume we know the AGN bolometric LF, $\Phi(L_{bol}, z)$, at all z. Assume also that the accumulation of BH mass is only via accretion with a mass-to-radiation conversion efficiency η. Thus $(1 - \eta)\Phi(L_{bol}, z)/\eta c^2$ gives the rate of accumulation of BH mass at redshift z per unit comoving volume. The integrated energy density from all AGNs is

$$u = \int_0^\infty dz \int_0^\infty \Phi(L, z) L dL \frac{dt}{dz}, \tag{9.24}$$

and the mass density of BH remnants is

$$\rho_{BH} = \int_0^\infty dz \int_0^\infty \Phi(L, z) \left(\frac{1-\eta}{\eta c^2} L\right) dL \frac{dt}{dz}, \tag{9.25}$$

where we made use of the relation $L = (\eta/(1 - \eta))c^2 \dot{M}$. Thus

$$\rho_{BH} = \frac{1-\eta}{\eta c^2} u. \tag{9.26}$$

There is no way to remove the dependence of ρ_{BH} on the $(1 - \eta)/\eta$ factor, which leads to a rather large uncertainty given the acceptable range of η inferred from accretion disk theory.

Obtaining the bolometric luminosity function of AGNs from the optical LF is not simple. The observations of the faint end of the luminosity function are not complete at high redshifts, and the optical–UV luminosity of type-II sources is uncertain. X-ray LFs that include type-II sources provide useful, less biased description of the population properties up to $z \sim 3$. As explained, the limitation of such LFs is at high redshift because of instrument sensitivity and the uncertainty associated with the unknown fraction of Compton thick sources.

The integration of Equation 9.25 is performed using $\Phi(L_X, z)$ in combination with bolometric correction factors like the ones described in Chapter 7. Earlier studies that used the Soltan argument assumed $\eta = 0.1$ and did not include corrections due to obscured AGNs, found $\rho_{BH}(z = 0) \sim 2 \times 10^5 \ M_\odot \ \text{Mpc}^{-3}$. Including obscured sources increases this density by a factor of about 2.

The $\rho_{BH}(z = 0)$ can also be estimated from the census of local BHs, that is, from the study of galaxies with bulges and the M–σ^* relationship (Chapter 8). The uncertainty here is partly due to the scatter in the M–σ^* relation and partly due to the difficulties in separating bulges from pseudo-bulges since the two correspond to different M_{BH} for the same $\sigma*$. Assuming the same relationship for bulges and pseudo-bulges gives $\rho_{BH} \sim 5 \times 10^5\ M_\odot\ \mathrm{Mpc}^{-3}$, in good agreement with the model.

9.7.4. The mass function of BHs

Having obtained redshift-dependent luminosity functions for AGNs, and the normalization of ρ_{BH} at $z = 0$, we proceed to determine a redshift-dependent BH mass MF, $\Phi(M, t)$. Two such functions will be considered: the MF of active BHs and the general MF of the entire BH population. It is convenient to use the time t rather than the redshift in such calculations; hence we follow the notation $f_{act}(M, t)$, $\Phi(L, t)$.

Assume a time t in the history of the universe when the comoving number density of BHs, active or quiescent, per unit comoving volume per unit mass M (the "number density mass function") is $N_{act}(M, t)$ for active BHs and $N(M, t)$ for all BHs. $N(M, t)$ can be interpreted as the first moment of a more specific distribution function, $g(M, \dot{M}, t)$, that depends on BH mass M, BH accretion rate, $\dot{M}$, and time:

$$N(M, t) = \int_0^\infty g(M, \dot{M}, t)d\dot{M}. \tag{9.27}$$

We can also use $g(M, \dot{M}, t)$ to derive the mean accretion rate for a given BH mass:

$$\langle \dot{M}(M, t) \rangle = \frac{1}{N(M, t)} \int_0^\infty \dot{M} g(M, \dot{M}, t)d\dot{M}. \tag{9.28}$$

Note again that the term MF used earlier is related to the number density MF by $\Phi(M, t) = N(M, t)M$.

It is simple to construct $N_{act}(M, t)$ from the bolometric LF of AGNs with measured BH mass. Such information is available for a very large number of type-I AGNs up to $z \simeq 2$ and for smaller samples at higher redshifts. In these type-I sources, we know both $g(M, \dot{M}, t)$ and $N(M, t)$. BH masses are not available for type-II AGNs with $z \gtrsim 0.3$ (for smaller redshifts, we can use the M–σ^* method) and for very faint type-I sources at high redshift. The situation is most problematic when approaching the flux limit of a given sample where the extension of the known MF to low luminosity is hampered by various selection effects.

A general solution for the number density mass function, $N(M, t)$, in cases where it cannot be constructed from individual sources can be obtained by using a continuity argument that relates the value of $N(M, t)$ at times prior to t to its value at time t. To follow this method we consider a general case where there is a source of BHs ("source function") at time t, $S(M, t)$, which is given in units of dN/dt and describes the change in time of the number of BHs due to mergers. The continuity equation is

$$\frac{\partial N(M, t)}{\partial t} + \frac{\partial}{\partial M}[N(M, t)\langle \dot{M}(M, t)\rangle] = S(M, t), \qquad (9.29)$$

and its solution provides the value of $N(M, t)$ at all times.

The source term, $S(M, t)$, represents cases where BHs are created or destroyed. This process takes place during BH mergers, where BHs of masses M_1 and M_2 merge to form a BH of mass M_3. In this case, the source terms $S(M_1, t)$ and $S(M_2, t)$ are negative, and $S(M_3, t)$ is positive. Given that most BH growth is by gas accretion, we can safely assume that at most redshifts, $S(M, t) = 0$.

The number density mass function is a derived quantity, and it is simpler, and more practical, to work with the observed LF. We can use the duty cycle introduced earlier, $f_{\text{act}}(M, t)$, to relate $N(M, t)$ to $\Phi(L, z)$:

$$\Phi(L, t)dL = f_{\text{act}}(M, t)N(M, t)dM. \qquad (9.30)$$

The two unknowns in this approach are $f_{\text{act}}(M, t)$ and its initial value, $f_{\text{act}}(M, t_0)$, where t_0 is an early time at redshift z_0 corresponding to the start of the BH growth process. A possible way to estimate $f_{\text{act}}(M, t_0)$ is to assume a period of continuous growth between the formation of the first (seed) BHs before t_0, and t_0. During this period, all BHs are active and thus $f_{\text{act}}(M, t_0) = 1$.

Equation 9.30 does not distinguish the duty cycles of sources with different $\dot{M}$. This is equivalent to assuming a constant $\Gamma = L/L_{\text{Edd}}$ for all sources with a given L. A more accurate treatment must therefore replace $f_{\text{act}}(M, t)$ by $p(\Gamma, t)f_{\text{act}}(M, t)$, where $p(\Gamma, t)$ is an additional function that specifies the distribution of Γ at all times. This is similar to what was discussed earlier regarding $g(M, \dot{M}, t)$. Such a general function has never been constructed for a large range of BH masses and redshifts, but there are several reasonable guesses that are consistent with the observations.

Assuming constant Γ, we can now perform a complete integration from $z = z_0$ to $z = 0$. The average accretion rate in Equation 9.29 can be written as

$$\langle \dot{M}(M, t)\rangle = f_{\text{act}}(M, t)\dot{M}(M, t), \qquad (9.31)$$

and the relationship between $N(M, t)$ and $\Phi(L, t)$ is given in Equation 9.30. We can thus use the known relationships between $\dot{M}$, t_{Edd} and L,[1]

$$L = \frac{\Gamma \eta c^2}{t_{\text{Edd}}} M = \frac{\eta c^2}{1 - \eta} \dot{M}, \qquad (9.32)$$

---

[1] Note again that in this book, the Eddington time t_{Edd} is not a constant since it includes the efficiency η.

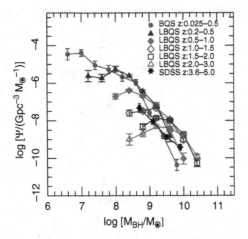

Figure 9.17. MFs of active BH at different redshifts derived from various AGN samples, as marked (adapted from Vestergaard and Osmer, 2009; reproduced by permission of the AAS).

to convert the continuity equation (9.29) to a modified equation involving L and $\Phi(L, t)$:

$$\frac{\partial N(M, t)}{\partial t} + \frac{\eta(1 - \eta)\Gamma^2 c^2}{t_{\text{Edd}}^2} \left[\frac{\partial}{\partial L}(L\Phi(L, t)) \right]_{L = \Gamma M \eta c^2 / t_{\text{Edd}}} = 0. \qquad (9.33)$$

This equation can be integrated over cosmic time starting from z_0, given the known LF,[2]

$$N(M, t) = N(M, t_0) + \frac{\eta(1 - \eta)\Gamma^2 c^2}{t_{\text{Edd}}^2} \int_z^{z_0} \left[\frac{\partial}{\partial L}(L\Phi(L, t)) \right]_{L = \Gamma M \eta c^2 / t_{\text{Edd}}} \frac{dt}{dz} dz. \qquad (9.34)$$

Going in steps from t_0 to the present time (from z_0 to $z = 0$) gives $N(M, t)$ at all times.

There are several realistic estimates of BH MFs that are based on the assumption of a constant L/L_{Edd}. The ones at lower redshift are based on the observations of many more sources and are hence more reliable. Obviously, the value of $N(M, z)$ at a certain z depends on its value at higher z, which introduces some uncertainties. Thus, much of the global uncertainty is due to incomplete information at high redshift. There are also more fundamental unknowns related to the possible time-dependent η (or BH spin) and the time and mass-dependent duty cycle. One solution which is in reasonable agreement with the observations at $z = 0$ assumes $\Gamma \sim 0.4$ and $\eta \sim 0.07$. Several examples are shown in Figure 9.17.

---

[2] Many papers use logarithmic expressions to describe the LF. The conversion between the systems is done by noting that $\Phi(L, t)dL = \Phi'(\log L, t)d \log L$.

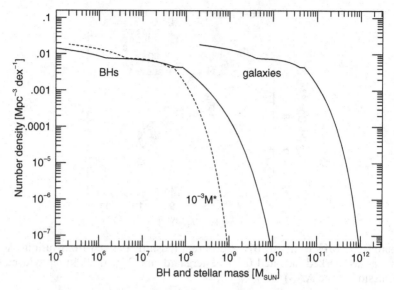

Figure 9.18. Observed galaxy MF in the local universe (adapted from data published in Li and White, 2009) and derived BH MF. A scaled-down galaxy MF, by a factor 1000, is shown for comparison (dashed line).

The value of $N(M, 0)$ can be checked against other estimates based on the local galaxy MF and the known relationship between M_{BH} and M_{bulge} (Chapter 8). Such a comparison is shown in Figure 9.18. The galaxy MF in this case was derived for a large number of low-redshift SDSS galaxies. The integrated stellar mass density in this case is $2.2 \times 10^8 M_\odot/Mpc^3$, assuming $H_0 = 70$ km/sec/Mpc. This mass density represents about 4 percent of the total baryonic mass in the universe; somewhat lower than other MFs. The *derived* BH mass function assumes $M_{BH} \propto M_{bulge}^{1.6}$, as measured for several local samples, and its integration suggests a local mass density of BH of $3.3 \times 10^5 M_\odot/Mpc^3$. This is somewhat smaller than several estimates based on the Soltan argument that were mentioned earlier ($\sim 5 \times 10^5 M_\odot/Mpc^3$) and is obviously a function of the exact $M_{BH}-M_{bulge}$ and $M_{BH}-M^*$ relationships. Given this, we can estimate the BH mass density relative to the critical density, $\Omega_{BH} \sim 3 \times 10^{-6}$.

9.8. Coevolution of galaxies and supermassive BHs

9.8.1. BH evolutionary models

BH evolutionary models that are based on redshift-dependent luminosity and mass functions can be used to follow the progression of BH mass with time. Despite the various limitations associated with such calculations, the general conclusions are robust. They support a scenario where the most massive BHs grew first while the less massive BHs, those with $M_{BH} \sim 10^7–10^8 M_\odot$, which are the majority of the local

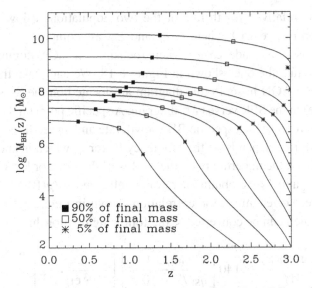

Figure 9.19. The growth history of BHs obtained from X-ray LFs and models of the XRB. The symbols indicate the points when a BH reaches a given fraction of its final mass. Note the much faster growth of the more massive BHs at earlier times (courtesy of A. Marconi).

active BH population, started their growth much later and are still growing at a fast rate today. Thus, BHs grow in an antihierarchical fashion. The results of one such model are shown in Figure 9.19. They are similar, in many ways, to the general correlation of M_{BH} with cosmic time shown in Figure 9.11.

The "downsizing" evolution of BHs is similar but not identical to the downsizing of galaxies discussed in § 9.2.2. This can be seen by comparing Figure 9.19 with Figure 9.4. In particular, it is not at all clear that the peak activities of the two populations (the highest SF density and the highest BH accretion density) occur at the same redshift. It is also not clear how the BH mass to stellar mass ratio evolves with time. Probing the AGN population to fainter luminosities, especially at large redshift, will no doubt improve our understanding in this area. Large deep surveys are being conducted, and new instruments are being built, to enable this challenging new research.

9.8.2. The relative growth rate of BH and stellar mass

Observations of active BHs, like the ones discussed earlier, provide information about their bolometric luminosity (L_{bol} or L_{AGN}) and the BH mass, M_{BH}. The values of M_{BH} and L_{AGN} can be compared with L_{SF} and M_*, which are known from other measurements. Thus, mass and luminosity for active BHs and for stars are known throughout the history of the universe. This information can be used to compare the relative growth rate of the two populations using diagrams like Figure 8.18 and Figure 8.19, where L_{AGN} and L_{SF} are compared.

The comparison is made easier when focusing on AGN-dominated sources ($L_{\mathrm{AGN}} > L_{\mathrm{SF}}$), like the ones shown in Figure 8.18. We can make the assumption that many of the AGN hosts in this diagram are located on the SF sequence (the main sequence) in the SFR versus M_* diagram. A specific point on the curve shown in this diagram represents a specific BH growth rate and a specific SFR in the host. The BH growth rate depends on the efficiency factor η, which is known to within a factor of ~ 2 and the uncertainty on the SFR is of the same order. We can use the measured L_{AGN} and L_{SF} to obtain the ratio g(stellar mass)$/g$(BH), where g stands for growth rate. For example, the correlation shown in Figure 8.18 suggests that $L_{\mathrm{SF}} \propto L_{\mathrm{bol}}^{0.7}$, which can be converted to the following relationship:

$$\frac{g(\text{stellar mass})}{g(\text{BH})} \simeq 140 \left[\frac{\eta_{\mathrm{BH}}/0.1}{\eta_{\mathrm{SF}}/7 \times 10^{-4}}\right] \left[\frac{L_{\mathrm{bol}}}{10^{43} \text{ erg s}^{-1}}\right]^{-0.3}, \qquad (9.35)$$

where η_{SF} is the SF efficiency in units of mc^2. Assuming the slope (-0.3) is not changing with redshift, we can use this expression to obtain an estimate of the relative growth times during episodes when the AGN dominates the emission of the system.

Modeling the entire population and the entire history requires three additional pieces of information. First, we need to know the fraction of AGN hosts that *are not* forming stars at a high rate because they are in hosts that are below the SF sequence. As explained in Chapter 8, such a census of AGN hosts is relatively easy to perform in the local universe, where counting all AGNs suggests that about two-thirds of high-ionization AGNs (high-ionization type-I and type-II AGNs but not LINERs) are found in hosts that are on the SF sequences. This fraction may be even higher at high redshifts, but such a census is currently unavailable. Second, we need to know the fraction of hosts that are *above* the SF sequence, perhaps due to mergers, since the constant in Equation 9.35 (140) depends on this assumption. Third, we need to take into account the growth rate during the SF-dominated part of the process, when $L_{\mathrm{SF}} > L_{\mathrm{AGN}}$ and g(stellar mass)$/g$(BH) is larger than in Equation 9.35 (see Figure 8.19). Various ingredients of such a complete model are still missing.

It is illuminating to estimate the relative duty cycles of BH growth and powerful SF events making the *wrong* assumption that $L_{\mathrm{AGN}} > L_{\mathrm{SF}}$ at all times. For this we use Equation 9.35 and the known M_*/M_{BH} in the local universe. As shown earlier (Figure 8.4), the ratio depends on the BH mass. For $M_{\mathrm{BH}} = 10^{7-8} M_\odot$, it is roughly 1500, and for the most massive BHs, with $M_{\mathrm{BH}} \sim 10^{10} M_\odot$, it is about 200. Since the numerical factor in Equation 9.35 is 140, the relative duty cycle at the low M_{BH} range is ~ 10 (1500/140). This means that a "typical" SF event, averaged over cosmic time, lasts about 10 times longer than the associated BH accretion events. For the most massive BHs, those producing $\sim 10^{47}$ erg s^{-1} while active, the ratio from Equation 9.35 is ~ 5, and the ratio of the two duty cycles is about $200/5 = 40$. Note again that we did not include in the calculation the SF-dominated phase of the growth, which, in principle, can be done by using diagrams like Figure 8.19.

9.8.3. Outflows and feedback

The significant correlation between BH mass and stellar (bulge) mass found in nearby galaxies suggests that BH evolution and galaxy evolution go hand in hand over a considerable fraction of the age of the universe. This raises questions about the nature of the physical processes that link the two, in particular, what BH-related processes can stop BH activity and shut down SF in the host galaxy. Such processes, if efficient enough, can explain the disagreement between observations and calculations of galaxy LFs, in particular, the lack of large massive spheroidal galaxies compared with simple theoretical predictions. Shutting off SF can also occur in different ways that are not related to BH activity. All such processes go under the name of "feedback."

Various modes of feedback

In the context considered here, feedback is a back reaction from galaxy formation or BH activity, on the host galaxy, on nearby systems, and on the BH itself. This can take one of three major routes.

BH-growth AGN feedback: This is the way by which the enhanced activity of the AGN results in a shutdown of mass supply to the BH (i.e., negative feedback). The process can operate in one of two ways. The strong ionizing radiation emitted by the AGN can ionize the gas in its surrounding increase its temperature and kinetic energy, and result in the termination of the accretion process. This can be very significant in cases of spherical accretion, and much less so if most of the accretion is via an accretion disk. The shut down of BH accretion can also occur if the BH activity is associated with strong outflows that carry much energy and/or momentum. Such outflows can sweep the accreted gas and stop the accretion.

SF AGN feedback: This process connects AGN growth, via accretion, and stellar mass growth, via star formation. The feedback can be either "positive" or "negative." The negative SF feedback is the process by which AGN-produced radiation, or AGN-induced outflows, interact with the cold star-forming gas in the galaxy, or the surrounding halo, and stop or slow down SF processes by ionizing and heating the gas, or by removing it from the system. Positive SF feedback operates in the opposite way. The AGN outflow interacts with a

low-density cold gas, below the Jeans mass, and increases its density until it
starts to form stars.

Stellar and supernova feedback: This process is not related to the central
AGN, but the basic physics is similar. Young massive stars in SF regions
produce intense ionizing radiation that can ionize neutral gas or photodisso-
ciate molecular gas in their vicinity. This can slow down, or even stop, SF
and accretion of cold gas from the halo. A powerful SN explosion, or stellar
winds, injects kinetic energy, which can heat the surrounding gas and eject it
from the galaxy or the halo. Such feedback, during the early stages of galaxy
formation, is thought to explain the lack of $L \ll L_*$ galaxies compared with
simple theoretical expectations that do not include feedback.

The physical of feedback

Focusing on AGN feedback, we note that there are several possible ways that
involve similar amounts of radiation or kinetic energy yet result in different effects
on their surroundings. Ionization-driven feedback is the simplest case. The strong
radiation field can ionize a large fraction of the surrounding gas, increasing its
temperature and preventing it from forming stars. The efficiency depends on the
level of ionization and the opacity of the absorbing gas, including its dust content.
Kinetic-type feedback can be the result of a quasi-spherical (large solid angle) flow
or a narrow relativistic jet. The first can have a large effect on most of the host
galaxy. The second can penetrate to large distances with little effect on the gas
in the galaxy because of its narrow solid angle. In fact, narrow jets are probably
important only as a way of heating up the surrounding halo gas or the intracluster
gas in large clusters of galaxies.

The exact nature of AGN feedback, its magnitude and the microphysics associated with it, is not well understood. The potential importance of the process can
be estimated by comparing the binding energy of a large galaxy with the total
electromagnetic energy radiated by a bright AGN over its lifetime. The binding
energy of a large galaxy is of order 10^{61} erg, and the radiated energy emitted by the
AGN is about $3 \times 10^{53} t_{yr} L_{bol,46}$ erg, where $L_{bol,46}$ is the bolometric luminosity in
units of 10^{46} erg s^{-1} and t_{yr} is the length of the active phase in years. A relatively
short period of activity, 10–100 Myr, is enough to release an amount of energy in
excess of the binding energy of the entire galaxy. An equivalent way to look at
this is to consider the total accreted mass that is required to release this amount of
energy. This translates to about $5 \times 10^6 \, M_\odot$, a small fraction of the mass of many
supermassive BHs. Thus, global energetics considerations suggest that feedback
may be of great importance.

The emission of an enormous amount of radiation is not, by itself, a signature of an efficient feedback. Most of this radiation can escape the galaxy and
never interacts with the surrounding material. Understanding the microphysics of

feedback is crucial for evaluating the importance of the process. For example, if the column density of the gas surrounding the BH is small, and the gas is close to the BH, it will be ionized and become transparent, and any further feedback will be negligible. If it is large and far enough from the BH, it must also be dusty, even in low-metallicity systems. In such a case, all the AGN radiation will be absorbed by the gas and dust, thus shielding the further away SF gas from the source of radiation. This, again, is a case of inefficient feedback.

There are several ways to convert the emitted radiation to mass outflows and winds that interact more efficiently with the surrounding gas. To estimate the feedback efficiency in such cases, we must consider the details of the process that create the wind, the wind kinetic energy, and its momentum.

As explained in Chapter 7, massive outflows are observed in a large number of AGNs. A likely mechanism to drive such outflows is radiation pressure force, and the likely launch location of the wind is in the vicinity of the BH: the accretion disk, the extended clumpy disk, the BLR, and the inner parts of the central torus. The velocity of such outflows must be close to the escape velocity at the point where the flow is launched. If this is the innermost part of the disk, the velocity is of order $0.1c$.

The interaction of AGN outflows with the surrounding gas, and the importance of negative feedback, can be evaluated by considering both the momentum and the kinetic energy associated with the flow. We can estimate the mass outflow rate using simple considerations of momentum conservation, assuming the optical depth of the wind is such that a fraction $\alpha(r)$ of the radiation is absorbed by the outflowing gas (this is the same factor used in Equation 5.75). We can write a momentum equation of the form

$$v_{\text{out}} \dot{M}_{\text{out}} \simeq \frac{\alpha(r) L_{\text{bol}}}{c}, \tag{9.36}$$

where $\dot{M}_{\text{out}}$ is the mass outflow rate and v_{out} is the flow velocity. A simple example of this type is the case where the outflow is Compton thick at all radii ($\alpha(r) = 1$) and $L_{\text{bol}} = L_{\text{Edd}}$. Such a situation may arise when the mass *inflow* rate through the disk is larger than $\dot{M}_{\text{Edd}}$, forcing a large fraction of this material to leave the disk in the form of a wind. The gas leaves the disk with a velocity that is close to the escape velocity. Given the large amount of gas, the Compton depth is large too. The momentum flux in this case is simply L_{bol}/c, and the fraction associated with feedback depends on the nature of the interaction with the surrounding gas.

An additional quantity of great importance is the total mechanical energy of the flow per unit time. For the case under study, this is given by

$$\frac{1}{2} \dot{M}_{\text{out}} v_{\text{out}}^2 \simeq \frac{1}{2} \frac{L_{\text{bol}}^2 \alpha^2}{\dot{M}_{\text{out}} c^2} \simeq \frac{1}{2} \frac{L_{\text{Edd}}^2}{\dot{M}_{\text{out}} c^2}. \tag{9.37}$$

Here, again, we need to consider what fraction of this energy can be delivered to the surrounding gas. This fraction depends on the physical states of the outflowing as well as the surrounding material.

The extreme case of outflowing Compton thick gas with a velocity that is close to the escape velocity and momentum flux of L_{bol}/c leads to additional interesting implications, especially in young galaxies. Consider a case of a self-gravitating isothermal sphere representing a proto-galaxy. Most of the material in this case is dark matter, and the fraction of gas is f_g. The asymptotic solution of the hydrostatic equation for an isothermal gas sphere with temperature T is

$$\rho = \frac{\sigma^2}{2\pi G r^2}, \tag{9.38}$$

where $\sigma = \sqrt{kT/m}$ is the particle velocity and ρ is the gas density at large distances from the BH.[3] The momentum of this wind can be transferred to the ISM, which results in a moving shell of swept-out material bounded by an inner shock. This scenario leads to specific correlations between the velocity of the outflowing gas, the velocity of the shell, and the accretion rate onto the BH.

The preceding scenario represents the most efficient way to transfer a large fraction of the momentum of the outflow to the surrounding material. In this case, all the momentum in the AGN radiation field can be transferred to the gas. This can result in an efficient negative feedback, which can halt the feeding of the AGN and affect, dramatically, the SF in its vicinity. In extreme cases, it can sweep the entire ISM and leave behind a gasless galaxy. This and similar processes have been invoked to explain the shutdown of stellar mass growth in the most massive galaxies.

Observational evidence for outflows and feedback

The study of AGN outflows that may indicate AGN or starburst feedback has become a major area of research. The earlier observations focused on the AGN-dominated aspects by means of optical, UV, and X-ray spectroscopy. They showed three distinct types of AGN winds identified by the properties of various absorption lines (Chapter 7): BAL outflows, NAL outflows, and HIG outflows. Another type of outflow was indicated by studying the narrow-emission-line profiles, in particular, the blue wing of the [O III] λ5007 line. While all such flows are common, estimating the mass outflow rate is extremely difficult and depends, in most of these cases, on unknown filling factors due to an unknown geometry. These unknowns can lead to very big uncertainties in estimating the mass outflow rate, and the associated kinetic energy and/or momentum, by up to an order of magnitude in some cases.

---

[3] Note that the formal solution diverges at $r = 0$.

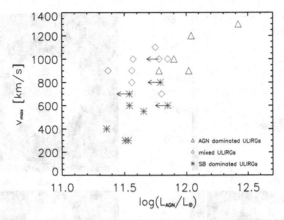

Figure 9.20. Maximum outflow velocities as a function of AGN luminosity in a sample of ULIRGs, as measured from the OH 119 P-Cygni profiles in Herschel PACS spectra (courtesy of E. Sturm).

Can any of the ideas related to AGN feedback be put to an observational test? Can we point to a case showing a clear connection between high-velocity flow, or strong radiation field, and the shutdown of SF, or the expulsion of SF gas from the galaxy? Perhaps the best evidence so far (2011) is coming from three types of observations: neutral atomic gas outflows in ULIRGs, molecular outflows in ULIRGs and LIRGs, and [O III] $\lambda5007$ outflows in high-redshift AGNs.

There are several well-studied cases of neutral atomic gas outflows in AGNs that reside in high-SFR-hosts. Most of them are detected via the NaI$\lambda\lambda5890$, 5896 D absorption lines. The general conclusion of such studies is that for fixed SFR, ULIRGs with higher AGN fractions have higher neutral gas outflow velocities, reaching values well above 1000 km s^{-1} in some broad-line AGNs.

The second line of evidence comes from FIR spectroscopy by Herschel, where molecular outflows are detected. The hydroxyl molecule (OH), in FIR spectra of many ULIRGs, provides the clearest evidence. In some of these objects, the (terminal) outflow velocities exceed 1000 km s^{-1}, and their outflow rates (up to ~ 1200 $M_\odot$/yr^{-1}) are several times larger than their SFRs. There is a clear connection between the outflow velocity and the presence of AGNs. The suggestion is that ULIRGs with a higher AGN luminosity have higher terminal velocities and shorter gas depletion time scales. The outflows in the observed ULIRGs are able to expel the cold gas reservoirs from the centers of these objects within 10^6–10^8 yr. A diagram summarizing these observations is shown in Figure 9.20. It shows the outflow velocity as a function of the SFR for objects that are dominated by the SF luminosity, objects that are dominated by the (buried) AGN luminosity, and intermediate-type objects where the two luminosities are comparable. The removal of cold molecular gas from the galaxies is a way to stop SF in a relatively short time. This can be considered as evidence for a negative AGN feedback.

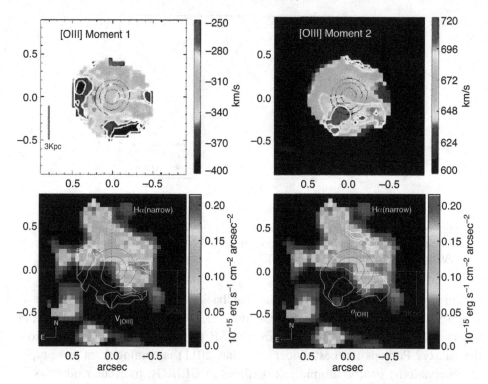

Figure 9.21. Top kinematical maps of the $z = 2.4$ AGN 2QZ20028-28 (adapted from Cano-Díaz et al., 2011) representing average [O III] λ5007 velocity (left) and velocity dispersion (right). (bottom) Narrow Hα flux with [O III] λ5007 velocity (left) and velocity dispersion (right) contours superimposed. The highest blue-shifted velocities and the largest velocity dispersions correspond to regions where narrow Hα emission is strongly suppressed. The [O III] λ5007/Hα line ratio and the upper limit on [N II] λ6584/Hα indicate that in these regions, Hα is powered by star formation, revealing the effect of AGN feedback. (See color plate)

There is a third, more direct line of evidence in a handful of sources like the one shown in Figure 9.21. Here the IFU maps show both high-velocity flow (in this case, in the [O III] λ5007 line) and a conical sector in the host galaxy, where the high-velocity region overlaps with a region where SFR, as indicated by the narrow Hα line, is considerably reduced. Unfortunately, the number of known cases of this type is so small that it is premature to draw general conclusions about the importance of such feedback.

9.9. Further reading

Galaxy evolution: A comprehensive description of all aspects is given in Mo et al. (2010). For galaxy stellar mass function, and luminosity function, see Pérez-Gonźales et al. (2008), Li and White (2009), and references therein.

Seed BHs: For a detailed review, see Volonteri (2010).

BH growth through mergers and secular evolution: There are dozens of papers discussing numerical simulations of mergers, including recoiling BHs, for example, Bernuzzi et al. (2011), Sijacki et al. (2009, 2011), Blecha et al. (2011), and references therein. Observational evidence for binary BHs is discussed in Tsalmantza et al. (2011), Shen et al. (2011), and references therein. Secular evolution in galaxies is reviewed in Kormendy and Kennicutt (2004) and in AGNs in Jogee (2006). For more references, see Cisternas et al. (2011).

The fundamental plane and kinetic power of AGN outflows: See Merloni et al. (2006), and Merloni and Heinz (2007).

BH growth via RIAF: See Cao (2007).

BH growth and evolution: A seminal paper regarding the accumulation of BH mass is Soltan (1982). For extensive reviews, see Marconi et al. (2004), Merloni and Heinz (2008), and Shankar (2009). For more detailed modeling, including spin evolution and downsizing, see Shankar et al. (2010), Fanidakis et al. (2011, 2012), and references therein. For metal and dust evolution at high redshift, see Valiante et al. (2012).

Luminosity and mass functions of BHs: Detailed articles that include many references to older works are Gilli et al. (2007), Croom et al. (2009), and Vestergaard and Osmer (2009). For the X-ray background, see Gilli et al. (2007), and references therein.

Feedback: Reviews as well as specific calculations based on various recipes can be found in Merloni and Heinz (2007), Hopkins et al. (2009), Hopkins and Elvis (2010), Fanidakis et al. (2011), and King (2010). For observational evidence of AGN outflows, see reviews by Veilleux et al. (2005) and Veilleux (2008) and additional information and references in Sturm et al. (2011) and Cano-Díaz et al. (2012).

10

Outstanding questions

Of the many questions addressed in this book, some are still open and require more attention. Answering these questions will lead to a much improved understanding of active and dormant BHs, of the physics in the various regions of AGNs, and of the complex evolutionary connections between massive BHs and their host galaxies. The questions that are considered to be more important are arranged subsequently in four large categories. Most of these issues were discussed in the previous chapters, and the list is only intended to serve as a reminder of the outstanding questions in this area of research and of the direction in which the field is going.

10.1. Questions related to the central power house

10.1.1. Black hole mass and spin

BH mass measurements in local type-II AGNs are based on the $M–\sigma^*$ relationship and the known relationships between bulge luminosity and BH size. For type-I AGNs, the estimates are based on RM-based measurements of R_{BLR} and the assumption of virialized BLRs. As of 2011, $H\beta$-based R_{BLR} estimates are available for about 40 sources, and it is not at all clear how well this sample represents the various types of AGNs, for example, radio loud versus radio quiet, high versus low luminosity, and high versus low L/L_{Edd}. The large, yet limited range of L_{bol} requires extrapolation to reach the highest-luminosity AGNs. This limits the accuracy of BH mass estimates in the most luminous AGNs. The slope of the relationship $R_{BLR} \propto L^{\alpha}$ is also not well established, with estimates ranging from 0.5 to 0.7. Moreover, it is not at all clear that a single slope applies to the entire range of L_{bol} (almost 5 orders of magnitude). There are various suggestions that $\alpha = 0.5$ is the "natural" slope because of the physics of the BLR gas. Such claims are not supported by photoionization modeling.

A more severe problem is the lack of a simple method to measure BH spin or the spin parameter a. A handful of objects show broad, relativistic Kα lines, which suggest $a \simeq 1$, but no direct way to measure a from optical or UV observations.

This has important consequences for BH evolution since the accretion efficiency η, and hence the growth rate of the BH (e.g., Equation 9.15), depends on the spin. There were several attempts to reverse the problem and use evolutionary models, which predict BH growth, to infer η and a. Such models leave much to be desired.

10.1.2. Accretion disk

Optically thick, geometrically thin accretion disks, mostly α-disks, are probably the major sources of optical–UV radiation near massive active BHs. While the very broad SED predicted for such objects is indeed typical of many AGNs, the agreement with specific spectra is not satisfactory. In particular, the observed optical–UV continuum slope is different from the one predicated by the models, and the slope of the extreme UV continuum, beyond the Lyman limit, is steeper than calculated, especially for sources with low M_{BH} and large L/L_{Edd}.

The variability of AGNs is well documented but not well understood. In particular, the optical structure functions introduced in Chapter 1 show no typical time scale and amplitudes that increase with time up to at least 10 years. It is also not clear why the variability amplitude is smaller in higher-luminosity and/or larger-BH mass sources. All this must be related to yet-to-be-understood accretion disk properties. For example, the variability time scale is considerably shorter than the one expected from changes in accretion rate (the viscous time scale) and is in better agreement with time scales that are typical of irradiated disks. However, simple energy conservation considerations show that X-ray-irradiated disks cannot explain the optical–UV radiation of most AGNs. Variability is likely to be related to disk instabilities and disk winds that are not included in most present-day disk models. The related X-ray variability may be due to different instabilities in the disk corona.

X-ray line and continuum emission from AGN disks are important yet not fully understood. The corona size and geometry is still an open issue, the soft X-ray (0.1–0.5 keV) part of the spectrum not fully modeled, and the level of ionization of the X-ray-emitting parts of the disk is not known. A related open issue is the complex interface between the disk and the X-ray-emitting corona. Present-day numerical disk models are not yet capable of answering such questions.

Broad relativistic iron lines have been in the news for years. However, their physics is still not fully understood. In particular, it is not clear why some (very few) sources show extremely broad iron lines while many others, with similar optical–UV properties, lack these features. There is also a possibility that spectral features attributed to relativistic lines are, in fact, due to complex absorption by large-column-density material in the vicinity of the central source.

Slim and thick accretion disks have been proposed to explain the changes in the disk spectrum and the reduced radiative efficiency due to advection expected in cases of very high accretion rates. Modeling such systems requires sophisticated numerical methods that are only partly available. An important observational issue is the expected deviation from the standard thin-disk spectrum in cases where $0.3 < \dot{M}/M_{\mathrm{Edd}} < 1$. This is the range where radiation pressure force dominates the shape of the disk, yet most SED fitting is still based on the (extrapolated) properties of thin accretion disks.

10.1.3. Radiatively inefficient flows and the nature of LINERs

RIAFs are likely to dominate the continuum emission in low accretion rate systems, yet such flows are not fully understood. In particular, the value of the limiting accretion rate that marks the transition between radiatively efficient and inefficient flows is not well determined (most of today's models assume $\dot{M}/M_{\mathrm{Edd}} \sim 0.01$). Several crucial numbers related to the energy balance in RIAFs, in particular, the way the released gravitational energy is divided between radiative and kinetic energy, are not known. These numbers are important for the understanding of AGN feedback and BH growth during the RIAF phase.

LINERs are the AGNs most associated with RIAFs. These objects outnumber the high-ionization AGNs by a very large factor, yet their inner physics is not fully understood. In particular, RIAF-powered LINERs are predicted to have extremely faint UV continua relative to the other parts of the SED. This is definitely not the case in some well-studied LINERs showing bright, pointlike UV continuum sources. Such objects tend to be similar to other LINERs in most of their other properties. RIAFs, if they exist in such objects, must have properties that are different than the "computer-produced" RIAFs of today.

10.1.4. The central radio and gamma-ray source

The most fundamental question in this category is probably related to the role of BH spin in producing radio and γ-ray jets. According to some models, this is the most important property that separates radio-loud from radio-quiet AGNs. There are more specific open questions regarding the physics of the jet related to the acceleration and radiation mechanisms. Several such questions are the role of MHD disk winds, the source of inverse Compton (IC) photons (accretion disk emission, BLR lines, or some combination of the two), and the differences between the various types of γ-ray jets in blazars. More accurate numerical modeling of magnetic AGN disks around fast-spinning BHs is likely to make a big improvement in this area.

10.2. Questions related to the broad and narrow emission lines

10.2.1. BLR physics and size

The size of the BLR has been measured by several successful reverberation mapping experiments. However, two-dimensional RM mappings, combining gas distribution and velocity, are available for only a handful of objects, and even these cases do not have the accuracy needed to distinguish between various kinematical models. High-quality location-velocity mapping of the BLR is still missing.

A more fundamental question is the origin and geometry of the BLR gas. In particular, it is not at all clear whether the system is made of individual cloudlets or whether it is more like a multiphase medium, with a large density range at every location. The first case requires the confinement of the clouds by external pressure, yet the origin of this pressure is not clear. Leading candidates are magnetic pressure and pressure by very hot intercloud gas. The second case involves winds and the LOC model. Here there can be a large density range at every location, and the various components are not in pressure equilibrium. Regarding the origin of the gas (or clouds), it is not clear whether the material is related to the central disk, perhaps through a disk wind, or is a remnant of a long-term, large-scale accretion episode.

Of the many spectroscopic issues related to the BLR gas, the questions related to the hydrogen line ratios, the intensity of the FeII lines, and the energy budget of the BLR are still unanswered.

10.2.2. Gas kinematics in the BLR

The gas kinematics in the BLR is not fully understood; in particular, the role of radiation pressure force in determining the gas motion is not understood. Such a force is likely to affect more the small-column-density material to the extent that the smaller clouds will be ejected from the system following large continuum outbursts. This material may or may not be replaced by disk-ejected winds. Related observational issues are the indications, if any, of inflow or outflow components in the emission-line profiles. In principle, this issue can be settled by a detailed, two-dimensional RM, but as explained, such results are only available for a handful of sources. An additional issue is the geometry and kinematics of the BLR gas in radio-loud AGNs that show broad, double-peak emission lines. The observed line profiles in such cases cannot be explained by simple inclination effects, which suggests that the radio properties and the gas motion are related in some way. Finally, the physics of the processes leading to various line profiles that are described by the eigenvector 1 sequence is not yet understood.

10.2.3. The narrow- and coronal-line regions

While the physics and kinematics of the NLR gas are better understood, there are several open issues that are related to the coronal-line gas and the general issue of dust in the NLR. A major problem is the reason for the lack of dust in the coronal-line region. This gas is way outside the dust sublimation distance, and the gas temperature, assuming photoionization, is not very high. Silicate and graphite dust grain can survive under such conditions. Another important issue is the role of internal dust in affecting the pressure profile of the NLR clouds. The reason for tendency of the NLR line intensities to decrease with increasing L_{bol} is not known. This must involve the covering factor of the NLR and, perhaps, the covering fraction of the central torus.

A large fraction of AGN hosts are SF galaxies. A practical problem related to NLR observations, and AGN classification, is the separation of AGN-excited gas from SF-excited gas in such objects, especially at high redshift, where most spectroscopic observations are not spatially resolved and include the entire galaxy.

10.2.4. Gas metallicity

The metallicity of the NLR and BLR gas is difficult to determine and requires photoionization modeling. Such models, when compared with BLR observations, can result in conflicting information about the gas metallicity. For example, the N/C abundance determined from the $N v \lambda 1240/C iv \lambda 1549$ line ratio tends to indicate a higher metallicity compared with the value deduced from the $[N iv]\lambda 1486/[C iii]\lambda 1750$ line ratio. There are also open questions related to the correlation of gas metallicity and other AGN properties such as source luminosity, the BH mass, and L/L_{Edd}.

10.3. Questions related to outflow and feedback

10.3.1. Outflow physics

The outstanding issues in this category are the relationship, if any, between atomic and molecular gas outflows in objects where both are observed, the driving mechanisms for both types of outflows, and the frequency of such winds in the AGN population. Mass outflow rates are difficult to estimate unless both density and column density (or filling factor) are known. Such information is only available for a handful of sources. Related to this is the unknown kinetic energy, or momentum, associated with the different types of flows. The exception are several well-observed cases of NAL systems obseved in the UV and X-ray parts of the spectrum.

The central accretion disk is a major mass reservoir, and disk winds were suggested to explain the most massive and energetic outflows. The physics of such winds, including the mechanisms lifting the gas from the surface of the disk; the role of magnetic fields; the central shielding required for efficient acceleration of BAL flows; and more are still unknown. The very elegant disk wind scenario that couples several AGN components (Chapter 7) is far from being a real physical model.

10.3.2. The physics of feedback

AGN negative feedback, which can suppress SF and further accretion onto the BH, is thought to be important especially in the final stages of evolution of massive galaxies. The feedback must be related to massive gas outflows and/or the strong AGN radiation field. Unfortunately, the interaction of the various types of AGN winds with the ISM and SF gas in the host is still lacking a concrete physical model. The microphysics involved in such processes is well understood, but it has never been incorporated into large-scale hydrodynamical calculations partly because of their limited numerical resolution. The result is that the AGN feedback efficiency, which is used in many galaxy evolutionary models, cannot be accurately calculated. Numbers used for the feedback efficiency in today's evolutionary calculations (of order 5% of the AGN luminosity) should be considered rough estimates and do not necessarily reflect the conditions in the systems in question. Of the many physical processes that are still missing from the calculations, the effect of the AGN-ionizing radiation is perhaps the easiest one to model. However, even this is not yet an integral part of most of today's models.

10.4. Questions related to AGN evolution

10.4.1. BH growth

The most important part of BH evolution is the way these objects grow over time. The details of the process must depend on the host galaxy and the interaction with other galaxies and are therefore better known in the local universe. Exponential growth has been adopted in most evolutionary models. The only real justification for this is if $\dot{M}/M_{\rm Edd} \sim 1$ and the Eddington limit imposes a natural upper limit on the rate. However, the great majority of AGNs are accreting at much slower rates. Accretion at a constant rate, over long periods of time, will result in smaller $M_{\rm BH}$ and $L/L_{\rm Edd}$ that is decreasing in time. Such models need to be considered.

As already mentioned, BH spin, and the related radiative efficiency η, is another important missing factor. The growth time is linear, with $\eta/(1-\eta)$, and thus the allowed range of spins is reflected in a large uncertainty on the duration of the process. Moreover, realistic evolutionary models must include the possibility of a long accretion episode that is broken into several shorter episodes, each with a different spin.

10.4.2. Duty cycle luminosity function and mass function

The fraction of BH lifetime that is spent in the active phase (BH duty cycle) must depend on BH mass and redshift. There is little information about this crucial parameter, and the available numbers are based on guesses rather than direct measurements or self-consistent models. Some of the important specific open questions in this area relate to the details of the creation of seed BHs in the early universe, their size distribution, their relationship to population III stars, and the details of their growth through mergers and accretion. Part of this evolution may be associated with ejection of intermediate-size BHs to the intergalactic medium due to high recoil velocity in the final stages of young galaxy mergers – a process with well-understood physics but less understood astronomical consequences.

The duty cycle of various size BHs at various redshifts is a crucial link in the construction of BH continuity equations (Chapter 9). The lack of knowledge in this case presents a major limitation in converting reasonably well measured LFs, at almost all redshifts, to almost completely unknown BH mass functions at all redshifts above about 0.3.

10.4.3. *Merger or secular evolution and completely obscured AGNs*

An additional unknown is the fraction of completely obscured active BHs in the early phases of their evolution and/or following large mergers. Several of today's models assume that a large part of the most rapid growth phase, with $L/L_{Edd} \sim 1$, is taking place while the accreting BH is obscured. The rate of accretion is going down once the BH blows out the cocoon and becomes visible. Like several other quantities, these BH properties are basically put in by hand in evolutionary calculations. It is not at all clear that the observations support this idea. In fact, there are many documented cases of "bare AGNs" showing very high accretion rates at all redshifts.

Additional open issues are related to the two different modes of BH growth: fast accretion during a merger and growth due to slower secular processes. If obscured AGNs are mostly the result of mergers, they must be more frequent in the early universe and not so frequent at later times, when secular evolution is more common. Details of these issues are still missing, and much of the lack of knowledge is, again, due to the limitations of present-day numerical simulations.

10.4.4. *M–σ* at high redshift*

The important correlations between BH mass, bulge mass, and stellar velocity dispersion in the local universe are the cornerstones of BH and galaxy evolutionary studies. The relationships are known, accurately, at low redshift, yet all attempts to extend them to higher redshifts ended with inconclusive results. There are claims that the correlations are identical to the ones observed in the local universe, at $z \sim 1$. There are equally supported claims of very different correlations. The main problem in this case is observational and is related to the great difficulties in determining M_* and σ_* in high-redshift, type-I AGNs. This field is awaiting more accurate observations, or perhaps other ideas, to overcome the difficulty.

10.4.5. The AGN–starburst connection

This area is progressing extremely fast due to a large increase in sample sizes and higher-quality observations due mostly to the superb Herschel observations. The result is that observations relating BH activity and SF in the host galaxy are now available all the way up to $z \sim 6$. Modeling the various suggested complex scenarios related to the AGN–starburst connection lags behind, and time is required to sort out the meaning of the various emerging relationships. An interesting open question is the exact relationship between L_{AGN} and L_{SF} (i.e., between BH accretion rate and SFR) at different redshifts; the differences of such relationships between mergers, where the host galaxy is probably situated above the SF sequence; and galaxies that form stars via more continuous secular processes. A related issue is the ability to distinguish between starburst and poststarburst AGN host galaxies. A more general issue is the similarity, if any, between the mass functions of galaxies and BHs at all redshifts.

References

Ackermann, M., Ajello, M., Allafort, A., et al. 2011, "The second catalog of active galactic nuclei detected by the Fermi large area telescope," ApJ, 743, 171.

Alonso-Herrero, A., Ramos Almeida, C., Mason, R., et al. 2011, "Torus and active galactic nucleus properties of nearby Seyfert galaxies: Results from fitting infrared spectral energy distributions and spectroscopy," ApJ 736, 82.

Antonucci, R. 1993, "Unified models for active galactic nuclei and quasars," AnnRevAstAp, 31, 473.

Arav, N., Moe, M., Costantini, E., et al. 2008, "Measuring column densities in quasar outflows: VLT observations of QSO 2359-1241," ApJ, 681, 954.

Assef, R. J., Denney, K. D., Kochanek, C. S., et al. 2011, "Black hole mass estimates based on C IV are consistent with those based on the Balmer lines," ApJ, 742, 93.

Baldwin, J. A. 1977, "Luminosity indicators in the spectra of quasi-stellar objects," ApJ, 214, 679.

Baldwin, J. A., Ferland, G. J., Korista, K. T., et al. 2004, "The origin of Fe II emission in active galactic nuclei," ApJ, 615, 610.

Barthel, P., & van Bemmel, I. 2003, "Radio galaxies: Unification and dust properties," NAR, 47, 199.

Becker, R. H., White, R. L., & Helfand, D. J., et al. 1995, "The FIRST survey: Faint images of the radio sky at twenty centimeters," ApJ 450, 559.

Bentz, M. C., Peterson, B. M., Netzer, H., et al. 2009, "The radius–luminosity relationship for active galactic nuclei: The effect of host-galaxy starlight on luminosity measurements. II. The full sample of reverberation-mapped AGNs," ApJ, 697, 160.

Bernuzzi, S., Nagar, A., & Zenginoğlu, A. 2011, "Binary black hole coalescence in the large-mass-ratio limit: The hyperboloidal layer method and waveforms at null infinity," PRD, 84, 084026.

Best, P. N., Kauffmann, G., Heckman, T. M., et al. 2005, "The host galaxies of radio-loud active galactic nuclei: Mass dependences, gas cooling and active galactic nuclei feedback," MNRAS, 362, 25.

Blaes, O., Hubeny, I., Agol, E., et al. 2001, "Non-LTE, relativistic accretion disk fits to 3C 273 and the origin of the Lyman limit spectral break," ApJ, 563, 560.

Blaes, O. 2007, "The central engine of active galactic nuclei," ASPC, 373, 75.

Blandford, R. D. 1990, "Physical processes in active galactic nuclei," *SAAS-FEE Advanced Course 20 on Active Galactic Nuclei*, Springer, 161.

Blanton, M. R., Hogg, D. W., Bahcall, N. A., et al. 2003, "The broadband optical properties of galaxies with redshifts 0.02 < z < 0.22," ApJ, 594, 186.

Blecha, L., Cox, T. J., Loeb, A., & Hernquist, L. 2011, "Recoiling black holes in merging galaxies: Relationship to active galactic nucleus lifetimes, starbursts and the MBH-* relation," MNRAS, 412, 2154.

Bonfield, D. G., Jarvis, M. J., Hardcastle, M. J., et al. 2011, "Herschel-ATLAS: The link between accretion luminosity and star formation in quasar host galaxies," MNRAS, 416, 13.

Boroson, T. A., & Green, R. F. 1992, "The emission-line properties of low-redshift quasi-stellar objects," ApJ, 80, 109.

Böttcher, M., Basu, S., Joshi, M., et al. 2007, "The WEBT campaign on the blazar 3C 279 in 2006," ApJ, 670, 968.

Bottorff, M. C., & Ferland, G. J. 2000, "Magnetic confinement, magnetohydrodynamic waves and smooth line profiles in active galactic nuclei," MNRAS, 316, 103.

Brammer, G. B., Whitaker, K. E., van Dokkum, P. G., et al. 2009, "The dead sequence: A clear bimodality in galaxy colors from $z = 0$ to $z = 2.5$," ApJL, 706, L173.

Cano-Díaz, M., Maiolino, R., Marconi, A., et al. 2012, "Observational evidence of quasar feedback quenching star formation at high redshift," A&A, 537, L8.

Cao, X. 2007, "Growth of massive black holes during radiatively inefficient accretion phases," ApJ, 659, 950.

Cao, X. 2009, "An accretion disc-corona model for X-ray spectra of active galactic nuclei," MNRAS, 394, 207.

Cao, X. 2010, "On the disappearance of the broad-line region in low-luminosity active galactic nuclei: The role of the outflows from advection dominated accretion flows," ApJ, 724, 855.

Capellupo, D. M., Hamann, F., Shields, J. C., et al. 2011, "Variability in quasar broad absorption line outflows – I. Trends in the short-term versus long-term data," MNRAS, 413, 908.

Cardiel, N., Elbaz, D., Schiavon, R. P., et al. 2003, "A multiwavelength approach to the star formation rate estimation in galaxies at intermediate redshifts," ApJ, 584, 76.

Chelouche, D., & Netzer, H. 2005, "Dynamical and spectral modeling of the ionized gas and nuclear environment in NGC 3783," ApJ, 625, 95.

Cisternas, M., Jahnke, K., Inskip, K. J., et al. 2011, "The bulk of the black hole growth since $z \sim 1$ occurs in a secular universe: No major merger-AGN connection," ApJ, 726, 57.

Cid Fernandes, R., Heckman, T., Schmitt, H., et al. 2001, "Empirical diagnostics of the starburst-AGN connection," ApJ, 558, 81.

Cid Fernandes, R., Stasińska, G., Mateus, A., et al. 2011, "A comprehensive classification of galaxies in the Sloan Digital Sky Survey: How to tell true from fake AGN?," MNRAS, 413, 1687.

Cisternas, M., Jahnke, K., Bongiorno, A., et al. 2011, "Secular evolution and a non-evolving black-hole-to-galaxy mass ratio in the last 7 Gyr," ApJL, 741, L11.

Clavel, J., Reichert, G. A., Alloin, D., et al. 1991, "Steps toward determination of the size and structure of the broad-line region in active galactic nuclei. I – an 8 month campaign of monitoring NGC 5548 with IUE," ApJ, 366, 64.

Collin, S., & Joly, M. 2000, "The Fe II problem in NLS1s," NAR, 44, 531.

Collin, S., & Kawaguchi, T. 2004, "Super-Eddington accretion rates in narrow line Seyfert 1 galaxies," A&A, 426, 797.

Croom, S. M., Smith, R. J., Boyle, B. J., et al. 2004, "The 2dF QSO redshift survey – XII. The spectroscopic catalogue and luminosity function," MNRAS, 349, 1397.

Croom, S. M., Smith, R. J., Boyle, B. J., et al. 2004, MNRAS, 349, 1397.

Croom, S. M., et al. 2008, "The 2dF-SDSS LRG and QSO survey: The spectroscopic QSO catalogue," MNRAS, 392, 19.

Croom, S. M., Richards, G. T., Shanks, T., et al. 2009, "The 2dF-SDSS LRG and QSO survey: The QSO luminosity function at 0.4 < z < 2.6," MNRAS, 399, 1755.

Croom, S. M. 2011, "Do quasar broad-line velocity widths add any information to virial black hole mass estimates?" ApJ 736, 161.

Crenshaw, D. M., Kraemer, S. B., & George, I. M. 2003, "Mass loss from the nuclei of active galaxies," ARAA, 41, 117.

Daddi, E., Dickinson, M., Morrison, G., et al. 2007, "Multiwavelength study of massive galaxies at z 2. I. Star formation and galaxy growth," ApJ, 670, 156.

Dasyra, K. M., Ho, L. C., Netzer, H., et al. 2011, "A view of the narrow-line region in the infrared: Active galactic nuclei with resolved fine-structure lines in the Spitzer archive," ApJ, 740, 94.

Dovčiak, M., Karas, V., & Yaqoob, T. 2004, "An extended scheme for fitting X-ray data with accretion disk spectra in the strong gravity regime," ApJS, 153, 205.

Davidson, K., & Netzer, H. 1979, "The emission lines of quasars and similar objects," RMP, 51, 715.

Davis, S. W., Woo, J.-H., & Blaes, O. M. 2007, "The UV continuum of quasars: Models and SDSS spectral slopes," ApJ, 668, 682.

Davis, S. W., & Laor, A. 2011, "The radiative efficiency of accretion flows in individual active galactic nuclei," ApJ, 728, 98.

Dopita, M. A., Groves, B. A., Sutherland, R. S., et al. 2002, "Are the narrow-line regions in active galaxies dusty and radiation pressure dominated?" ApJ 572, 753.

Dutton, A. A., van den Bosch, F. C., & Dekel, A. 2010, "On the origin of the galaxy star-formation-rate sequence: Evolution and scatter," MNRAS, 405, 1690.

Emerson, D. 1996, *Interpreting Astronomical Spectra*, John Wiley.

Elitzur, M. 2008, "The toroidal obscuration of active galactic nuclei," NAR, 52, 274.

Elitzur, M., & Ho, L. C. 2009. "On the disappearance of the broad line region in low luminosity active galactic nuclei," ApJL, *ApJL* 701, L91–L94.

Elvis, M., Wilkes, B. J., McDowell, J. C., et al., 1994, "Atlas of quasar energy distributions," ApJS, 95, 1.

Elvis, M. 2000, "A structure for quasars," ApJ, 545, 63.

Emmering, R. T., Blandford, R. D., & Shlosman, I. 1992, "Magnetic acceleration of broad emission-line clouds in active galactic nuclei," ApJ, 385, 460.

Eracleous, M., & Halpern, J. P. 2003, "Completion of a survey and detailed study of double-peaked emission lines in radio-loud active galactic nuclei," ApJ, 599, 886.

Fan, X. 2006, "Evolution of high-redshift quasars," NAR, 50, 665–671.

Fan, X., Carilli, C. L., & Keating, B. 2006, "Observational constraints on cosmic reionization," ARAA, 44, 415.

Fanidakis, N., Baugh, C. M., Benson, A. J., et al. 2011, "Grand unification of AGN activity in the CDM cosmology," MNRAS, 410, 53.

Fanidakis, N., Baugh, C. M., Benson, A. J., et al. 2012, "The evolution of active galactic nuclei across cosmic time: What is downsizing?," MNRAS, 419, 2797.

Fine, S., Croom, S. M., Bland-Hawthorn, J., et al. 2010. "The CIV linewidth distribution for quasars and its implications for broad-line region dynamics and virial mass estimation," MNRAS, 409, 591.

Ferrarese, L., & Merritt, D. A. 2000, "A fundamental relation between supermassive black holes and their host galaxies," ApJ, 539, L9L12.

Fosbury, R. 2006, "AGN beyond the 100pc scale," *Physics of Active Galactic Nuclei at all Scales*, edited by Danielle Alloin, Rachel Johnson, and Paulina Lira. LNP, 693, 121.

Frank, J., King, A., & Raine, D. J. 2002, *Accretion Power in Astrophysics: Cambridge University Press, Third Edition.*

Fritz, J., Franceschini, A., & Hatziminaoglou, E. 2006, "Revisiting the infrared spectra of active galactic nuclei with a new torus emission model," MNRAS, 366, 767.

Gebhardt, K., Bender, R., Bower, G., et al. 2000, "A relationship between nuclear black hole mass and galaxy velocity dispersion," ApJ, 539, L13.

Gilli, R., Comastri, A., & Hasinger, G. 2007, "The synthesis of the cosmic X-ray background in the Chandra and XMM-Newton era," A&A, 463, 79.

Greene, J. E., & Ho, L. C. 2005, "A comparison of stellar and gaseous kinematics in the nuclei of active galaxies," ApJ, 627, 721.

Groves, B., Kewley, L., Kauffmann, G., et al. 2006, "An SDSS view of type-2 AGN classification," NAR, 50, 743.

Grupe, D., Komossa, S., Leighly, K. M., & Page, K. L. 2010, "The simultaneous optical-to-X-ray spectral energy distribution of soft X-ray selected active galactic nuclei observed by swift," ApJS, 187, 64.

Gültekin, K., Richstone, D. O., Gebhardt, K., et al. 2009, "The M-σ and M-L relations in galactic bulges, and determinations of their intrinsic scatter," ApJ, 698, 198.

Haas, M., Siebenmorgen, R., Schulz, B., et al. 2005, "Spitzer IRS spectroscopy of 3CR radio galaxies and quasars: Testing the unified schemes," A&A, 442, L39.

Hamann, F., & Ferland, G. 1999, "Elemental abundances in quasistellar objects: Star formation and galactic nuclear evolution at high redshifts," ARAA, 37, 487.

Hasinger, G., Miyaji, T., & Schmidt, M. 2005, "Luminosity-dependent evolution of soft X-ray selected AGN. New Chandra and XMM-Newton surveys," A&A, 441, 417.

Heckman, T. M., & Kauffmann, G. 2006, "The host galaxies of AGN in the Sloan Digital Sky Survey," NAR, 50, 677.

Herrnstein, J. R., Moran, J. M., Greenhill, L. J., et al. 1999, "A geometric distance to the galaxy NGC4258 from orbital motions in a nuclear gas disk," Nature, 400, 539.

Hirashita, H., Buat, V., & Inoue, A. K. 2003, "Star formation rate in galaxies from UV, IR, and H estimators," A&A, 410, 83.

Ho, L. C. 2008, "Nuclear activity in nearby galaxies," ARAA, 46, 475.

Hopkins, P. F., Murray, N., & Thompson, T. A. 2009, "The small scatter in BH-host correlations and the case for self-regulated BH growth," MNRAS, 398, 303.

Hopkins, P. F., & Elvis, M. 2010, "Quasar feedback: More bang for your buck," MNRAS, 401, 7.

Hubeny, I., Blaes, O., Krolik, J. H., et al. 2001, "Non-LTE models and theoretical spectra of accretion disks in active galactic nuclei. IV. Effects of Compton scattering and metal opacities," ApJ, 559, 680.

Jogee, S. R. 2006, "The fueling and evolution of AGN: Internal and external triggers," *Physics of Active Galactic Nuclei at all Scales*, edited by Danielle Alloin, Rachel Johnson, and Paulina Lira, LNP, 693, 143.

Kennicutt, R. C., Jr. 1998, "Star formation in galaxies along the Hubble sequence," ARAA, 36, 189.

Kennicutt, R. C., Jr., Hao, C.-N., Calzetti, D., et al. 2009, "Dust-corrected star formation rates of galaxies. I. Combinations of Hα and infrared tracers," ApJ, 703, 1672.

Kormendy, J., & Richstone, D. 1995, "Inward bound the search for supermassive black holes in galactic nuclei," ARAA, 33, 581.

Kormendy, J., & Kennicutt, R. C., Jr. 2004, "Secular evolution and the formation of pseudobulges in disk galaxies," ARAA, 42, 603.

Kormendy, J., Fisher, D. B., Cornell, M. E., et al. 2009, "Structure and formation of elliptical and spheroidal galaxies," ApJS, 182, 216.

Krolik, J. H., 1999, *Active Galactic Nuclei*, Princeton University Press.

Kuo, C. Y., Braatz, J. A., Condon, J. J., et al. 2011, "The megamaser cosmology project. III. Accurate masses of seven supermassive black holes in active galaxies with circumnuclear megamaser disks," ApJ, 727, 20.

Kaastra, J. S. 2008, "High spectral resolution X-ray observations of AGN," AN, 329, 162.

Kauffmann, G., Heckman, T. M., Tremonti, C., et al. 2003, "The host galaxies of active galactic nuclei," MNRAS, 346, 1055.

Kauffmann, G., Heckman, T. M., & Best, P. N. 2008, "Radio jets in galaxies with actively accreting black holes: New insights from the SDSS," MNRAS, 384, 953.

Kauffmann, G., & Heckman, T. M. 2009, "Feast and famine: Regulation of black hole growth in low-redshift galaxies," MNRAS, 397, 135.

Kaspi, S., & Netzer, H. 1999, "Modeling variable emission lines in active galactic nuclei: Method and application to NGC 5548," ApJ, 524, 71.

Kaspi, S., Smith, P. S., Netzer, H., et al. 2000, "Reverberation measurements for 17 quasars and the size-mass-luminosity relations in active galactic nuclei," ApJ, 533, 631.

Kaspi, S., Maoz, D., Netzer, H., et al. 2005, "The relationship between luminosity and broad-line region size in active galactic nuclei," ApJ, 629, 61.

Kaspi, S., Brandt, W. N., Maoz, D., et al. 2007, "Reverberation mapping of high-luminosity quasars: First results," ApJ, 659, 997.

Kawaguchi, T., Shimura, T., & Mineshige, S. 2001, "Broadband spectral energy distributions of active galactic nuclei from an accretion disk with advective coronal flow," ApJ, 546, 966.

Kewley, L. J., Groves, B., Kauffmann, G., et al. 2006, "The host galaxies and classification of active galactic nuclei," MNRAS, 372, 961.

King, A. R. 2010, "Black hole outflows," MNRAS, 402, 1516.

Kishimoto, M., Antonucci, R., Blaes, O., et al. 2008, "The characteristic blue spectra of accretion disks in quasars as uncovered in the infrared," Nature, 454, 492.

Korista, K. T., & Goad, M. R. 2000, "Locally optimally emitting clouds and the variable broad emission line spectrum of NGC 5548," ApJ, 536, 284.

Korista, K. T., & Goad, M. R. 2004, "What the optical recombination lines can tell us about the broad-line regions of active galactic nuclei," ApJ, 606, 749.

Lawrence, A., & Elvis, M. 2010, "Misaligned disks as obscurers in active galaxies," ApJ, 714, 561.

Li, C., & White, S. D. M. 2009, "The distribution of stellar mass in the low-redshift universe," MNRAS, 398, 2177.

Lister, M. L., Aller, M., Aller, H., et al. 2011, "γ-Ray and parsec-scale jet properties of a complete sample of blazars from the MOJAVE program," ApJ, 742, 27.

Madau, P. 1988, "Thick accretion disks around black holes and the UV/soft X-ray excess in quasars," ApJ, 327, 116.

Maiolino, R., Risaliti, G., Salvati, M., et al. 2010, "Comets orbiting a black hole," A&A, 517, A47.

Maoz, D. 2007, "Low-luminosity active galactic nuclei: Are they UV faint and radio loud?," MNRAS, 377, 1696.

Marconi, A., Risaliti, G., Gilli, R., et al. 2004, "Local supermassive black holes, relics of active galactic nuclei and the X-ray background," MNRAS, 351, 169.

Marconi, A., Axon, D. J., Maiolino, R., et al. 2008, "The effect of radiation pressure on virial black hole mass estimates and the case of narrow-line Seyfert 1 galaxies," ApJ, 678, 693.

Marziani, P., & Sulentic, J. W. 2012, "Quasar outflows in the 4D eigenvector 1 context," AstRv, 7, 33.

Matsuoka, K., Nagao, T., Marconi, A., et al. 2011, "The mass-metallicity relation of SDSS quasars," A&A, 527, A100.

McHardy, I. 2010, "X-ray variability of AGN and relationship to galactic black hole binary systems," LNP, 794, 203.

McLure, R. J., & Dunlop, J. S. 2004, "The cosmological evolution of quasar black hole masses," MNRAS, 352, 1390.

McConnell, N. J., Ma, C.-P., Gebhardt, K., et al. 2011, "Two ten-billion-solar-mass black holes at the centres of giant elliptical galaxies," Nature, 480, 215.

Merloni, A., Körding, E., Heinz, S., et al. 2006, "Why the fundamental plane of black hole activity is not simply a distance driven artifact," NAR, 11, 567.

Merloni, A., & Heinz, S. 2007, "Measuring the kinetic power of active galactic nuclei in the radio mode," MNRAS, 381, 589.

Merloni, A., & Heinz, S. 2008, "A synthesis model for AGN evolution: Supermassive black holes growth and feedback modes," MNRAS, 388, 1011.

Miller, L., Turner, T. J., & Reeves, J. N. 2008. An absorption origin for the X-ray spectral variability of MCG-6-30-15. *A&A* 483, 437–452.

Miller, L., Turner, T. J., & Reeves, J. N. 2009, "The absorption-dominated model for the X-ray spectra of typeI active galaxies: MCG-6-30-15," MNRAS, 399, L69.

Mo, H., van den Bosch, F., & White, S. 2010, *Galaxy formation and evolution*, Cambridge University Press.

Müller-Sánchez, F., Prieto, M. A., Hicks, E. K. S., et al. 2011, "Outflows from active galactic nuclei: Kinematics of the narrow-line and coronal-line regions in Seyfert galaxies," ApJ, 739, 69.

Mor, R., & Trakhtenbrot, B. 2011, "Hot-dust clouds with pure-graphite composition around type-I active galactic nuclei," ApJL, 737, L36.

Mor, R., & Netzer, H. 2012, "Hot graphite dust and the infrared spectral energy distribution of active galactic nuclei," MNRAS, 420, 526.

Narayan, R., & Quataert, E. 2005, "Black hole accretion," Science, 307, 77.

Narayan, R., & McClintock, J. E. 2008, "Advection-dominated accretion and the black hole event horizon," NAR, 51, 733.

Negrete, C. A., Dultzin, D., Marziani, P., & Sulentic, J. W. 2012. "Broad-line region physical conditions in extreme population a quasars: A method to estimate central black hole mass at high redshift," ApJ, 757, 62.

Netzer, H. 1985, "Quasar discs. I – The Baldwin effect," MNRAS, 216, 63.

Netzer, H. 1990, "AGN emission lines," *SAAS-FEE advanced course 20 on active galactic nuclei*, Springer, 57.

Netzer, H., Laor, A., & Gondhalekar, P. M. 1992, "Quasar discs. III – Line and continuum correlations," MNRAS, 254, 15.

Netzer, H., & Peterson, B. M. 1997, "Reverberation mapping and the physics of active galactic nuclei," ATS, 218, 85.

Netzer, H., Kaspi, S., Behar, E., et al. 2003, "The ionized gas and nuclear environment in NGC 3783. IV. Variability and modeling of the 900 kilosecond chandra spectrum," ApJ, 599, 933.

Netzer, H., & Trakhtenbrot, B. 2007, "Cosmic evolution of mass accretion rate and metallicity in active galactic nuclei," ApJ, 654, 754.

Netzer, H., Lutz, D., Schweitzer, M., et al. 2007a, "Spitzer quasar and ULIRG evolution study (QUEST). II. The spectral energy distributions of palomar-green quasars," ApJ, 666, 806.

Netzer, H., Lira, P., Trakhtenbrot, B., et al. 2007b, "Black hole mass and growth rate at high redshift," ApJ, 671, 1256.

Netzer, H. 2008, "Ionized gas in active galactic nuclei," NAR, 52, 257.

Netzer, H. 2009, "Accretion and star formation rates in low-redshift type II active galactic nuclei," MNRAS, 399, 1907.

Netzer, H., & Marziani, P. 2010, "The effect of radiation pressure on emission-line profiles and black hole mass determination in active galactic nuclei," ApJ, 724, 318.

Osterbrock, D. E., & Ferland, G. J. 2006, *Astrophysics of gaseous nebulae and active galactic nuclei*, University Science Books.

Pérez-González, P. G., Rieke, G. H., Villar, V., et al. 2008, "The stellar mass assembly of galaxies from $z = 0$ to $z = 4$: Analysis of a sample selected in the rest-frame near-infrared with spitzer," ApJ, 675, 234.

Peterson, B. M. 1997, *An introduction to active galactic nuclei*, Cambridge University Press.

Peterson, B. M., & Bentz, M. C. 2006, "Black hole masses from reverberation mapping," NewAsRev, 50, 796.

Peterson, B. M. 2008, "The central black hole and relationships with the host galaxy," NewAsRev, 52, 240.

Proga, D., & Kallman, T. R. 2004, "Dynamics of line-driven disk winds in active galactic nuclei. II. Effects of disk radiation," ApJ, 616, 688.

Quintilio, R., & Viegas, S. M. 1997, "Theoretical emission-line profiles of active galactic nuclei and the unified model. I. The face-on torus," ApJ, 474, 616.

Reynolds, C. S., & Nowak, M. A. 2003, "Fluorescent iron lines as a probe of astrophysical black hole systems," Physics Reports, 377, 389.

Richards, G. T., Lacy, M., Storrie-Lombardi, et al. 2006. "Spectral energy distributions and multiwavelength selection of type 1 quasars," ApJS Supp, 166, 470–497.

Richards, G. T., Nichol, R. C., Gray, A. G., et al. 2008, "Efficient photometric selection of quasars from the sloan digital sky survey: 100,000 z < 3 quasars from data release one," ApJS, 155, 257.

Richards, G. T., Myers, A. D., Gray, A. G., et al. 2009. "Efficient photometric selection of quasars from the sloan digital sky survey. II. ~1,000,000 quasars from data release 6," ApJS, 180, 67.

Risaliti, G., Salvati, M., & Marconi, A. 2011, "[O III] equivalent width and orientation effects in quasars," MNRAS, 411, 2223.

Robinson, A. 1995, "The profiles and response functions of broad emission lines in active galactic nuclei," MNRAS, 276, 933.

Robson, I. 1996, *Active Galactic Nuclei*, John Wiley.

Ross, R. R., & Fabian, A. C. 2005, "A comprehensive range of X-ray ionized-reflection models," MNRAS, 358, 211.

Rosario, D. J., Santini, P., Lutz, D., et al. 2012, "The mean star formation rate of X-ray selected active galaxies and its evolution from z ~ 2.5: Results from PEP-Herschel," A&A, 545, A45.

Różańska, A., & Madej, J. 2008, "Models of the iron Kα fluorescent line and the Compton Shoulder in irradiated accretion disc spectra," MNRAS, 386, 1872.

Rybicky, G. B., & Lightman, A. P. 1979, *Radiative processes in astrophysics*, John Wiley.

Salim, S., Rich, R. M., Charlot, S., et al. 2007, "UV star formation rates in the local universe," ApJS, 173, 267.

Sani, E., Lutz, D., Risaliti, G., et al. 2010, "Enhanced star formation in narrow-line Seyfert 1 active galactic nuclei revealed by Spitzer," MNRAS, 403, 1246.

Sani, E., Marconi, A., Hunt, L. K., et al. 2011, "The Spitzer/IRAC view of black hole-bulge scaling relations," MNRAS, 413, 1479.

Santini, P., Rosario, D. J., Shao, L., et al. 2012, "Enhanced star formation rates in AGN hosts with respect to inactive galaxies from PEP–Herschel observations," A&A, 540, A109.

Schawinski, K., Urry, C. M., Virani, S., et al. 2010, "Galaxy Zoo: The fundamentally different co-evolution of supermassive black holes and their early- and late-type host galaxies," ApJ, 711, 284.

Shakura, N. I., & Sunyaev, R. A. 1973, "Black holes in binary systems. Observational appearance," A&A, 24, 337.

Shang, Z., Brotherton, M. S., Wills, B. J., et al. 2011, "The next generation atlas of quasar spectral energy distributions from radio to X-rays," ApJS, 196, 2.

Shankar, F. 2009, "The demography of supermassive black holes: Growing monsters at the heart of galaxies," NAR, 53, 57.

Shankar, F., Crocce, M., Miralda-Escudé, J., et al. 2010, "On the radiative efficiencies, Eddington ratios, and duty cycles of luminous high-redshift quasars," ApJ, 718, 231.

Schmidt, M., et al. 2012, "The color variability of quasars," ApJ, 744, 147.

Schweitzer, M., Lutz, D., Sturm, E., et al. 2006, "Spitzer Quasar and ULIRG Evolution Study (QUEST). I. The origin of the far-infrared continuum of QSOs," ApJ, 649, 79.

Shemmer, O., Netzer, H., Maiolino, R., et al. 2004, "Near-infrared spectroscopy of high-redshift active galactic nuclei. I. A metallicity-accretion rate relationship," ApJ, 614, 547.

Shen, Y., Greene, J. E., Strauss, M. A., et al. 2008, "Biases in virial black hole masses: An SDSS perspective," ApJ, 680, 169.

Shen, Y., Liu, X., Greene, J. E., et al. 2011, "Type 2 active galactic nuclei with double-peaked [O III] lines. II. Single AGNs with complex narrow-line region kinematics are more common than binary AGNs," ApJ, 735, 48.

Sigut, T. A. A., & Pradhan, A. K. 2003, "Predicted Fe II emission-line strengths from active galactic nuclei," ApJS, 145, 15.

Sijacki, D., Springel, V., & Haehnelt, M. G. 2009, "Growing the first bright quasars in cosmological simulations of structure formation," MNRAS, 400, 100.

Sijacki, D., Springel, V., & Haehnelt, M. G. 2011, "Gravitational recoils of supermassive black holes in hydrodynamical simulations of gas-rich galaxies," MNRAS, 414, 3656.

Sikora, M., Stawarz, L., & Lasota, J.-P. 2007, "Radio loudness of active galactic nuclei: Observational facts and theoretical implications," ApJ, 658, 815.

Sikora, M., Stawarz, L., & Lasota, J.-P. 2008, "Radio-loudness of active galaxies and the black hole evolution," NAR, 51, 891.

Soltan, A. 1982, "Masses of quasars," MNRAS, 200, 115.

Stamerra, A., J. Becerra, G. Bonnoli, L. Maraschi, F. Tavecchio, D. Mazin, K. Saito, for the MAGIC Collaboration, Y. Tanaka, D. Wood, and for the Fermi/LAT Collaboration 2011. Challenging the high-energy emission zone in FSRQs. *ArXiv e-prints*.

Stern, J., & Laor, A. 2012, "Type 1 AGN at low z − II. The relative strength of narrow lines and the nature of intermediate type AGN," MNRAS, 426, 2703.

Sturm, E., González-Alfonso, E., Veilleux, S., et al. 2011, "Massive molecular outflows and negative feedback in ULIRGs observed by Herschel–PACS," ApJL, 733, L16.

Sulentic, J. W., Marziani, P., & Dultzin-Hacyan, D. 2000, "Phenomenology of broad emission lines in active galactic nuclei," ARAA, 38, 521.

Sulentic, J. W., Marziani, P., & Zamfir, S. 2009, "Comparing Hβ line profiles in the 4D Eigenvector 1 context," NAR, 53, 198.

Tadhunter, C. 2008, "An introduction to active galactic nuclei: Classification and unification," NAR, 52, 227.

Tavecchio, F., Maraschi, L., Wolter, A., et al. 2007, "Chandra and Hubble Space Telescope observations of gamma-ray blazars: Comparing jet emission at small and large scales," ApJ, 662, 900.

Tombesi, F., Cappi, M., Reeves, J. N., et al. 2010, "Evidence for ultra-fast outflows in radio-quiet AGNs. I. Detection and statistical incidence of Fe K-shell absorption lines," A&A, 521, A57.

Tommasin, S., Spinoglio, L., Malkan, M. A., et al. 2010, "Spitzer–IRS high-resolution spectroscopy of the 12 μm Seyfert galaxies. II. Results for the complete data set," ApJ, 709, 1257.

Tommasin, S., Netzer, H., Sternberg, A., et al. 2012, "Star formation in LINER host galaxies at z $\sim$ 0.3," arXiv:1201.3792.

Trakhtenbrot, B., Netzer, H., Lira, P., et al. 2011, "Black hole mass and growth rate at z $\sim$ 4.8: A short episode of fast growth followed by short duty cycle activity. ApJ, 730, 7.

Trakhtenbrot, B., and H. Netzer 2012. Black Hole Growth to $z = 2-$ I: Improved Virial Methods for Measuring M_BH and L/L_Edd. *ArXiv e-prints*.

Tran, H. D. 2010, "Hidden double-peaked emitters in seyfert 2 galaxies," ApJ, 711, 1174.

Trichas, M., Georgakakis, A., Rowan-Robinson, M., et al. 2009, "Testing the starburst/AGN connection with SWIRE X-ray/70 μm sources," MNRAS, 399, 663.

Trump, J. R., Hall, P. B., Reichard, T. A., et al. 2006, "A catalog of broad absorption line quasars from the Sloan Digital Sky Survey Third Data Release," ApJS, 165, 1.

Trump, J. R., Impey, C. D., Taniguchi, Y., et al. 2009, "The nature of optically dull active galactic nuclei in COSMOS," ApJ, 706, 797.

Tsalmantza, P., Decarli, R., Dotti, M., et al. 2011, "A systematic search for massive black hole binaries in the Sloan Digital Sky Survey spectroscopic sample," ApJ, 738, 20.

Turner, T. J., & Miller, L. 2009. X-ray absorption and reflection in active galactic nuclei. AstApRv, 17, 47–104.

Urry, C. M., Scarpa, R., O'Dowd, M., et al. 2002, "Host galaxies and the unification of radio-loud AGN," NAR, 46, 349.

Urry, M. 2011, "Gamma-ray and multiwavelength emission from blazars," ApJ&A, 32, 139.

Vanden Berk, D. E., Richards, G. T., Bauer, A., et al. 2001, "Composite quasar spectra from the sloan digital sky survey," AsJ, 122, 549.

Valiante, R., Schneider, R., Salvadori, S., et al. 2011, "The origin of the dust in high-redshift quasars: The case of SDSS J1148+5251," MNRAS, 416, 1916.

Vasudevan, R. V., & Fabian, A. C. 2007, "Piecing together the X-ray background: Bolometric corrections for active galactic nuclei," MNRAS, 381, 1235.

Veilleux, S., & Osterbrock, D. E. 1987, "Spectral classification of emission-line galaxies," ApJS, 63, 295 1G.

Veilleux, S., Cecil, G., & Bland-Hawthorn, J. 2005, "Galactic winds," AnnRevAstAp, 43, 769.

Veilleux, S. 2008, "AGN host galaxies," NAR, 52, 289.

Vestergaard, M., & Peterson, B. M. 2006, "Determining central black hole masses in distant active galaxies and quasars. II. Improved optical and UV scaling relationships," ApJ, 641, 689.

Vestergaard, M., & Osmer, P. S. 2009, "Mass functions of the active black holes in distant quasars from the Large Bright Quasar Survey, the Bright Quasar Survey, and the color-selected sample of the SDSS Fall Equatorial Stripe," ApJ, 699, 800.

Villforth, C., et al. 2010, "A new extensive catalog of variable active galactic nuceli in the GOODS fields and a new statistical approach to variability selection," ApJ, 723, 737.

Volonteri, M. 2010, "Formation of supermassive black holes," AApRev, 18, 279.

Véron-Cetty, M. P., & Véron, P. 2000, "The emission line spectrum of active galactic nuclei and the unifying scheme," 1GAApRev, 10, 81.

Wang, J.-M., & Netzer, H. 2003, "Extreme slim accretion disks and narrow line Seyfert 1 galaxies: The nature of the soft X-ray hump," A&A, 398, 927.

Wild, V., Heckman, T., & Charlot, S. 2010, "Timing the starburst–AGN connection," MNRAS, 405, 933.

Wills, B. J., Netzer, H., & Wills, D. 1985, "Broad emission features in QSOs and active galactic nuclei. II – New observations and theory of Fe II and H I emission," ApJ, 288, 94.

Winter, L. M., Mushotzky, R. F., Terashima, Y., et al. 2009, "The suzaku view of the swift/bat active galactic nuclei. II. Time variability and spectra of five 'hidden' active galactic nuclei," ApJ, 701, 1644.

Worrall, D. M. 2009, "The X-ray jets of active galaxies," AApRev 17, 1.

Wuyts, S., Förster Schreiber, N. M., van der Wel, A., et al. 2011, "Galaxy structure and mode of star formation in the SFR-mass plane from $z \sim 2.5$ to $z \sim 0.1$," ApJ, 742, 96.

Xu, Y., Bian, W.-H., Yuan, Q.-R., et al. 2008, "The origin and evolution of CIV Baldwin effect in QSOs from the Sloan Digital Sky Survey," MNRAS, 389, 1703.

Index

Printed in the United States
by Baker & Taylor Publisher Services

Printed in the United States
by Baker & Taylor Publisher Services